ABOUT THE AUTHOR

Patrick A. Domenico is currently an associate professor of geology at the University of Illinois, Urbana. He received his B.S. degree in geology and M.S. degree in engineering geology from Syracuse University and earned his Ph.D. degree in hydrology from the University of Nevada. In addition to teaching, Professor Domenico has worked with the California Department of Water Resources and the Desert Research Institute; he is a registered engineer in the State of Nevada and an associate editor of *Water Resources Research*. He has written several technical papers and reports on water resources and has been active as a part-time consulting hydrologist for various firms and individuals.

CONCEPTS AND MODELS IN GROUNDWATER HYDROLOGY

McGRAW-HILL INTERNATIONAL SERIES
IN THE EARTH AND PLANETARY SCIENCES

FRANK PRESS, *Consulting Editor*

EDITORIAL BOARD

KENNETH O. EMERY, Woods Hole Oceanographic Institution
KONRAD KRAUSKOPF, Stanford University
BRUCE MURRAY, California Institute of Technology
RAYMOND SIEVER, Harvard University
EX OFFICIO MEMBER
ALBERT E. J. ENGEL, University of California, San Diego
SVERRE PETTERSSEN, emeritus—University of Chicago

AGER · Principles of Paleoecology
BERNER · Principles of Chemical Sedimentology
BROECKER and OVERSBY · Chemical Equilibria in the Earth
DE SITTER · Structural Geology
DOMENICO · Concepts and Models in Groundwater Hydrology
EWING, JARDETZKY, and PRESS · Elastic Waves in Layered Media
GRANT and WEST · Interpretation Theory in Applied Geophysics
GRIFFITHS · Scientific Method in Analysis of Sediments
GRIM · Applied Clay Mineralogy
GRIM · Clay Mineralogy
HOWELL · Introduction to Geophysics
JACOBS, RUSSELL, and WILSON · Physics and Geology
KRAUSKOF · Introduction to Geochemistry
KRUMBEIN and GRAYBILL · An Introduction to Statistical Models in Geology
LEGGET · Geology and Engineering
MENARD · Marine Geology of the Pacific
MILLER · Photogeology
OFFICER · Introduction to the Theory of Sound Transmission
RAMSAY · Folding and Fracturing of Rocks
ROBERTSON · The Nature of the Solid Earth
SHROCK and TWENHOFEL · Principles of Invertebrate Paleontology
STANTON · Ore Petrology
TURNER · Metamorphic Petrology
TURNER and VERHOOGEN · Igneous and Metamorphic Petrology
TURNER and WEISS · Structural Analysis of Metamorphic Tectonites
WHITE · Seismic Waves: Radiation, Transmission, and Attenuation

CONCEPTS AND MODELS IN GROUNDWATER HYDROLOGY

PATRICK A. DOMENICO
Associate Professor of Geology
University of Illinois

McGraw-Hill Book Company
New York   St. Louis   San Francisco   Düsseldorf   Johannesburg
Kuala Lumpur   London   Mexico   Montreal   New Delhi
Panama   Rio de Janeiro   Singapore   Sydney   Toronto

*To my wife Delores,
and to my children
Daniel, Philip, Trina, and Patrick*

**CONCEPTS AND MODELS IN
GROUNDWATER HYDROLOGY**

Copyright © 1972 by McGraw-Hill, Inc. All rights reserved. Printed in the United States of America. No part of this publication may be reproduced, stored in a retrieval system, or transmitted, in any form or by any means, electronic, mechanical, photocopying, recording, or otherwise, without the prior written permission of the publisher.

Library of Congress Catalog Card Number 79-168751

07-017535-7

1 2 3 4 5 6 7 8 9 0 K P K P 7 9 8 7 6 5 4 3 2

This book was set in Times Roman, and was printed and bound by Kingsport Press, Inc. The designer was Richard Paul Kluga; the drawings were done by Oxford Illustrated Limited. The editors were Bradford Bayne and Barry Benjamin. Matt Martino supervised production.

Contents

preface   vii

1   Introduction to a System Concept in Groundwater Hydrology   1

   1.1   Notion of a System   2
   1.2   Systems and Modeling Concepts   4
   1.3   Hydrologic and Water-resource Systems   9
   1.4   Major Subsystems and Their Investigation   22
   1.5   Organization and Scope of the Book   30

part one   The Groundwater Basin as a Lumped Parameter System   39

2   Hydrologic Models   41

   2.1   Conventional Input-Output Analysis   42
   2.2   Basin Assemblages and Regression Analysis   53
   2.3   Information Theory and Hydrologic Variables   57
   2.4   Concluding Statement   72

3   Optimization Models   78

   3.1   Overview of Objectives and Concepts in Groundwater Management   79
   3.2   Basic Economic and Hydrologic Concepts   91
   3.3   Strategies for Isolated Groundwater Basins   93
   3.4   Related Problems in Valuation   107
   3.5   Mathematical Programming   113
   3.6   Concluding Statement   135

part two   Conservation Principles and Applications   143

4   Energy and Its Transformations   145

   4.1   Mechanical Energy and Fluid Flow   146
   4.2   Measurements of Potential Energy and Interpretations   164
   4.3   Other Forms of Energy   176
   4.4   Extensions for Two Miscible Fluids: Sea-Water Intrusion in Coastal Aquifers   183
   4.5   Energy Concepts Applied to Chemical Thermodynamics   191
   4.6   Concluding Statement   202

5   Compressibility, Elasticity, and Main Equations of Flow   209
    5.1 Effective Stress and Natural Body Forces   210
    5.2 Storativity of Elastic Aquifers   216
    5.3 Elastic Compression and Water-level Fluctuations   223
    5.4 Extensions for Low-permeability Units   228
    5.5 Effective Stress and Resistance to Shear   235
    5.6 Main Equations of Flow   239
    5.7 Concluding Statement   246

part three   The Groundwater Basin as a Distributed-Parameter System   251

6   Regional Groundwater Flow: Theoretical Models of the Steady-state and Related-field Observations   253
    6.1 Theoretical Models: Solutions of Laplace's Equation   254
    6.2 Surface Features of Groundwater Flow   273
    6.3 Temperature of Groundwater   279
    6.4 Chemistry of Groundwater   283
    6.5 Extensions for Environmental, Engineering, and Geological Applications   296
    6.6 Concluding Statement   309

7   Theoretical Models of the Unsteady State: Well Flow, Consolidation Theory, Simulation   315
    A. Well Flow   316
    7.1 The Heat-flow Analogy: Radial Flow to a Well in an Infinite Aquifer   317
    7.2 The Steady Response   327
    7.3 Extensions and Modifications of the Nonequilibrium Equation   332
    7.4 A More General Theory of Radial Flow   338
    7.5 Radial Flow to a Well in a Finite Aquifer   347
    7.6 Water-level Response at the Pumped Well   351
    B. Consolidation Theory   356
    7.7 Theoretical Development for One-dimensional Consolidation   356
    7.8 Mathematical Analysis   360
    C. Simulation   367
    7.9 Analog Simulation   368
    7.10 Digital Simulation   381
    7.11 Simulation and Objectives Functions   385

author index   396

subject index   400

Preface

Basically, this book is an attempt to present the main ideas of groundwater hydrology as an organized entity rather than a collection of facts and formulas with no underlying organization. Central to this theme is the organization of various hydrologic models and concepts into logical groupings based on mathematical and physical principles. Chapter 1 sets the stage for this grouping, and establishes a classification of models and methods that proves to be useful in the conceptualization of most hydrologic problems. From this chapter, two main problems of analysis emerge: lumped-parameter and distributed-parameter representations of groundwater systems. These are treated subsequently in Part One and Part Three, respectively.

Briefly, a lumped-parameter representation means that the spatial variations in the properties, behavior, or response surface of a system are ignored. A distributed-parameter representation takes into account the detailed spatial variations in properties, behavior, or response surface. All groundwater systems are of course distributed in that the properties are fundamentally geologic, with inevitable variations in space. When the variations are small, they may be ignored and the system may be "lumped." Geology plays a most important role in the analysis of distributed-parameter problems.

I have felt, too, that the student is entitled to know something of the interrelations between the subject matter of groundwater hydrology and those of the long list of supporting sciences from which it derives its strength, such as geology, physics, chemistry, and several branches of engineering. In order to emphasize this point, the organization of the text is intended to mitigate the natural tendency to isolate phenomena and principles into narrowly confined channels. This is accomplished (or at least attempted) by the expansion of a number of related basic ideas rather than by a survey of a multitude of separate topics. Hence Chapter 4 in Part Two works within the broad theme of energy and its transformations, which permits an integrated discussion of the first and second laws of thermodynamics as they pertain to free and potential energies, to entropy production in irreversible groundwater flow and in chemical reactions, and to concepts put forth in mechanical energy, two-phase systems, and chemical thermodynamics. Chapter 5 has compressibility of rock units as a main theme and permits us to identify the analogies between topics treated independently in groundwater hydrology and in soil mechanics. In particular, it is demonstrated that these disciplines deal in large part not only with a similar set of transient-flow problems, but with an identical set of parameters which are merely called by different names.

This same unifying theme is carried over to Part Three, which treats of the distributed-parameter system. Chapter 6 deals with a broad theory of regional groundwater flow, which basically develops the concepts of potential theory, the investigation of the regional movements of groundwater and, from these movements, the demonstration of the other phenomena. The "other phenomena" include temperature and chemical distributions, and surficial manifestations of flow. These are treated within the context of a single system of flow, rather than as independent measurements and observations.

Chapter 7 is more conventional in approach, dealing with the better-known models of the unsteady state, with total emphasis on movement of water under hydraulic gradients far in excess of those characteristic of the natural flow regime. Main features of this chapter include (1) the relation between equations used to describe the flow of water in both high- and low-permeability materials, and (2) the interrelation between these equations and others which are identical in form and principle but deal with the flow of heat and electricity. For analogous initial and boundary conditions, the solution to one of these equations is a solution to all of them.

Two chapters in this book are devoted to the lumped-parameter system. Much of the material in these chapters might be new to many readers. In particular, the concepts of optimization in Chapter 3 are not generally familiar to physical scientists, and they are included for the sole purpose of explaining the power of a set of techniques and ideas that may usher in a whole new era of groundwater-resource management. This chapter can of course be excluded from any serious study of the physical aspects of groundwater hydrology, but it is my hope that it is at least read by those who are totally unfamiliar with these ideas.

In retrospect, this book is written by a geologist with the primary aim of reaching an audience of geologists and engineers interested in groundwater. It is intended for students with little previous experience in the subject, and its purpose is to instruct the student how to think about hydrologic phenomenon. It is not intended to teach him mathematics. However, calculus is an essential prerequisite for most of the chapters. Additional training in statistics, some methods of operations research, computer science, and differential equations would be helpful, but is not necessary. Although most of the mathematical formulations are introduced from basic principles, there is no attempt at an exhaustive treatment of existence theorems or mathematical methods of solution. In this manner, the mathematics that is employed is strongly intuitive and is seen mostly as the language and logic of the discipline, with equations as a symbolic form for physical laws. In addition, mathematics is used as a precise mode of expressing ideas about relations, as a technique for manipulating ideas and data to get answers, and—most often—as a precise statement of a problem.

Because of the integration of ideas from a wide variety of disciplines, an attempt to employ a consistently familiar system of symbols for hydrologic notation has been difficult. Where one symbol is generally accepted for two different meanings or parameters, ambiguity is avoided by retaining that symbol and using an asterisk or some other distinctive sign for one of them, and by restating—time and again—the meaning of a given symbol whenever it appears.

My primary acknowledgments are to the imaginative and creative scientists of many nations who, over the past several decades, have shaped and reshaped the ideas of hydrology as we now know them. More immediately, I was fortunate to obtain the views, opinions, and corrections from a number of reviewers, especially John D. Bredehoeft, U.S. Geological Survey, and Leo A. Heindl, National Research Council, and Jacob Bear, Technion-Israel Institute of Technology. In addition, I am indebted to William Back, U.S. Geological Survey, who read Sections 4.5 and 6.4 to their benefit; to Petar Kokotovic, Coordinated Science Laboratory, University of Illinois, and Dale Meredith, Department of Civil Engineering, University of Illinois, who read Chapter 3; to Victor Palciauskas, Department of Geology, University of Illinois, who read parts of Chapters 4 through 7; and to Frank Schwartz, graduate student at the University of Illinois, who read and contributed to Sections 4.5 and 6.4. In my giving these men sincere thanks, they are exempt from responsibility for whatever omissions or errors may be present; nor does this acknowledgment necessarily imply their complete agreement with the approach and material covered.

This work represents, of course, an individual's view of groundwater hydrology, and it is inevitable that I should have missed or omitted material which others may consider important. Of the material covered, whenever a source was known, credit has been given. If omissions have occurred, they have been unintentional. I should be pleased if readers would be kind enough to bring to my attention such work or credits that they may consider significant, but which have been omitted, and any of the errors which inevitably occur in the course of writing and rewriting.

Patrick A. Domenico

1
Introduction to a System Concept in Groundwater Hydrology

The problems in groundwater hydrology, or discipline dealing with the occurrence, origin, movement, quality, recovery, and use of groundwater, appear to fall naturally into three classes:

1. *Problems studied from a scientific point of view in order to promote a better description of hydrologic phenomena* In numerous cases, the observer notes inputs and outputs in nature or in the laboratory, and attempts to formulate verbal explanations or construct models to describe the relations observed. As the methods of natural science are based on the hypothesis that all processes and phenomena are interrelated, there is little doubt that the problems under study involve systems—here defined as an assemblage of objects joined in regular interaction to which certain laws may apply. Science is interested in determining these laws.

2. *Problems studied from an engineering point of view, which are largely the converse of those studied by science* Given a natural system, engineering is concerned with its use to achieve certain objectives. In more specific terms, engineering is interested in groundwater systems only to the extent that they

achieve some utility or purpose. To this end, the engineer is concerned with the laws discovered by science to predict performance, reliability, cost of development, maintainability, or life expectancy. Again, there is little doubt that the problems under study involve systems—here defined as diverse parts of a whole, which accomplish an operational process.

3. *Problems studied from the point of view of management and planning, who share incentives and objectives with the engineer, with the additional incentive of control* Control is defined as the monitored state of the system. *State variables* define the condition of the major components of the system, whereas *decision variables* act to modify the state. Examples of state variables include water quality, water levels, and number of wells and their distribution. Such information, when compared with established criteria, defines the discrepancy between the existing state of the resource and some proposed state; that is, it defines the *management problem*. Other than detailed monitoring, management is interested in the establishment of output models that describe the consequences if a water-resource system is developed in an unregulated manner, and intervention models that describe, a priori, the probable result of intervention. Problems in management and planning clearly involve systems—here defined as diverse parts of a whole, which are subject to a common plan or serve a common purpose.

Two things are evident from the above discussion. First, a system implies interrelated objects or procedures, rather than a collection of parts, so that the three definitions represent, in fact, three different ways of saying the same thing. Second, study of the problem types cited is closely aligned with the study of models of some sort. A model can be defined as a representation of reality that attempts to explain the behavior of some aspect of it and is always less complex than the real system it represents. Largely because of the complex interdisciplinary interests in groundwater, the models differ markedly in purpose, information requirements, assumptions, usefulness, and even the kind of mathematics employed.

Since it can be argued that groundwater hydrology, as practiced in modern times, is a combination of concepts and perspectives that pertain to the scientific, engineering, and management aspects of groundwater in the hydrologic cycle, there is need for a general approach to the subject. Such an approach, if tied to the notion of a system—somehow defined—may provide an objective standard by which the literature in the field can be most effectively processed and organized for study and new problems recognized and organized for investigation and solution.

1.1 NOTION OF A SYSTEM

The term *system* as introduced above is commonly associated with several different meanings. The breadth of these definitions permits an entire field under consideration to be taken as the system: two or more basins characterized by interflow, and sharing a common physiographic boundary; a conjunctively operated groundwater supply and surface reservoir or pipeline; the Colorado River, including surface storage and tributaries; the California Water Project. Indeed, this concept permits consideration of isolated assemblages ranging from two parallel flow lines traced from source to sink to the entire hydrologic cycle.

The system in the cases cited above is identified with a real or abstract object. Thus, the groundwater geologist may refer to a "groundwater system," meaning an aggregation of rocks with certain properties which freely admit and transmit water bounded by other rocks that do not. The hydrodynamicist, on the other hand, may refer to the distribution of fluid potential within these same rocks as a "groundwater flow system." In the case of the former, the system is the whole thing as it exists, and encompasses many variables. The sheer bulk of such a system precludes absolute knowledge concerning it. In the latter case, the system is referred to as one or more sets of variables (or states) which are adequate for the purpose being served. As the concept cannot be applied to any whole incapable of being studied, it is often better to concentrate on parts of the whole that are complete within themselves and are sufficient for the practical purpose being served.

A groundwater system is of special interest if it is characterized by one or both of the following:

1. It reacts to excitation in a deterministic or probabilistic manner.
2. It can be supplied with a set of rules to govern its operation.

A *deterministic system* is one that is defined by definite cause-and-effect relations. Cause is generally referred to as excitation, while effect refers to reaction, or response. In some studies, a natural system is thought to be sufficiently understood if the cause-and-effect relations within it are understood. Certainly there is nothing wrong with this level of understanding. For groundwater systems, there is just something missing. That something is the factor relating cause and effect, invariably the geology of the system. Consider, for example, the abstract nature of a map showing pumpage and its distribution (a cause) related to water-level change (an effect), or water-level change (a cause) related to the amount and distribution of land subsidence (an effect). These cause-and-effect relations will remain abstract in their meaning until they are related to the geology of the system. Unless, of course, the geology is of the simplest type conceivable.

This leads to another useful function of deterministic systems, especially for engineering purposes, that of predicting effects arising from causes. Indeed, quantitative hydrology exists because deterministic cause-and-effect relations can be measured and modeled.

A probabilistic system allows no precise prediction, but may provide expected values within the limitations of the probability terms which define its behavior. A stochastic, or random, variable is a variable quantity with a definite range of values, each one of which, depending on chance, can be attained with a definite probability. An expected value is the mean value of a random variable. If any one aspect of an otherwise deterministic relation is random in nature, the whole relation is rendered stochastic. One example is provided by the accepted equation of continuity used for inventory studies,

$$\Delta S = R_N + Q_i - X - Q_o \tag{1.1}$$

where ΔS is change in groundwater storage, R_N is natural replenishment by precipitation, Q_i is inflow, X is pumpage, and Q_o is outflow. This relation is generally treated as deterministic. However, if natural replenishment by precipitation is random with an associated probability distribution, Eq. (1.1) is stochastic regardless of its deterministic elements. In practice, the concept of determinism prevails among groundwater hydrologists, and stochastic problems are seldom recognized or treated as such.

From statement two given earlier, a system may consist not only of an assemblage of interrelated elements, but also of a set of rules for its operation. In broader terms, a system is any set of elements which can be usefully considered related in the accomplishment of a common goal. An inherent assumption here is that the system contains certain elements which are within the control of man and, in combination with the uncontrollable elements, is investigated to seek answers of an optimum, or "best value," class. Hence, it is possible to ask (and often ascertain) what is the best way of operating a groundwater basin or integrated groundwater–surface-water supply. Clearly, any concept of best is relative to a concisely formulated statement of objectives, which may be economical, hydrological, social, political, or legal. The idea of objectives, or an objective function, is thus fundamental to a best-value class of problems.

1.2 SYSTEMS AND MODELING CONCEPTS

The system concept is used differently by various disciplines, primarily because of the different types of problems investigated. Karplus (1958), for example, is concerned with the engineering analysis of physical systems and defined a system as an assemblage of elements ranging in number from one to infinity, in either lumped or continuous form, which reacts to excitation in a known or predictable manner. Streeter (1966) identified system with a particular quantity of matter (lumped) or a specified region in space (distributed). Ashby (1961), on the other hand, recognized that modern-day science is concerned with systems other than those of mass and energy and defined system engineering as the science of controlling systems of high complexity. The ambiguity of the term "high complexity" is removed when it is understood to mean that the quantities of information handled in the control process are large relative to the channel capacity of the controlling system.

Getting closer to hydrologic systems, Dooge (1967) was concerned with water in its natural state and defined a system as "any structure, device, scheme, or procedure, real or abstract, that interrelates in a given time reference, an input, cause, or stimulus of matter, energy, or information and an output, effect, or response of information, energy, or matter." This is essentially the definition adopted earlier in the text, and emphasizes the interrelations between the parts of the whole independent of the tools available to study the system. This suggests that a systems approach to hydrology is as old as concern for the nature of the water regime itself, and only the terminology has changed. Nevertheless, there are a few good reasons to retain such an approach. First, as a teaching tool, it brings the main principles into focus, which allows reasonable ties to be made between principles that are closely interdependent. This provides a framework that is easier to deal with than a variety of theories, explanations, and mathematical methods that have no underlying organization. Second, as a research strategy, it forces one to look at an entire problem as a whole, to formulate objectives, and to construct and operate within a model of the system—a real or abstract simplification of the situation appropriate to the problem at hand. The emphasis is clearly on the formulation of models, and in the final analysis, the accuracy and utility of the model depend on how much of the real system remains after one simplifies and cuts problems to manageable size.

A more precise definition of a groundwater system would take cognizance of the fact that hydrologists are concerned with the transport of matter, energy, and information, suggesting that a system is an interrelation of parts, which behaves according to some description and whose function is to operate on matter, energy, or information, or on any two or all three. A groundwater system, which is concerned with water in its natural state, may be defined in accordance with this definition. This is shown schematically in Fig. 1.1.

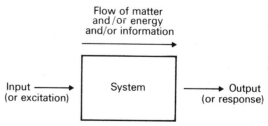

Fig. 1.1 Schematic representation of a system.

OPEN SYSTEMS AND CLOSED SYSTEMS

Within the context of the last definition, it is possible to distinguish between absolutely closed, relatively closed, and open systems: An absolutely closed (isolated) system exchanges neither energy nor matter with its environment, a relatively closed system exchanges energy but not matter with its environment, and an open system exchanges both energy and matter with its environment.

Classical physics is concerned mainly with the study of closed systems (Prigogine, 1967). Hence, physical chemistry deals with reactions, their rates, and the chemical equilibria established in closed vessels where reactants have been brought together. The basic characteristic of closed systems is that, given a certain amount of free or potential energy, they develop toward states of maximum entropy (Von Bertalanffy, 1950), the term entropy being used to express the degree to which energy becomes unavailable to perform work. This, of course, is the essence of the second law of thermodynamics, which states that entropy must increase to a maximum and, eventually, the process grinds to a halt when equilibrium is achieved. Most of the models employed in chemical thermodynamics are based on closed system concepts.

By contrast, an open system is maintained and preserved by a constant supply and removal of material and energy, and may manifest one important property denied the closed system. An open system may regulate or adapt itself in response to changes in the supply of energy and matter from the outside. Thus, it may attain a "dynamic equilibrium," or steady state, wherein the import and export of matter and energy are equal; more precisely, the tendency is to strive toward equilibrium through internal adjustments within the system. The self-adjusting behavior of open systems has a direct analogy with fluvial processes in geomorphology (Leopold and Langbein, 1962; Chorley, 1962), as well as with practically all dynamic aspects of groundwater hydrology. From a mathematical point of view, the widely used concept of a control (differential) volume in the derivation of flow equations, with inputs and outputs accounted for by conservation principles, is perhaps the most self-explanatory example of an open system. It follows that most of the models employed in groundwater hydrology are based on open-system concepts.

A fundamental example of open-system behavior has been presented by Theis (1938; 1940), who examined the operation of a groundwater basin in response to a pumping stress. Prior to such development, groundwater basins are in a condition of dynamic equilibrium, with recharge being discharged by natural means. Discharge by wells is an additional stress put on the groundwater body, which must be balanced by (1) an increase in recharge; (2) a decrease in natural discharge; (3) a loss of storage in the groundwater body; or (4) a combination of these factors. The essential physical features that determine which of these factors dominate are (Theis, 1940):

INTRODUCTION TO A SYSTEM CONCEPT IN GROUNDWATER HYDROLOGY

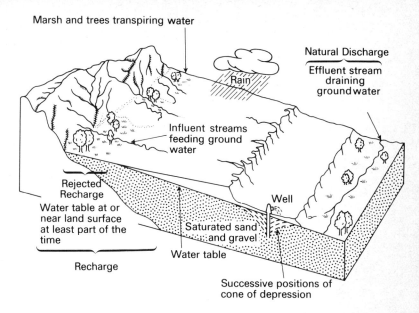

Fig. 1.2 Factors controlling the response of a groundwater basin to discharge by wells. (*After Theis*, 1940.)

1. The distance to, and character of, the recharge
2. The distance to the area of natural discharge
3. The manner in which the groundwater basin responds to pumping

With reference to 1, if recharge was rejected prior to pumping, there is a strong possibility of balancing well discharge by increased recharge. This is shown diagrammatically in Fig. 1.2, in which there is rejected recharge at the edge of the mountain area. Pumping in this area will allow more water to infiltrate the underground portions of the basin. The well shown in the figure will also affect the nearby area of natural discharge after some water is taken from storage. In order to do so, the water table must be depressed in that area. Pumping in areas of rejected recharge, or of natural discharge, permits the basin to act as a pirating agent to procure water from other components of the hydrologic cycle.

The factors controlling the manner in which a groundwater basin responds to pumping will be discussed in detail elsewhere. It is clear, however, that the response surface must ultimately include areas of rejected recharge or natural discharge, or both, in order to promote a new equilibrium in the system. Other examples of open-system behavior to be dealt with in later chapters include the fluctuation of the position of a freshwater-saltwater interface in coastal basins and land subsidence in response to pumping.

LUMPED-PARAMETER SYSTEMS AND DISTRIBUTED-PARAMETER SYSTEMS

When attempting to model hydrologic or water-resource systems, two fundamental problems of analysis are encountered:

1. Problems which apply to the system regarded as lumped
2. Problems which apply to distributed-parameter systems

A *black box* is a term used for a lumped-parameter system in which inputs and outputs can be measured or estimated although the processes which interrelate them are not often observable. Figure 1.1 may serve as a black box, complete with inputs and outputs. Input terminals represent the excitation variables, which influence system behavior, and are identified by arrows pointing toward the box. Output terminals represent the response variables, representing those aspects of system behavior which are of interest, and are identified by arrows pointing away from the box. Intermediate variables, which are neither excitation nor response variables, are assumed to be embedded in the box, and are important because of their combined effect on the relation between input and output. However, detailed knowledge of the internal mechanism relating input and output is not necessary in so-called black-box analyses.

The distinctive feature of a lumped-parameter, or black-box, problem in groundwater hydrology is that a space coordinate system is not required in problem formulation and solution. As a practical example, a rise in water levels in wells over a certain time interval (a response variable) may be converted to recharge without any regard to the location of wells in the field or to their spacing (or to the manner in which recharge reaches the water table or even to the amount of rainfall). A similar case can be made for some streamflow-recharge and precipitation-discharge models (Duckstein and Kisiel, 1968), recession studies dealing with groundwater discharge or declining water levels, evapotranspiration estimates by empirical methods, and inventory investigations utilizing Eq. (1.1). For most of these examples, time, but not space, is an important variable. Indeed, the hydrologic cycle itself is often presented as a black-box abstraction of lumped elements, and it follows that most of the classic methods employed in its investigation are of a black-box nature. It follows that the system parameters are expressed in lumped form when the total system is regarded as located at a single point in space.

In addition, black-box analyses are used rather extensively in optimization studies, where emphasis is on extreme-value problems, such as maximizing a time flow of economic returns associated with groundwater utilization. A severe limitation of these studies is the lack of suitable means to describe the spatial interdependencies of the pumping-response surface within a specified region.

The distinctive feature of the distributed-parameter problem is that the internal space of the system is described by a distribution of points, each of which requires examination. For a mathematical solution, the initial data must include not only the values of the properties at all points within the system, but also the location of the boundaries. The largest class of problems of this type deals with solution to partial differential equations describing the flow of groundwater through porous media. It follows that a space coordinate system comprises a necessary part of problem formulation and solution.

An important distinction has to be made between steady (space variables only) and unsteady (both space and time variables) applications. Steady-flow applications have permitted a long-desired relinquishing of the black-box approach in the analysis of the natural flow of groundwater in the hydrologic cycle. Unsteady applications have (at least until more recent times) been more useful for studies of smaller scale, as in the vicinity of pumping wells. Meyboom (1966) has commented on this observation, identifying time and scale as the main difference in the approach to groundwater flow.

LINEAR SYSTEMS AND NONLINEAR SYSTEMS

The concept of linearity can be presented as a property of a certain class of differential equations or a property of systems. A linear differential equation consists of a sum of linear terms, a linear term being first degree in the dependent variables and their derivatives. Linearity in differential equations is synonymous with the principle of superposition, which states that the derivative of a sum of terms is equal to the sum of the derivatives of the individual terms. Expressed in terms of system response, the total effect resulting from several stresses acting simultaneously is equal to the sum of the effects caused by each of the stresses acting separately.

A linear system has the property that if (Distefano, Stubberud, and Williams, 1967) (1) an input $X_1(t)$ produces an output $Y_1(t)$ and (2) an input $X_2(t)$ produces an output $Y_2(t)$, then (3) an input $b_1 X_1(t) + b_2 X_2(t)$ produces an output $b_1 Y_1(t) + b_2 Y_2(t)$ for all pairs of inputs $X_1(t)$ and $X_2(t)$ and all pairs of constants b_1 and b_2. It follows that linear systems may be represented by linear equations, differential or otherwise.

1.3 HYDROLOGIC AND WATER–RESOURCE SYSTEMS

Several of the above-cited concepts can be applied to the study of water in the natural environment, and these may be classified by their content, or subject matter. Included here are the open-closed and lumped-parameter–distributed-parameter distinctions to be discussed throughout the book. Other concepts are primarily descriptive, and may be classified by the features which they attempt to describe. These classes include at least the following:

1. Concepts which distinguish between subsystems within a given system
2. Concepts which relate to the interaction between various subsystems
3. Concepts which distinguish between different kinds of systems.

These features are already interwoven in the fabric of classical and modern groundwater hydrology, and will be uncoupled and examined in a descriptive manner in the following pages.

THE HYDROLOGIC CYCLE AS A LUMPED-PARAMETER SYSTEM

Hydrology is defined as the science that deals with processes governing the depletion and replenishment of water resources of the land areas of the earth (Wisler and Brater, 1949). The oceans, from which water vapor is carried into the air and then transferred to the land, constitute the ultimate source and sink of this water. Precipitated water may run off over the land to streams, may be intercepted by lakes, swamps, and surface depressions, may infiltrate the ground only to be transpired by plants, or may replenish a groundwater supply only to turn up later in a spring or stream. The march of events marking the progress of a particle of water from the atmosphere to the land masses and oceans and its return to the atmosphere is termed the *hydrologic cycle*.

The hydrologic cycle is often illustrated schematically by diagrams that lump the atmosphere, land, and ocean areas into single components (Fig. 1.3). This type of presentation masks the fact that components of the cycle may often be evaluated somewhat independently of the complete cycle, and that each component may be thought of as having subcomponents. A more informative method of presentation is by rectangular boxes and connecting lines that portray various moisture inputs and outputs (Fig. 1.4). The atmosphere gains moisture from the oceans and land areas, E, and releases it back in the form of precipitation P. Precipitation is disposed of by evaporation to the atmosphere, E, runoff to the channel network of streams, Q_o, and infiltration through the soil, F. Water in the soil is subject to transpiration T; outflow to the channel network, Q_o, referred to as *interflow*; and downward percolation to the groundwater, R_N. The groundwater reservoir may receive water Q_i from, and release water Q_o to, the channel network and the atmosphere, T. In some instances, water of volcanic, or magmatic, origin M from the lithosphere may be added to the groundwater or to the atmosphere. The ultimate destination of waters not returned directly to the atmosphere is the oceanic areas R_O, with numerous short circuits along the way.

The hydrologic cycle shown in Fig. 1.4 is a complex unit formed of

INTRODUCTION TO A SYSTEM CONCEPT IN GROUNDWATER HYDROLOGY

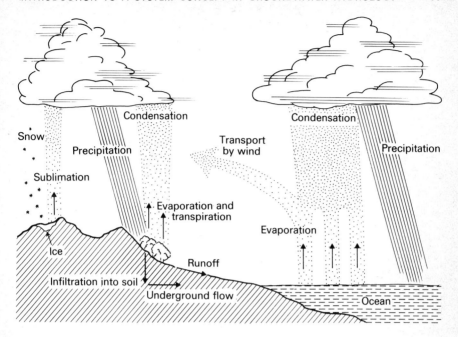

Fig. 1.3 Schematic representation of the hydrologic cycle.

many diverse parts serving the common purpose of renewing and disposing of water of the land area. Interaction between the parts is the basis for treating the cycle as a hydrologic system. As the cycle is solar powered, energy but not matter (water) is exchanged with the outside environment, and the hydrologic cycle is a relatively closed system. Among the processes involved are precipitation, evaporation, infiltration, and surface and subsurface runoff.

As mentioned previously, it sometimes is possible to isolate parts of the whole that are sufficiently descriptive for the practical purpose being served. The meteorologist and atmospheric physicist, for example, are concerned with the atmosphere (Fig. 1.5a), the soil physicist with the soil component (Fig. 1.5b), and the surface-water hydrologist with rainfall-runoff relations in individual basins (Fig. 1.5c). These figures demonstrate that isolation of subsystems from the larger system cannot exclude the lines of moisture transport connecting the subsystem with the outside environment. The individual systems shown, therefore, are subsystems of yet a larger system, and are open in the sense that water and energy within them are interchanged with water and energy from other components of the hydrologic cycle. A subsystem may be regarded as closed only when it can be evaluated independently of the total environment. In practice, this means enlarging the

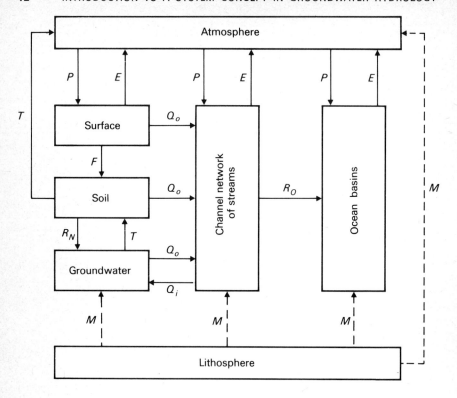

Fig. 1.4 Elements of the hydrologic cycle. (*Modified, after Dooge*, 1967.)

system under study by linking various subsystems. For lumped-parameter systems, the linkages may often be established by merely accounting for the additional inflows and outflows over time. For distributed-parameter systems, a subsystem may be isolated by introducing the effect of its environment through boundary conditions.

Identification and isolation of subsystems not only aid in defining the categories of interest to scientists investigating various components of the hydrologic cycle, but define the network of inputs and outputs in communication with the total environment. This network can often be visualized and sometimes evaluated by examining the hydrologic equation formulated from continuity considerations

$$I - O = \Delta S \tag{1.2}$$

where I is inflow during a given period of time, O is outflow, and ΔS is the change of water in storage. For the soil physicist (Fig. 1.5b), this equation may take the form

$$F - (T + R_N + Q_o) = \Delta S \tag{1.3}$$

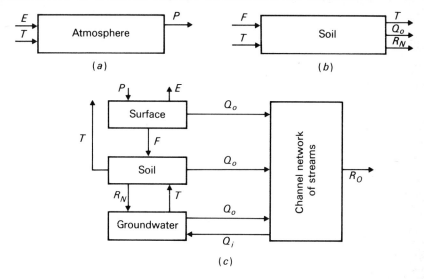

Fig. 1.5 Subsystems of the hydrologic cycle of interest to (*a*) atmospheric physicists and meteorologists, (*b*) soil physicists, and (*c*) surface-water hydrologists. (*Part (c) after Dooge*, 1967.)

whereas the form of continuity corresponding to the hydrologist's interest may be

$$P - (E + T + R_o) = \Delta S \tag{1.4}$$

where ΔS refers to the lumped change in soil, ground, and surface-water storage. Each item of the equations cited has the units of a discharge, or volume per unit time. The equations may be expanded, or abbreviated, depending on circumstances. Most important, they relay information on what components of the hydrologic cycle affect the various disciplines concerned with its study. The inclusion of the time element is required in that the individual subsystems are not closed and can only be treated as such if an accounting is taken of the inflow and outflow over a period of record.

In its most elaborate form, the groundwater subsystem includes the soil, groundwater, and lithosphere components, and encompasses the largest source of fresh water in the hydrologic cycle. This is illustrated by the diagram showing the vertical distribution of groundwater, or the water profile (Fig. 1.6). In this diagram, the vadose zone corresponds to the unsaturated zone with its characteristic inputs and outputs, the phreatic zone corresponds to the groundwater component, and the lithosphere corresponds to all subsurface water below the phreatic zone. The intermediate zone separates the saturated phreatic zone from soil water. It can be absent in areas of high

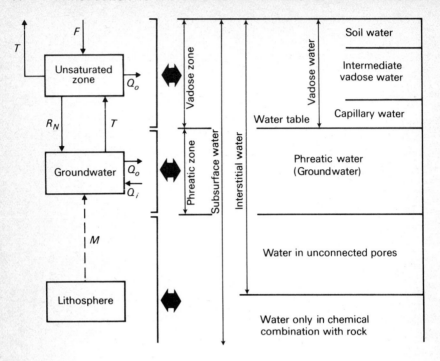

Fig. 1.6 The groundwater subsystem and the water profile.

precipitation, and hundreds of feet thick in arid areas. The water table marks the bottom of the capillary water and the beginning of the saturated zone, or groundwater.

The vadose zone occupies the attention of many hydrologists, whereas subsurface water below the phreatic zone is considered only infrequently, and then largely by a few geochemists or geophysicists. Most groundwater hydrologists are concerned largely with the phreatic zone (Fig. 1.7). A form of continuity that applies to the single component of Fig. 1.7 is

$$R_N + Q_i - (T + Q_o) = \Delta S \tag{1.5}$$

The equations cited above may be expressed in the form of an ordinary differential equation

$$I(t) - O(t) = \frac{dS}{dt} \tag{1.6}$$

Equation (1.6), termed the *hydrologic* or *storage equation*, may also be referred to as the lumped-parameter, deterministic performance equation. Its application—as also the application of any lumped-parameter equation—always requires that the pertinent variables be represented by "averages" over the

INTRODUCTION TO A SYSTEM CONCEPT IN GROUNDWATER HYDROLOGY

space of interest. From a mathematical point of view, the term *lumped-parameter* is generally used to signify ordinary differential equations, whereas the term *distributed-parameter* is used to signify partial differential equations.

A variety of instruments, techniques, and empirical formulas have been developed to measure or estimate the quantity of water in various components of the hydrologic cycle. The primary measurements include precipitation, evaporation, and runoff, and have been discussed at length by Linsley and others (1958) and Kazmann (1965). Empirical relations are used extensively to estimate total water loss, or evapotranspiration (Thornthwaite, 1948; Blaney, 1952; Penman, 1956).

THE HYDROLOGIC CYCLE AS A DISTRIBUTED-PARAMETER SYSTEM

In the lumped-parameter approach to the movement of water within the hydrologic cycle, the detailed processes and actual mechanics of movement are ignored. This is somewhat disconcerting to some hydrologists, especially groundwater hydrologists, who are interested in the details of movement in saturated environments. It is known, for example, that the potential energy of a fluid is the dominating quantity in a mechanical flow process, and that spatial changes in this quantity indicate the directional movements of the fluid. In order to formulate this problem in a satisfactory manner, the watershed or groundwater basin must be treated as a distributed-parameter system. This requires formulation and solution of partial differential equations based on conservation principles of both mass and energy. With few noteworthy exceptions (Chen and Chow, 1968), the complicated physics of surface runoff renders this type of formulation either intractable or overly frustrating for most surface-water problems. On the other hand, possibly due to justifiably ignoring interactions between the subsystems of the hydrologic cycle, distributed-parameter models are at the forefront of groundwater hydrology. To discuss these models at this time would preempt the main considerations of this book.

WATER-RESOURCE SYSTEMS

The traditional hydrologic cycle depicting water in its natural state is not only overly simplified, but gives an erroneous picture of the true state of affairs.

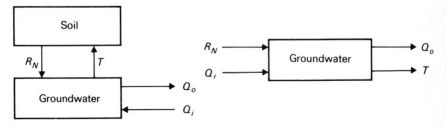

Fig. 1.7 Subsystems of the groundwater system.

Serious departures from the natural setting are the result of interaction between man and the hydrologic environment: few river systems are unregulated by surface-storage facilities; many developed groundwater basins are characterized by pumping outputs that exceed natural inputs; recycling of water is common in areas where nonconsumptive uses allow the return of water to the cycle; water transfers from areas of excess to areas of deficiency are becoming more common; atmospheric scientists are actively pursuing the goal of augmenting and controlling precipitation. In short, the watershed or groundwater basin in many cases can no longer be described as a hydrologic unit suitable for investigation of natural hydrologic phenomena, and it has elements that interact with the interests of biological and social scientists (Ackermann, 1969).

The present-day hydrologic cycle might well be represented by a large number of interconnected water-resource subsystems, each of which serves a different function, but all of which are intended for a common goal (Fig. 1.8). The degree of success in achieving this goal is a measure of the effectiveness of

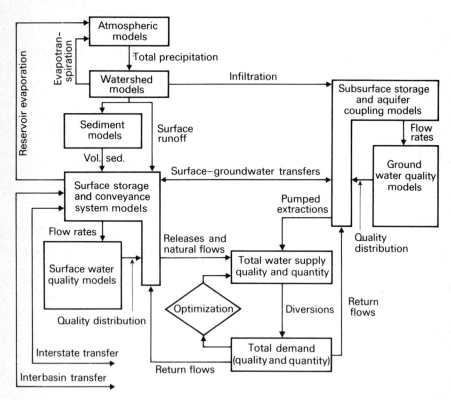

Fig. 1.8 Water-resource systems. (*After Ackermann*, 1969.)

the system. As in the hydrologic cycle, the total system is closed in a hydrologic sense, but open subsystems are recognized and may be isolated from the total environment. Atmospheric models are the domain of atmospheric scientists; watershed models are the domain of the surface-water hydrologist; sediment and surface-storage models are generally dealt with by hydraulic engineers; groundwater models fall within the competence of the groundwater hydrologist. The water quality aspects of the subsystems may attract the sanitary engineer as well as the biologist's interest, whereas the optimization of supply and demand is of sufficient difficulty to attract the attention of engineers, economists, and scientists in operations research. Although water-resource systems operate within the hydrologic cycle, modifying its component parts, the environment outside the physical components of such systems is a complex of constraints involving social, economic, legal, and political factors, and thereby attracts social and political scientists and members of the legal profession. In this sense, man-made constraints must be considered an additional input to a water-resource system.

The complex interdisciplinary interests in water-resource systems suggests that certain subsystems within the hydrologic cycle may easily be optimized to the detriment of others with which they are interconnected. The modern-day approach to water-resource systems recognizes this drawback and fosters the theme that subsystems should complement rather than hinder each other. Operationally, emphasis is placed on the following items: (1) system identification, including the recognition of subsystems to be dealt with and their interaction with other system components; (2) formulation of objectives within the given hydrologic, social, political, and legal constraints that are operative; (3) the development of alternative approaches to achieve these objectives; and (4) comparison of the effectiveness of each alternative on a rational basis. These items constitute much of the subject matter of Chapter 3.

A primary unit in groundwater-resource studies is the aquifer, a lithologic unit or combination of units capable of yielding water to pumped wells or springs. An aquifer may be coextensive with a geologic formation, a group of formations, or a part of a formation; or it may cut across formations so diversely as to be essentially independent of any geologic unit. Units of low permeability that either bound the aquifer or are interstratified with it are referred to as *confining beds*. An aquifer, a confining unit, or a combination of aquifers and confining units that comprise a reasonably distinct source of groundwater forms an aquifer system.

The importance of aquifers as sources of water is reflected in the various classification schemes that have been proposed over the years. The first exhaustive attempt at classification was by Meinzer (1923), who classified the water-yielding capabilities of formations in accordance with geologic age. His classification system does not suggest a statistically significant correlation between age and yield, but merely that knowledge of the succession of formations is essential to an understanding of the circulation of water through them. Meinzer also classified the formations according to 21 physiographic provinces long accepted by the Association of American Geographers. The detailed description of aquifers within physiographic provinces has provided considerable aid to the exploration for groundwater.

Thomas (1951) pointed out that the geology alone is not a sufficient basis for a classification of aquifers because it does not provide information on the problems associated with their development. In a comprehensive study of over 70 areas, he recognized 3 main types of problems:

1. Reservoir problems, or problems in replenishment
2. Pipeline problems, or problems in movement
3. Water-course problems, where groundwater is in hydraulic connection with surface water and the development of one affects the other.

The criteria selected by Thomas that best delineates these problems include the proved ability of sediments to yield water and the manner in which they are replenished. Three types of groundwater areas are shown in Fig. 1.9:

1. Consolidated water-bearing rocks whose recharge areas generally coincide with areas of outcrop
2. Loose water-bearing materials, chiefly sand and gravel, rechargeable directly by precipitation
3. Water courses consisting of a channel occupied by a perennial stream which may act as a source of replenishment.

It is possible to combine and utilize the work of Meinzer (1923) and Thomas (1951) in such a way that at least preliminary information concerning the availability of water can be obtained for any region in the United States. It is known, for example, that the unconsolidated deposits in Fig. 1.9 constitute the largest and most productive aquifers. Of the consolidated rocks, sandstones and conglomerates are the most important, followed by carbonate and volcanic rocks. Crystalline, igneous, and metamorphic rocks are the least productive and only serve as a minor source of water. Further, Fig. 1.9 indicates that only about one-half of the area in the United States is underlain by aquifers capable of yielding moderate to large quantities of water. The comprehensiveness of the data presented by Meinzer (1923), Tolman (1937), and Thomas (1951), and supplemented by the more recent survey of McGuinness (1963), suggests that the main exploratory stage for groundwater in the United States is in the past.

Kazmann (1965) classified aquifers in accordance with their primary function. He recognized:

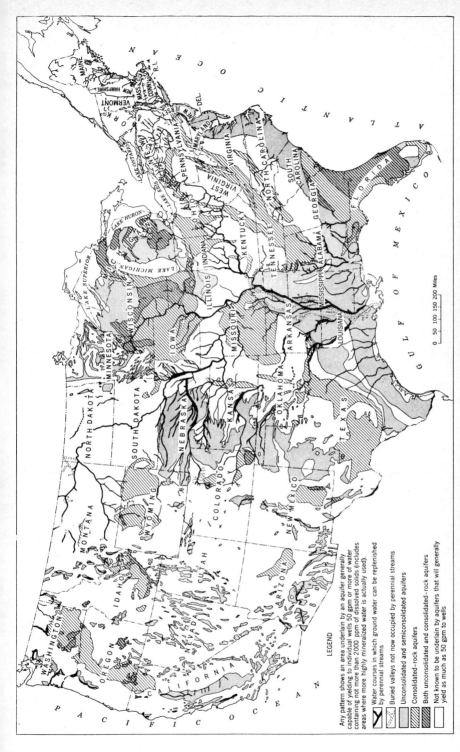

Fig. 1.9 Principal aquifers in the United States. (*After Thomas, 1951, by permission of McGraw-Hill Book Co.*)

1. The aquifer as a filter plant, where the intake of water is important, as in stream-aquifer systems
2. The aquifer as a reservoir, where recharge-discharge relations are significant to its development
3. The aquifer as a mine, where replenishment is small or negligible with respect to pumping

The distinction between filter plants and reservoirs is found in the presence or absence of induced infiltration from streams. The distinction between reservoirs and groundwater mines lies in the nature and rate of recharge.

Hall and Dracup (1967) recognized that aquifers are more than sources of water, but storage reservoirs that require proper management for efficient use. With respect to their management, aquifers may be classified as:

1. Reservoirs to maintain sustained yield, that is, to ensure a balance between inflows and withdrawals
2. Mines that may be exploited at various rates, but ultimately to exhaustion
3. Reservoirs for long-term storage, artificially produced
4. Transmission systems to eliminate or reduce the need for surface pipelines and canals for distribution
5. Energy sources in that pumping lifts are modified by management
6. Water-quality control tools through use of the filtering characteristics of aquifers, especially the artificial recharge of waste water.

This classification refers not only to managed aquifer use, but to manageable resources for alternative and potentially feasible uses. Items 1 and 2 refer to groundwater basins characterized by a lack of adequate surface-water resources. Under these two classes, manageable resources are limited to the annual volume of natural recharge and the minable one-time volume of groundwater storage. The remaining items refer to integrated groundwater–surface-water systems. Manageable resources include underground reservoir space, 3; transmission characteristics, 4; water-level fluctuations, 5; and water-quality aspects, 6.

Items 3 through 6 in the management classification of Hall and Dracup cannot be considered in the absence of artificial recharge. Artificial recharge has been defined by Todd (1959a) as augmenting the natural infiltration of precipitation or surface water into underground formations by some method of construction, spreading of water, or a change in natural conditions. Artificial recharge is often used to (1) replenish depleted supplies, (2) prevent or retard salt-water intrusion, or (3) store water underground where surface-storage facilities are inadequate to supply seasonal demands. California has long been the leader in artificial-recharge operations, generally for the reasons cited in (1) and (2) and, in more recent years, for reason (3). Several methods of getting water underground have been developed, including recharge pits, ponds, and wells and water-spreading grounds. Literature in this field is voluminous and includes the publications of the Ground Water Recharge Center in California, which deal primarily with infiltration and water spreading (Schiff, 1954, 1955; Behnke and Bianchi, 1965); the extensive work of Baumann, dealing primarily with the theoretical aspects of recharge through wells (1952; 1957; 1963; 1965); the experience on Long Island (Brashears, 1946; Johnson, 1948; Cohen and others, 1968); and a comprehensive annotated bibliography prepared by Todd for the United States Geological Survey (1959b). In recent years, scientists in Israel have contributed much theoretical and experimental research in this area (Symposium of Haifa, 1967).

The placement of water underground makes use of geologic formations as storage facilities and is an integral part of the planned utilization of groundwater basins in the coastal plain of Los Angeles County (California Department of Water Resources, 1966) and of the Israel water plan (Buras, 1967). Underground storage will undoubtedly play a large role in the future of other areas. The need for such storage is emphasized not only by the increasing use of water, but by problems of surface storage, transportation, and use of water from all sources to meet future expanding requirements. Underground storage may be an efficient method for meeting these requirements, provided it is accompanied by the coordinated operation of both surface and subsurface supplies. The potential of underground basins as storage facilities is well demonstrated in the San Joaquin Valley, where the storage capacity has been estimated to be nine times the capacity of both presently constructed surface-water reservoirs and those proposed for construction under the California water plan (Davis and others, 1959).

The emphasis on the discharge of groundwater by wells is central to the classification of aquifers. In the late 1950s, this discharge provided about 15 percent of the nation's water demand. At that time, as is currently the case, there was no national water problem, only individual problems distributed throughout the states. One of the problems, groundwater mining, was examined by Bagley (1961), who investigated pumpage in excess of replenishment in western United States. According to Bagley, nearly 60 percent of the total groundwater withdrawals in 1955 were for irrigation in the 17 Western states. California, Arizona, New Mexico, and Texas pumped for irrigation about one-half of the total groundwater withdrawals for all purposes in the United States. California, the largest user of groundwater, had an estimated overdraft, or pumpage in excess of replenishment, equal to one-half of the total groundwater withdrawals in the state. On a nationwide basis, Bagley estimated that over one-fourth of all groundwater withdrawals were mined.

Estimates made of the quantities of water used in 1965 indicate a total

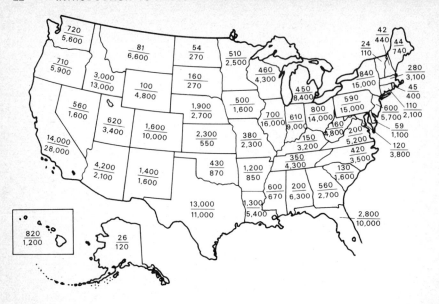

Fig. 1.10 Total water withdrawals by states (excluding that for hydroelectric power). Lower number designates withdrawals from surface-water sources, and upper number designates withdrawals from groundwater sources. (*After Murray*, 1968.)

use of about 310 billion gallons per day (gpd), approximately 20 percent of which was supplied by groundwater (Murray, 1968). Of the total water use, about 54 percent was used by industry, 38 percent for irrigation, 7 percent for public supplies, and 1 percent for rural domestic and stock uses. Figure 1.10 shows the total water withdrawn for individual states.

1.4 MAJOR SUBSYSTEMS AND THEIR INVESTIGATION

Although both surface water and groundwater are interrelated components of one global hydrologic system, the methods and tools applied to the study of each differ considerably. Many modern researchers in surface-water hydrology are interested in precipitation-to-runoff relations and are inclined to treat the watershed as lumped, either deterministically or stochastically. Their interest in the groundwater component of runoff logically centers around the hydrological equation, or some other lumped-parameter description of groundwater flow. On the other hand, many groundwater hydrologists are preoccupied with the details of groundwater flow and are concerned with the groundwater basin as a distributed-parameter system of the deterministic variety. This is a fundamental difference in approach and does not mean that one is good in some sense and the other is bad. Rather, depending largely on the type of problem addressed, one approach is generally favored

INTRODUCTION TO A SYSTEM CONCEPT IN GROUNDWATER HYDROLOGY

over the other because of predictive reliability, information demands, or mathematical tractability. A comparison of these methods should provide a clearer picture of model formulation and a starting point for the study of groundwater as presented in this book.

THE SURFACE–WATER SUBSYSTEM

The study of surface-water hydrology is an important area for practice and research in engineering that has been in a state of flux over the past three or four decades. During this time, practicing surface-water hydrologists have witnessed the transition from an art based largely on unsubstantiated empirical formulas and methods to a full-fledged engineering system designed to arrive at workable relations between pertinent variables of the hydrologic cycle. Current topics of study have been reviewed by Amorocho and Hart (1964) and include physical hydrology and systems investigations (Fig. 1.11). Regardless of which approach is examined, the primary role of the system is that of generating output from input (that is, runoff from precipitation), or of interrelating input and output. The partial achievement of this objective

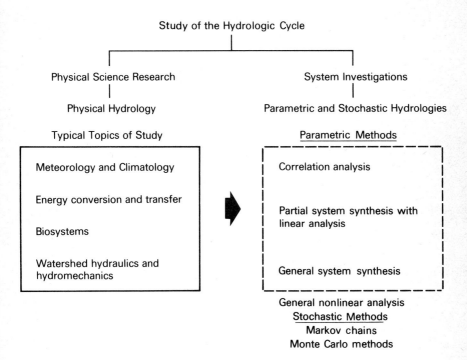

Fig. 1.11 Topics and methods of hydrological study. (*Modified, after Amorocho and Hart*, 1964.)

is due to an improved understanding of the runoff component of the hydrologic cycle on a higher conceptual level than previously thought necessary.

The methodology of physical hydrology is identified as a scientific inquiry into the basic operations of each of the components of the hydrologic cycle for the purpose of obtaining a full understanding of their interaction. The prevailing concept is that output depends on the nature of input, the physical laws involved, and the nature of the system. With the system approach, emphasis is on measurements of pertinent variables and procedures from which explicit algebraic relations may be derived. The complex detail of the physical laws may be ignored in this procedure.

Parametric hydrology has been defined by the Committee on Surfacewater Hydrology of the American Society of Civil Engineers (1963) and paraphrased by Amorocho and Hart (1964) as the development of relations among physical parameters in hydrologic events, and the use of these relations to generate, or synthesize, nonrecorded hydrologic sequences. The authors describe in detail the methods of parametric hydrology and designate by shading those that depend on physical knowledge of the hydrologic cycle (Fig. 1.11). Of the methods cited, only general systems synthesis (simulation) as a method of investigation has a serious parallel in groundwater hydrology, and is one of the better-known methods in surface-water hydrology. This is due to the efforts of Linsley and his co-workers (Linsley and Crawford, 1960; Crawford and Linsley, 1964). As applied to a typical watershed, the synthesis model incorporates assumed or empirical functional relations between components of the hydrologic cycle which explicitly dictate the disposition of input to the system. The basis for the model is the lumped-parameter hydrological equation introduced earlier. As shown in Fig. 1.12, model output (runoff) is compared with actual runoff data. If the agreement is satisfactory, that is, if the model faithfully reproduces output in agreement with actual recorded data, the synthesis is acceptable. If model output and actual runoff data are not compatible, the functional relations between system components are modified until an acceptable correspondence is achieved. The nonuniqueness of this process, as with all attempts at synthesis, is apparent.

Figure 1.13 shows more clearly the workings of the hydrologic cycle in general systems synthesis as opposed to a purely descriptive approach to the physical components of runoff. Input to the system includes average values of precipitation and potential evapotranspiration. Output includes actual evapotranspiration and simulated runoff. Inputs and outputs may be taken over any convenient unit of time, such as monthly, weekly, or hourly. Those processes which require functional relations are clearly depicted. Channel inflow attributed to runoff from the watershed impervious area is generally a simple linear relation between the fraction of impervious area in the watershed (a constant) and rainfall. Precipitated water available after adjustment

INTRODUCTION TO A SYSTEM CONCEPT IN GROUNDWATER HYDROLOGY

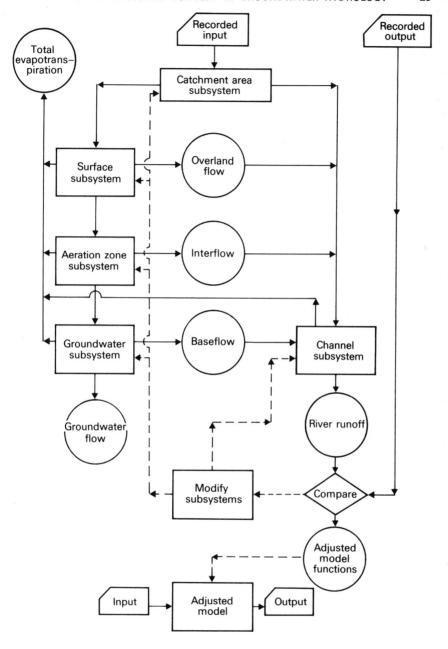

Fig 1.12 Flow chart of a runoff synthesis (*After Amorocho and Hart*, 1964.)

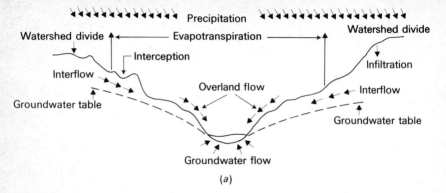

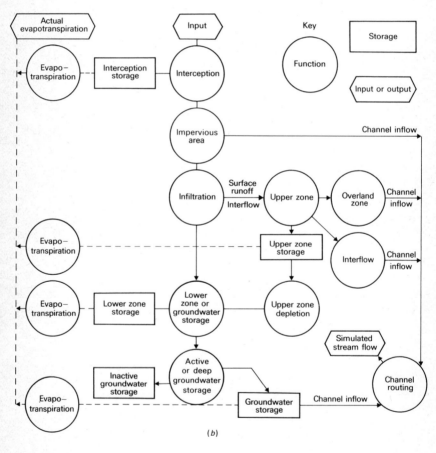

Fig. 1.13 Diagrams showing (a) the physical components of runoff, and (b) a general systems synthesis of runoff. (*Part (b) after Linsley, 1969*.)

for impervious runoff and evapotranspiration is recorded in the upper zone, and becomes available for overland flow, interflow, and upper-zone storage. Infiltration to the lower zone, on the other hand, is a function of the maximum and minimum infiltration rates, the maximum capacity of the lower-zone storage (a constant), and antecedent lower-zone moisture. Groundwater runoff may be a simple relation between some constant determined from subsurface runoff to streams and groundwater storage. Interaction between the processes of infiltration, surface runoff, groundwater storage, etc., is taken into account in arriving at the functional relations. These relations constitute much of the subject matter in conventional courses in surface-water hydrology. Indeed, the flow chart as a functional description of the lumped-parameter hydrologic cycle provides the basis for organization of instructional material in such courses (Linsley, 1969).

The stochastic problem has been described by Amorocho and Hart (1964): Given a historic sequence of events, what inferences can be derived from their statistical distribution so that probabilities of future sequences can be assessed? Stochastic hydrology deals with uncertainty, and attempts to allow for the fact that everything is not known. Of the methods cited in Fig. 1.11, monte carlo may be employed when the events are purely random, that is, when the state (flow) of the system at any given time is independent of the state at any other time and depends only on the underlying probability distribution. Markov chains may be employed when the individual events depend, in part, on previous events, that is, when nature's strategy in the selection of any given state depends on its selection of one or more previous states. In general, there is no serious parallel between these methods and the methods employed to study groundwater in the hydrologic cycle.

Some explanatory remarks treating Markov processes as a class of models between the extremes of deterministic and random models may be instructive at this point (Krumbein, 1967). The synthesis operation shown in Fig. 1.12 is a typical example of a deterministic model. The explicit cause-and-effect relation permits an exact prediction of any state of the system, given a starting state and the underlying descriptive process. For the random model, the state at any point in time is independent of the state at any other time and depends only on the underlying probability distribution. The degree of dependency of a state on a previous state, then, ranges from complete dependency in the case of deterministic models to complete independence in the case of random models. Interpreted in terms of memory of the process, deterministic models have a long memory, extending clear back to the initial conditions, whereas random models have no memory at all. The Markov model is intermediate between these extremes, characterized by a limited memory. The order of the model (first, second, third, etc.) establishes the extent of the memory, or the dependency of the process on preceding states. The element of probability in Markov models, however, precludes exact prediction of future states, in this respect resembling the random model.

Example 1.1 Recorded surface flows for an ephemeral stream in an arid area subject to occasional bursts of rainfall have been classified according to the following states:

State I: 0 units of flow
State II: $>0 < 1{,}000$ units of flow
State III: $>1{,}000 < 2{,}000$ units of flow

Eighteen consecutive periods of runoff have been recorded and are given as follows:

Period	1	2	3	4	5	6	7	8	9	10	11	12	13	14	15	16	17	18
State	I	I	III	II	I	I	I	I	II	I	III	II	I	I	I	III	II	I

If these data are treated as random events, the probabilities of occurrence may be obtained by dividing the number of occurrences of each state by the total number of possible occurrences:

State I: $\frac{11}{18} = 0.611$

State II: $\frac{4}{18} = 0.222$

State III: $\frac{3}{18} = 0.166$

If it is suspected that nature's strategy in the selection of a given state depends on its selection of a previous state, one-step markovian dependencies may be examined by preparing a stochastic matrix. A stochastic matrix is prepared by observing the frequencies of all possible transitions—that is, the number of times a given state i is followed by state j, where i equals 1, 2, 3; j equals 1, 2, 3. As only three states are recognized,

$$\begin{array}{c} \text{To state } j \rightarrow \\ \text{From} \\ \text{state } i \\ \downarrow \end{array} \begin{array}{c} \\ \text{I} \quad \text{II} \quad \text{III} \\ \text{I} \\ \text{II} \\ \text{III} \end{array} \begin{bmatrix} p_{11} & p_{12} & p_{13} \\ p_{21} & p_{22} & p_{23} \\ p_{31} & p_{32} & p_{33} \end{bmatrix} \quad \begin{bmatrix} \text{III} & \text{I} & \text{III} \\ \text{II} & 0 & 0 \\ 0 & \text{III} & 0 \end{bmatrix} \quad \begin{bmatrix} 0.6 & 0.1 & 0.3 \\ 1.0 & 0 & 0 \\ 0 & 1.0 & 0 \end{bmatrix}$$

where the first matrix designates the probability of a transition to state j during the following time period, given that the system now occupies state i; the second matrix records such transitions, as determined from the data above; and the third matrix gives the transition probabilities, which are found by dividing the number of transitions to a given state j by the total number of transitions out of a given state i. For example, the first row is obtained by noting that there were 10 transitions out of state I, or 6/10, 1/10, 3/10. Hence, the probability of a transition to state I, II, or III, given that the system now occupies state I, is 0.6, 0.1, and 0.3, respectively.

INTRODUCTION TO A SYSTEM CONCEPT IN GROUNDWATER HYDROLOGY

THE GROUNDWATER SUBSYSTEM

An appreciation of the value of operating within a systematic classification of models and methods such as presented in Fig. 1.11 is not lacking in hydrologists concerned with groundwater. However, a classification system based on statistical methods and models is only necessary when the interpretation of data depends as much upon the character of the model adopted as upon the quality of the data itself (Chorley, 1962). In general, the concept of determinism is so deeply ingrained in groundwater hydrology that the uncertainty aspects of prediction are seldom raised.

In spite of these differences, the physical and systems approach discussed

Physical Science Research

Physical Hydrology

Topics of Interest

- Influence of geology on an enclosed dynamic fluid
- Geologic occurrence and exploration for groundwater
- Water quality and geochemical investigations
- Macroscopic evaluation of hydrologic and simple medium properties

Systems Investigations

Lumped and Distributed Parameter Hydrology

Lumped – Parameter Systems

Hydrologic Models
- Application of the hydrologic equation
- Recession characteristics
- Regression analysis
- Information theory and hydrologic variables

Optimization Models
- Strategies for isolated groundwater basins and related problems in valuation
- Mathematical programming

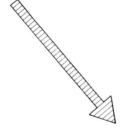

Distributed – Parameter Systems

- Regional (steady) groundwater flow and its relationship with other system properties, including distribution of chemical constituents, temperature distribution, and surface manifestations of flow.

- Unsteady flow in response to external excitation, including flow to wells, consolidation theory, and simulation of aquifers.

Fig. 1.14 Investigations in groundwater hydrology.

by Amorocho and Hart (1964) may be applied to groundwater research in general. A topic of interest in surface-water hydrology is the relation between precipitation and runoff; a parallel lumped-parameter topic in groundwater hydrology might be the relation between precipitation and recharge. Physical science research has its counterpart in groundwater research. Geology is the main theme here, with emphasis placed on determining the influence of geology on an enclosed fluid in motion. Main topics of interest include the geologic occurrence of water, water quality, and evaluation and tabulation of hydrologic and simple medium properties (Fig. 1.14). The results of such investigations are very often of a descriptive nature.

The terms *parametric* and *stochastic* have little or no meaning that can be substantiated by the main stream of literature in groundwater hydrology. On the other hand, for classification purposes, the terms *lumped-* and *distributed-parameter* appear broad enough to cover most of the quantitative methods of interest (Fig. 1.14). The arrow in Fig. 1.14 indicates that investigation of problems of the distributed-parameter type depends a great deal on knowledge of geology, whereas geology plays a relatively unimportant role in lumped-parameter investigations.

Finally, in favorable comparison with Fig. 1.11, one aspect of the science emphasizes topics of study, whereas another deals with methods and models.

1.5 ORGANIZATION AND SCOPE OF THE BOOK

The areas of groundwater investigations as given in Fig. 1.14 make up the main theme of this book. Chapters 2 and 3 deal with the groundwater basin as a lumped-parameter system. Chapter 2 is a brief treatment of some conventional input-output methods, and some basic ideas organized around the concept of point measurements over time in basins. Chapter 3 is concerned with the systematic determination of optimal strategies for isolated groundwater basins and integrated water-resource systems. In this chapter, the physical system establishes the physical constraints within which a set of economic operating rules may function. This point of view appears appropriate in that increasing water demand often transforms physical problems into economic problems. The thesis that the pertinent response variables are most often treated in lumped form is established early in the chapter, and demonstrates that an optimization problem, so conceived, does not constitute a field problem. Once this becomes fully understood, it is possible to deal with the optimization problem on a completely different level than the field problem. Emphasis is on the goals of management and the derivation of optimal decision rules and strategies that reflect these goals, notwithstanding the lumped-system response.

Implicit in the idea of a distributed-parameter system is the realization that the mathematical problem must be formulated in partial differential equations, with the space variables x, y, z, and the time variable t designated as independent variables. Chapters 4 and 5 are concerned primarily with the identification and interpretation of system parameters and their formulation in partial differential equations. In addition, as this presentation deals with the main theme of energy and mass conservation laws, opportunities arise to examine related theories and phenomena whose fundamentals share this same theme. Hence, discussions on two-phase systems and chemical thermodynamics may be appropriately considered under the main banner of conservation principles. It is hoped that this approach to the subject helps in tying certain theories with the principles which they share in common, and does not distract too much from the main purpose of the chapters.

Chapters 6 and 7 deal exclusively with distributed-parameter systems. Chapter 6 attempts to describe a flow system through a chosen dynamic property, and to relate this property to other system parameters. The chosen property is groundwater potential, and related system properties include chemical character, temperature, and surface manifestations of flow. The system is assumed to be in a steady state, and Laplace's equation applies. Chapter 7 deals with the individual unit within a groundwater basin, its compressibility, and its suitability for water supply. Topics dealt with include well flow, consolidation theory, and simulation. The system is assumed to be in an unsteady state, and the diffusion equation applies. Geology is recognized and treated as an important input to the study of distributed-parameter systems.

PROBLEMS AND DISCUSSION QUESTIONS

1.1 Open systems have been described as adaptive, i.e., capable of conforming to environmental jolts by attempting to achieve a new dynamic equilibrium. Discuss the manner in which this adjustment manifests itself in the following readings:

a. Theis, C. V.: The significance and nature of a cone of depression in groundwater bodies, *Econ. Geol.*, vol. 38, pp. 889–902, 1938 (or C. V. Theis: The source of water derived from wells—Essential factors controlling the response of an aquifer to development, *Civ. Eng. (N.Y.)*, pub. by Amer. Soc. Civ. Engrs., pp. 277–280, May, 1940).

b. Banks, H. O., and R. C. Richter: Sea water intrusion into groundwater basins bordering the California coast and inland bays, *Trans. Amer. Geophys. Union*, vol. 34, pp. 575–582, 1953.

c. Kohout, F. A., and H. Klein: Effect of pulse recharge on the zone of diffusion in the Biscayne aquifer, *Int. Ass. Sci. Hydrol.*, Symp. Haifa, Publ. 72, pp. 252–270, 1967.

d. Poland, J. F., and G. H. Davis: Subsidence of the land surface in Tulare-Wasco (Delano) and Los Banos–Kettleman City areas, San Joaquin Valley, California, *Trans. Amer. Geophys. Union*, vol. 37, pp. 287–296, 1956.

1.2 From Fig. 1.4, explain how input for a given time period (such as precipitation) may be output from a preceding time period (such as evapotranspiration).

1.3 Consider the following moisture-accounting procedure for simulation in accordance with the schematic of Fig. P1.3. The upper zone corresponds to water on the surface, the bottom zone to the soil horizon, and the groundwater zone to water below the water table.

Constants

C_{imp} = fraction of impervious area

UZ_{max} = maximum upper-zone moisture capacity

BZ_{max} = maximum bottom-zone moisture capacity

F_{max} = maximum infiltration rate of the soil

F_{min} = minimum infiltration rate of soil = zero

CGW = groundwater discharge constant

K = time period

Functions

$P_{net\,1} = RF - Q_{imp}$

$Q_{imp} = C_{imp} RF(K)$

$UZ(K) = UZ(K-1) + P_{net\,1} - PET(K)$, where PET = potential evapotranspiration, or amount of water loss from a surface if water is available in unlimited supply

$F = F_{max} - \dfrac{F_{max} - F_{min}}{BZ_{max}} BZ(K-1)$

If $F \geq P_{net\,2}$, then $Q_{over} = 0$.

If $F < P_{net\,2}$, then $Q_{over} = P_{net\,2} - F$.

$P_{net\,3} = P_{net\,2} - Q_{over}$

$AET(BZ) = \dfrac{BZ(K-1)}{BZ_{max}} PET$

$P_{net\,4} = P_{net\,3} \dfrac{BZ(K-1)}{BZ_{max}}$

$QGW = CGW\ GWS(K-1)$

a. Explain the reasoning or logic for each of the functions cited above.
b. Under what conditions will the infiltration rate and actual evapotranspiration from bottom-zone storage take on their maximum and minimum rates? Demonstrate these functions by simple linear graphs.
c. Explain the logic of the following operations, which describe the runoff process (simplified) when the moisture available to the upper zone is not sufficient to satisfy the demands of potential evapotranspiration:

   (a) $Q_{imp} = C_{imp} RF(K)$
   (b) $P_{net\,1} = RF(K) - Q_{imp}$

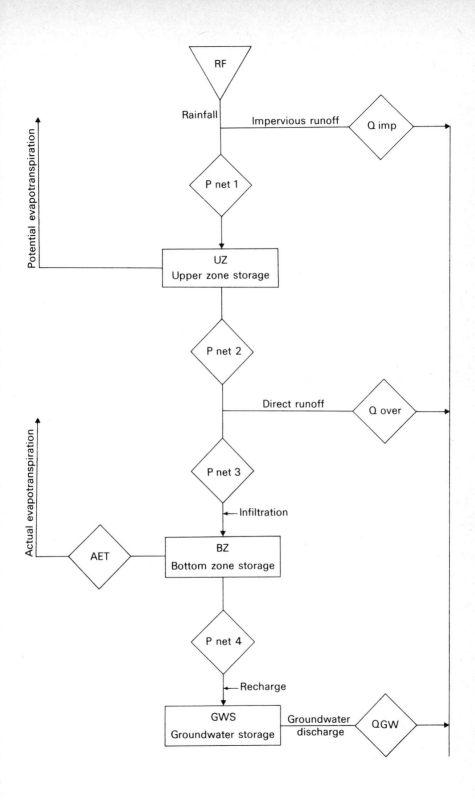

Statement: $UZ(K) = UZ(K-1) + P_{net\,1} - PET(K)$
If $UZ(K)$ is negative, $PET(K) > [UZ(K-1) + P_{net\,1}]$.

(c) $AET(UZ) = UZ(K-1) + P_{net\,1}$

(d) $UZ(K) = 0$

(e) $Q_{over} = 0$

Statement: $AET(BZ) = \dfrac{BZ(K-1)}{BZ_{max}} PET(K)$

If $AET(BZ) \geq PET(K) - [UZ(K-1) + P_{net\,1}]$, then $AET(BZ) = PET(K) - [UZ(K-1) + P_{net\,1}]$.
If $AET(BZ) < PET(K) - [UZ(K-1) + P_{net\,1}]$, then $AET(BZ) = \dfrac{BZ(K-1)}{BZ_{max}} PET(K)$.

(f) $BZ(K) = BZ(K-1) - AET(BZ)$

(g) $P_{net\,4} = 0$

(h) $QGW(K) = CGW\ GWS(K-1)$

(i) $GWS(K) = GWS(K-1) - QGW(K)$

(j) $Q_{total} = Q_{imp} + QGW(K)$

d. Explain the logic of the following operations, which describe the runoff process (simplified) when the moisture available to the upper zone is just sufficient to satisfy the demands of potential evapotranspiration:

(a) $Q_{imp} = C_{imp} RF(K)$

(b) $P_{net\,1} = RF(K) - Q_{imp}$

Statement: $UZ(K) = UZ(K-1) + P_{net\,1} - PET(K)$
If $UZ(K) = 0$, then $PET(K) = UZ(K-1) + P_{net\,1}$.

(c) $AET(UZ) = PET(K)$

(d) $UZ(K) = 0$

(e) $Q_{over} = 0$

(f) $BZ(K) = BZ(K-1)$

(g) $QGW(K) = CGW\ GWS(K-1)$

(h) $GWS(K) = GWS(K-1) - QGW(K)$

(i) $Q_{total} = Q_{imp} + QGW(K)$

e. Explain the logic of the following operations, which describe the runoff process (simplified) when the moisture available to the upper zone is more than sufficient to satisfy the demands of potential evapotranspiration:

(a) $Q_{imp} = C_{imp}\ RF(K)$

(b) $P_{net\,1} = RF(K) - Q_{imp}$

Statement: $UZ(K) = UZ(K-1) + P_{net\,1} - PET(K)$
If $UZ(K)$ is positive, then $[UZ(K-1) + P_{net\,1}] > PET(K)$.

INTRODUCTION TO A SYSTEM CONCEPT IN GROUNDWATER HYDROLOGY

Statement: $P_{net\,2} = UZ(K) - UZ_{max}$
If $P_{net\,2}$ is negative, or zero:

(c) $UZ(K) = UZ(K)$

(d) Continue with items (e) through (i) in part d.

If $P_{net\,2}$ is positive:

(e) $\qquad UZ(K) = UZ_{max}$

If $F \geq P_{net\,2}$:

(f) $Q_{over} = 0$

(g) $P_{net\,3} = P_{net\,2}$

If $F < P_{net\,2}$:

(f) $Q_{over} = P_{net\,2} - F$

(g) $P_{net\,3} = P_{net\,2} - Q_{over}$

(h) $\qquad P_{net\,4} = P_{net\,3} \dfrac{BZ(K-1)}{BZ_{max}}$

(i) $BZ(K) = BZ(K-1) + P_{net\,3} - P_{net\,4}$

(j) $QGW(K) = CGW\ GWS(K-1)$

(k) $GWS(K) = GWS(K-1) + P_{net\,4} - QGW(K)$

(l) $Q_{total} = Q_{imp} + Q_{over} + QGW(K)$

f. *Class project:* Formulate the simplified runoff process as described in c, d, and e for instructions to be given to a digital computer.

1.4 Two common parameters used to measure the degree of urban development in a watershed are the percentage of land occupied by impervious cover, and the percentage of the area serviced by sewers (L. B. Leopold: Hydrology for urban planning—A guidebook on the hydrologic effects of urban land use, *U.S. Geol. Surv. Circ.* 554, 1968). Discuss the most likely effects of each of these parameters on the relation between rainfall and runoff and rainfall and recharge as you now understand them. How can simple models such as described in Problem 1.3 be used to study the effects of watershed alteration?

1.5 Do you detect any general relation between climate (arid, semiarid, humid) and the ratio of groundwater to surface-water use as depicted in Fig. 1.10? With water pollution and increased industrialization in eastern United States, do you envision any long-term changes in this ratio?

1.6 Using the paper by Meinzer (1923) or the book by Tolman (1937), ascertain the main aquifers in your state. Using the paper by McGuinness (1963), determine the major problems and problem areas. Of the various ways to classify aquifers, which most nearly reflects the problem areas you discover?

1.7 Prepare a flow chart of similar form to that in Fig. 1.12 with precipitation as recorded input, but with:

(a) Spring discharge as the simulated (recorded) output
(b) Groundwater storage as the simulated (recorded) output

If you were to model these systems with a digital computer, what significant functional relations would be required?

1.8 a. Comment on the reason that a summation of each of the row values of the stochastic matrix given in Example 1.1 always equals 1.

b. From this matrix, is there any reason to suspect the partial dependency of a given state supplied almost completely by groundwater discharge on a previous state consisting largely of surface runoff? Explain.

c. A water-well driller is continually submitting bids for contract awards. We recognize two states, a successful bid (state I) and an unsuccessful bid (state II). The consecutive appearance of these states for the past few years is as follows:

State: I, I, II, II, I, II, I, II, I, II, I, II, I, II, II, I, I, II, I, II

Is the occurrence of a given state random in nature, or do you suspect some dependencies on a previous state? Explain.

d. A certain amount of uncertainty is involved in predicting any future flow. Which of the models of Example 1.1 most effectively reduces at least part of this uncertainty? Explain.

REFERENCES

Ackermann, W. C.: Scientific hydrology in the United States, The Progress of Hydrology, *Intern. Seminar for Hydrol. Professors, 1st,* Univ. of Illinois, Champaign, pp. 563–571, 1969.

American Society of Civil Engineers, Hydraulics Division: Communication of the committee on surface-water hydrology, 1963.

Amorocho, J., and W. E. Hart: A critique of current methods in hydrologic systems investigations, *Trans. Amer. Geophys. Union,* vol. 45, no. 2, pp. 307–321, 1964.

Ashby, W. R., Defining the field of systems engineering—its relationship to the biological and medical sciences, *Proc. Natl. Symp. Impact Systems Eng. Concepts Eng. Ed. Res.,* The Foundation for Instrumentation, Education, and Research, Inc., New York, pp. 25–27, 1961.

Bagley, E. S.: Water rights law and public policies relating to groundwater "mining" in the southwestern states, *J. Law and Econ.,* vol. 4, pp. 144–174, 1961.

Baumann, P.: Groundwater movement controlled through spreading, *Trans. Amer. Soc. Civil Engrs.,* vol. 117, pp. 1024–1060, 1952.

———: Basin recharge, in Groundwater Development, A symposium, *Trans. Am. Soc. Civil Engrs.,* vol. 122, pp. 458–473, 1957.

———: Theoretical and practical aspects of well recharge, *Trans. Am. Soc. Civil Engrs.,* vol. 128, pp. 739–764, 1963.

———: Technical development in groundwater recharge, *Advan. Hydrosc.,* ed. by V. T. Chow, vol. 2, pp. 209–279, 1965.

Behnke, J., and W. Bianchi: Pressure distributions in layered sand columns during transient and steady-state flows, *Water Resources Res.,* vol. 1, no. 4, pp. 557–562, 1965.

Blaney, H. F.: Consumptive use of water, *Trans. Amer. Soc. Civil Engrs.,* vol. 117, pp. 949–973, 1952.

Brashears, M. L.: Artificial recharge of groundwater on Long Island, New York, *Econ. Geol.,* vol. 41, pp. 503–516, 1946.

Buras, N.: Systems engineering and aquifer management, *Intern. Assoc. Sci. Hydrol., Symp. Haifa,* Publ. 72, pp. 466–473, 1967.

California Department of Water Resources: Planned utilization of groundwater basins: Coastal plain of Los Angeles County, Appendix C, Operations and economics, 1966.

Chen, C. L., and V. T. Chow: Hydrodynamics of mathematically simulated surface runoff, *Hydraul. Eng. Series,* no. 8, Dept. Civil Eng., Univ. Illinois, 1968.

Chorley, R. J.: Geomorphology and general systems theory, *U.S. Geol. Surv., Profess. Papers* 500-B, 1962.

Cohen, P., O. Frank, and B. Foxworthy: An atlas of Long Island's water resources, *N.Y. State Water Resources Comm. Bull.* GW62, 1968.

Crawford, N. H., and R. K. Linsley: A conceptual model of the hydrologic cycle, *Intern. Assoc. Sci. Hydrol., Publ.* 63, pp. 573–587, 1964.

Davis, G. H., and others: Groundwater conditions and storage capacity in the San Joaquin Valley, California, *U.S. Geol. Surv., Water Supply Papers* 1469, 1959.

Distefano, J. J., A. R. Stubberud, and I. J. Williams: "Theory and Problems of Feedback and Control Systems," Schaum's Outline Series, McGraw-Hill Book Company, New York, 1967.

Dooge, J. C. I.: The hydrologic cycle as a closed system, in *Proc. Intern. Hydrol. Symp., Fort Collins, Colo.,* pp. 98–113, Sept. 6–8, 1967.

Duckstein, L., and C. C. Kisiel: General systems approach to groundwater problems, *Proc. Natl. Symp. Analysis Water Resource Systems, Denver, Colo.,* Amer. Water Resources Assoc., pp. 100–115, 1968.

Hall, W. A., and J. A. Dracup: The optimum management of groundwater resources, Paper presented at Intern. Conf. Water for Peace, Washington, D.C., May 23–31, 1967.

Johnson, A. H.: Groundwater recharge on Long Island, *J. Amer. Water Works Assoc.,* vol. 49, no. 11, pp. 1159–1166, 1948.

Karplus, W. J.: "Analog Simulation," McGraw-Hill Book Company, New York, 1958.

Kazmann, R. G.: "Modern Hydrology," Harper and Row, Publishers, Incorporated, New York, 1965.

Krumbein, W. C.: Fortran IV computer programs for Markov chain experiments in geology, Computer Contribution 16, *State Geol. Surv. Kansas, Univ. Kansas Publ.,* Lawrence, 1967.

Leopold, L. B., and W. B. Langbein: The concept of entropy in landscape evolution, *U.S. Geol. Surv., Profess. Papers* 500-A, 1962.

Linsley, Jr., R. K.: Digital simulation in hydrology and its role in hydrologic education, *Prog. Hydrol., Intern. Seminar Hydrol. Professors, 1st,* Univ. Illinois, Champaign, pp. 563–571, 1969.

———, M. A. Kohler, and J. L. H. Paulhus: "Hydrology for Engineers," McGraw-Hill Book Company, New York, 1958.

——— and N. H. Crawford, Computation of a synthetic streamflow record on a digital computer, *Intern. Assoc. Sci. Hydrol.,* Publ. 51, pp. 526–538, 1960.

McGuinness, C. L.: The role of groundwater in the National Water Situation, *U.S. Geol. Surv. Water Supply Papers,* 1800, 1963.

Meinzer, O. E.: The occurrence of groundwater in the United States, with a discussion of principles, *U.S. Geol. Surv., Water Supply Papers,* 489, 1923.

Meyboom, P.: Current trends in hydrogeology, *Earth Sci. Revs.,* vol. 2, no. 4, pp. 345–364, 1966.

Murray, C. R.: Estimated use of water in the United States: *U.S. Geol. Surv., Circ.* 556, 1968.

Penman, H. L.: Estimating evaporation, *Trans. Amer. Geophys. Union,* vol. 37, pp. 43–46, February, 1956.

Prigogine, I.: "Introduction to Thermodynamics of Irreversible Processes," Interscience Publishers, a division of John Wiley & Sons, Inc., New York, 1967.

Schiff, L.: Water spreading for storage underground, *Agr. Eng.,* vol. 35, pp. 794–800, 1954.

———: The status of water spreading for groundwater replenishment, *Trans. Amer. Geophys. Union,* vol. 36, pp. 1009–1020, 1955.

Streeter, V. L.: "Fluid Mechanics," 4th ed., McGraw-Hill Book Company, New York, 1966.

Symposium of Haifa, *Intern. Assoc. Sci. Hydrol.*, Publ. 72, 1967.
Theis, C. V.: The significance and nature of a cone of depression in groundwater bodies, *Econ. Geol.*, vol. 33, pp. 889–902, 1938.
――――: The source of water derived from wells—Essential factors controlling the response of an aquifer to development, *Civil Eng. (N.Y.)*, Publ. Amer. Soc. Civil Engrs., pp. 277–280, May, 1940.
Thomas, H. E.: "The Conservation of Groundwater," McGraw-Hill Book Company, New York, 1951.
Thornthwaite, C. W.: An approach toward a rational classification of climate, *Geograph. Rev.*, vol. 38, pp. 55–94, 1948.
Todd, D. K.: "Groundwater Hydrology," John Wiley & Sons, Inc., New York, 1959a.
――――: Annotated bibliography on artificial recharge of groundwater through 1954, *U.S. Geol. Surv., Water Supply Papers,* 1477, 1959b.
Tolman, C. F.: "Ground Water," McGraw-Hill Book Company, New York, 1937.
Von Bertalanffy, L.: The theory of open systems in physics and biology, *Science,* vol. 111, pp. 23–29, 1950.
Wisler, C. O., and E. F. Brater: "Hydrology," John Wiley & Sons, Inc., New York, 1949.

part one
The Groundwater Basin as a Lumped-parameter System

2
Hydrologic Models

The lumped approximation of the characteristics of a watershed was pointed out as being suited for the investigation and solution of many problems in surface-water hydrology. A similar case may be made to justify certain investigations of groundwater in the hydrologic cycle, especially when dealing with inflow-outflow-storage relations. These relations are particularly difficult to describe because of the dampening effect of storage on inflow, the lag, or delay, between the time water enters the system and the time this same water exits, the variable rates and diffuse manner by which inflow and outflow can occur, and the heterogeneity of the geology. A quantitative analysis is facilitated by considering the whole groundwater basin as occupying a single point in space. In this manner, the spatial variations in the pertinent parameters may be ignored.

The description of any lumped-parameter system for analytic evaluation involves specification of the following items:

1. The nature of input
2. The nature of output
3. A descriptive model relating inputs, outputs, and system states, in time

This chapter introduces the reader to a few techniques that may be used to relate input and output in hydrologic studies. The field tasks that require attention generally fall into one of the following two classes:

1. Ascertain the average value of some hydrologic variable or variables at one or several points in a basin for several values of time.
2. Ascertain the average value of some hydrologic variable or variables at one or several points in an assemblage of basins for several values of time.

Data of the type described in 1 are generally employed in conventional inventory studies, such as those dealing with yield, recharge, and recession characteristics, all of which are reasonably straightforward. Data of the type described in 2 may be utilized in regression equations to determine the degree that each of several independent variables is responsible for variations in a given dependent variable. Data of the type described in 1 or 2 may be employed as input to models developed in modern communication theory, which deal with uncertainty and its measurement.

2.1 CONVENTIONAL INPUT–OUTPUT ANALYSIS

Inputs and outputs for conventional hydrological models are generally water volumes per unit time, such as recharge and discharge and surface inflows and outflows. The fundamental idea common to the variety of situations to be discussed is that the hydrological equation or some other simple equation, empirically derived, is usually employed. For purposes of this section, the hydrological equation is given in general terms,

$$I(t) - O(t) = \frac{dS}{dt} \qquad (2.1)$$

Several applications are possible:

1. The term *inventory* is generally reserved for investigations in which a detailed accounting of inflow, storage, and outflow is attempted for time intervals, such as years or other units of time, during a period of observation (Meinzer, 1932). As noted earlier, inclusion of the time element is required in that the groundwater subsystem is not closed and can only be treated as such if an accounting is taken of inflow and outflow over a period of record. In the Pomperaug basin, Connecticut, for example, available data were organized on a monthly basis for a 3-year period (Meinzer and Stearns, 1928). The hydrologic equation is the basis for such studies.
2. Where inflow terms are balanced by outflow terms, the change in storage

is zero. This provides the necessary conditions to arrive at safe yield estimates and to predict recharge from precipitation.
3. When outflow occurs in the absence of inflow, a general recession model may be formulated. This permits an evaluation of the outflow quantities, the effects on groundwater storage, or the inflow that takes place following the recession.

YIELD ANALYSIS

The objective of many groundwater-resource studies is the determination of how much water is available for pumping, that is, determination of the maximum possible pumping compatible with stability of the groundwater supply. The term *safe yield* as an indicator of this maximum use rate has had an interesting evolution since first introduced. Lee (1915) first defined safe yield as

> ... the limit to the quantity of water which can be withdrawn regularly and permanently without dangerous depletion of the storage reserve.

Meinzer (1923) defined safe yield as

> ... the rate at which water can be withdrawn from an aquifer for human use without depleting the supply to the extent that withdrawal at this rate is no longer economically feasible.

Hence, the "dangerous depletion" spoken of by Lee takes on economic overtones in Meinzer's definition, and both speak of permanency of withdrawals.

Meinzer's definition was expanded by Conkling (1946), who described safe yield as an annual extraction of water which does not:

1. Exceed average annual recharge
2. Lower the water table so that the permissible cost of pumping is exceeded.
3. Lower the water table so as to permit intrusion of water of undesirable quality.

A fourth condition, the protection of existing rights, was added by Banks (1953).

The single-valued concept of safe yield, as now understood, ambiguously encompasses hydrologic, economic, quality, and legal considerations. The controversial nature of the concept is clearly demonstrated in 43 pages of discussion of Conkling's (1946) 28-page paper, by no less than 10 authorities. Todd's (1959) compact definition of safe yield as

> ... the amount of water which can be withdrawn from (a groundwater basin) annually without producing an undesired result

adds nothing to clarify the situation in that the "undesired results" include concern for available water supply, economics of pumping, quality, and water rights. Thomas (1951) and Kazmann (1956) have suggested abandonment of the term because of its indefiniteness.

Methods of determining safe yield are generally based on conservation considerations expressed in the hydrologic equation and include (Conkling, 1946; Todd, 1959) (1) the Hill method, which is merely a plot of annual pumping versus average water-level change, allowing identification of the pumping draft associated with zero water-level change (Fig. 2.1a); (2) the Harding method, which is a plot of retained flow (surface inflow minus outflow) versus average water-level change, the zero-change in water level again designating safe yield (Fig. 2.1b); and (3) the zero water-level change method, which is based on the premise that if the groundwater storage elevation is the same at the beginning and end of a long period of pumping, the average net draft over this period is an estimate of safe yield (Fig. 2.1c).

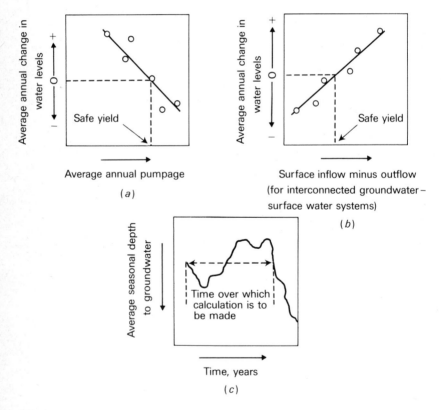

Fig. 2.1 Safe-yield determinations by (a) the Hill method, (b) the Harding method, and (c) the zero water-level change method.

HYDROLOGIC MODELS

It follows the safe yield is interpreted as that part of basin replenishment that finds its way to pumping wells such that perennial pumping at this rate results in an annual zero net change in water levels. In principle, a satisfactory method would rank all possible pumping rates in accordance with the criterion of zero net water-level change, selecting the rate that satisfies this objective. In practice, such an approach is impossible in that safe yield has no unique or constant value, its value at any time depending on the spacing and location of wells and their influence on the dynamics of interchange between groundwater and other elements of the hydrologic cycle. That is to say, the variables in question are evaluated on historical data, and any balance which exists is valid only for the period of record examined. A withdrawal in excess of the pumpage recorded during the period of record will, in general, increase inflow, or decrease outflow, and so will lead to a new equilibrium at a lower storage level. Thus, methods proposed for determination of safe yield typically reflect the framing of a difficult problem of the distributed type (distribution of wells, physical properties, areas of recharge and discharge) into one of a lumped nature. In concept, therefore, the idea of a safe yield encompasses a great deal more and is considerably more sophisticated than the methods proposed to ascertain its value. Optimal yield, a more dynamic concept, is discussed in Chapter 3.

PRECIPITATION–RECHARGE RELATIONS

Long-term fluctuations of water levels in wells in response to changes in precipitation have been observed in several areas (Gleason, 1942; Fischel, 1956). In attempts to correlate the observed relations, departures from average water levels have been compared with cumulative departures from average precipitation (Fig. 2.2). The water-level departure curve differs from the actual hydrograph in that it indicates the relative position of the water level at any time, the zero line giving the long-term average. The distance above or below the line gives the distance that the water level is above or below its average position and is often a better indicator of recharge from precipitation than the actual hydrograph. For example, the period of deficient precipitation from August, 1932, to March, 1933, is accompanied by a rise in water levels, as demonstrated by the hydrograph, but a decline in the departure curve. Hence, the rise was less than the average over this period (Wenzel, 1936).

Jacob (1943) utilized the idea of cumulating departures from a moving rather than a constant average precipitation and compared this result with an average water-level graph. As groundwater levels are a function of precipitation during the preceding few years rather than precipitation during still earlier years (Leggette, 1942), moving averages have the property of progressively eliminating the weight given to early time periods.

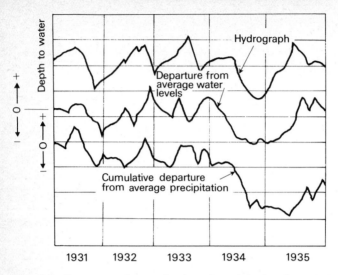

Fig 2.2 Departure of water levels and precipitation from their average values. (*After Wenzel*, 1936.)

Precipitation-recharge models have been used to estimate recharge from precipitation and water-level data, the estimated value used in safe-yield computations (Hantush, 1957; Malmberg, 1965). The method centers on ascertaining exactly what periods of record a given basin is in a state of dynamic equilibrium, which reduces the hydrologic equation to

$$\text{Inflow} = \text{outflow} = X + Q_o \tag{2.2}$$

where X is pumping, Q_o is natural discharge, and the lumped or average change in groundwater storage is zero for the period examined. This occurs when declines in mean annual water levels in certain parts of a basin are offset by rises in mean annual water levels in other parts of the basin. Under these conditions, inflow to the basin may be described by

$$\text{Inflow} = C(\bar{R}_{N_n}) \tag{2.3}$$

where $\bar{R}_{N_n}$ is the effective average rate of precipitation, and C is a function of the extent of the recharge area and the percentage of effective rainfall that reaches the water table.

The effective average rate of precipitation was defined by Jacob (1943) as

> ... the rate of precipitation which, had it been maintained throughout the past, would have produced the same water table profile as actually existed at that particular time.

HYDROLOGIC MODELS

The value of effective average rate of precipitation is computed by

$$\bar{R}_{N_n} = \sum_{i=1}^{k} \frac{2(k + 1 - i)}{k(k + 1)} R_{n+1-i} \tag{2.4}$$

where $\bar{R}_N$ equals the effective average precipitation at the end of the nth period, R equals the precipitation during any period, and k equals the number of periods the rainfall of a given period is effective. The value of k is found empirically by examining well hydrographs. For example, hydrographs of observations wells at the edge of the recharge area in the Roswell Basin show that after heavy rains in 1941, the water levels continued to rise through 1943 despite a decrease in annual rainfall and increase in pumping during the years 1942 and 1943 (Hantush, 1957). Hence, if k is found to equal 3 years, and a given year (say, 1935) appears to be one in which dynamic equilibrium conditions prevailed, the 3-year effective average rate of precipitation is

$$\bar{R}_{N_{35}} = \frac{R_{33}}{2} + \frac{R_{34}}{3} + \frac{R_{35}}{6} \tag{2.5}$$

For this condition

$$C\bar{R}_{N_n} = X + Q_o \tag{2.6}$$

It follows that three or more values of $\bar{R}_{N_n}$ determined under conditions of steady flow can be plotted against pumpage for the corresponding years to yield a straight line, provided natural discharge is the same for these periods. The slope of the line $\Delta X/\Delta \bar{R}_{N_n}$ gives the value of C, and the intercept of the line on the $\bar{R}_{N_n}$ scale gives the value of Q_o/C. Hence, this black-box method precludes direct knowledge of the percentage of precipitation that reaches the water table and the extent of the recharge area, these factors being already incorporated in the empirical constant which is found indirectly.

Jacob (1944) gave an extended application of this method for the purpose of correlating average groundwater levels and precipitation.

The emphasis in recent years given to stochastic methods in surface-water studies suggests that the renewability of the hydrologic cycle may be of a random character, and one may question the validity of a deterministic approach to recharge and the significance of average values. Chapman (1964) has examined this question, and concluded that the importance of random or deterministic input for groundwater basins (at least as far as their exploitation is concerned) depends largely on the relation between storage volume and inflows. According to Chapman, when the volume of water in storage is considerably greater than annual (or seasonal) replenishment, the net effect of variation in replenishment is negligible. Chapman quantified this relation by a storage/flow ratio, which he defined as the ratio of storage upstream from a cross-sectional area to the flow past the section. The ratio has the units of time. If the ratio is large, random fluctuations in recharge are dampened, and specific problems can be analyzed with regular inputs and outputs. A ratio of at least 50 years in arid areas and considerably less in humid areas is suggested as sufficient for this purpose. Where the storage/flow ratio is insufficient to provide an adequate stabilizing influence, the problem must be treated as one with irregular inputs and outputs.

RECESSION CHARACTERISTICS: A GENERAL EXPONENTIAL MODEL

The term *recession* refers to the decline of natural outflow from a system in response to the absence of inflow and is assumed or known from experience to follow an exponential decay law. Applications in groundwater hydrology generally deal with the recession characteristics of stream hydrographs (baseflow component of streams) and the downward trend of water levels in wells or spring discharge in the absence of recharge. Recession characteristics have also been found useful in empirical studies which attempt to relate the geology of a watershed to streamflow parameters.

The baseflow component of streams represents the withdrawal of groundwater from storage, termed *groundwater recession*. However, because baseflow is generally studied by examination of stream hydrographs, three subsystems of the hydrologic cycle are involved: (1) the subsystem of direct surface runoff, (2) the subsystem of the soil component, and (3) the subsystem of groundwater. This follows from the fact that streamflow consists of three components which reflect three different ways in which streams receive water (Barnes, 1939; Meyer, 1940): (1) surface runoff, which consists of water that flows over the land surface; (2) interflow, which consists of water that flows part of the way underground, but does not become part of the main groundwater body; and (3) baseflow, which is natural groundwater discharge. The problem faced by the hydrologist is, then, separation of the stream hydrograph into its various components and interpretation of the baseflow component. Components of a typical hydrograph and the source and magnitude of the baseflow component are shown schematically in Fig. 2.3.

Barnes (1939) suggested that the recession of each of the components of a typical hydrograph can be approximated with the empirical regression equation

$$Q_t = Q_o K_r^t \tag{2.7}$$

where Q_o is discharge at any given time, Q_t is discharge t time units after Q_o, K_r is the recession factor, and t is the time interval. It follows that the numerical value of K_r is a function of the time interval selected. A semilogarithmic plot of stream discharge versus time with discharge on the logarithmic scale gives a straight line, with the slope's defining K_r (Fig. 2.4).

HYDROLOGIC MODELS

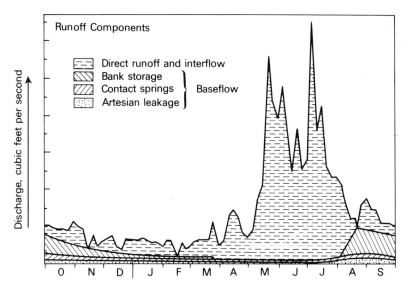

Fig 2.3 Components of runoff.

Butler (1957) gave the equation

$$Q = \frac{K_1}{10^{t/K_2}} \tag{2.8}$$

where K_1 equals Q if t equals zero [equivalent to Q_o of Eq. (2.7)], and K_2 equals t if Q equals $0.1K$. In other words, K_2 is equal to the time increment corresponding to a log cycle change in Q. The volume of discharge corresponding to a given recession is found by integrating the flow rate of Eq. (2.8),

$$\text{Vol} = \int_{t_1}^{t_2} Q \, dt = \frac{-K_1 K_2/2.3}{10^{t/K_2}} \bigg|_{t_1}^{t_2} \tag{2.9}$$

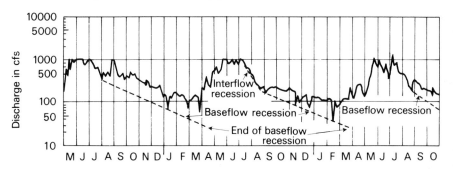

Fig 2.4 Stream hydrograph showing baseflow recession. (*After Meyboom*, 1961.)

Equation (2.9) is valid for any time period of interest on the baseflow recession part of the hydrograph and represents the total volume of groundwater discharge over the interval selected. Total potential groundwater discharge Q_{tp} is defined as the total volume of groundwater that would be discharged during an entire recession if complete depletion were to take place uninterruptedly (Meyboom, 1961), and it is determined by evaluating Eq. (2.9) over the limits t equals zero to t equals infinity. This gives

$$Q_{tp} = \frac{-K_1 K_2/2.3}{10^{\infty/K_2}} - \frac{-K_1 K_2/2.3}{10^{0/K_2}} \tag{2.10}$$

or

$$Q_{tp} = \frac{K_1 K_2}{2.3} \tag{2.11}$$

The difference between the remaining potential groundwater discharge at the end of a given recession and the total potential groundwater discharge at the beginning of the next recession is a measure of the recharge that takes place between recessions (Meyboom, 1961).

Example 2.1 Determine the approximate recharge volume between the first two recessions of Fig. 2.4.

The first recession has an initial value of 500 ft³/sec, K_1, and takes about 7.5 months to complete a log cycle of discharge, K_2. Total potential discharge is calculated from Eq. (2.11)

$$Q_{tp} = \frac{K_1 K_2}{2.3}$$

$$= \frac{500 \text{ ft}^3/\text{sec} \times 7.5 \text{ months} \times 30 \text{ days/months} \times 1{,}440 \text{ min/day} \times 60 \text{ sec/min}}{2.3}$$

$$= 4{,}222 \times 10^6 \text{ ft}^3$$

The groundwater volume discharged through the total recession lasting approximately 8 months is determined by evaluating Eq. (2.9) over the limits t equals zero to t equals 8 months

$$\frac{K_1 K_2}{2.3} - \frac{K_1 K_2/2.3}{10^{t/K_2}} = 4{,}222 \times 10^6 \text{ ft}^3 - \frac{4{,}222 \times 10^6 \text{ ft}^3}{10^{8/7.5}}$$

or about 3,800 ft³. Baseflow storage still remaining at the end of the recession can be determined by evaluating Eq. (2.9) from t equals 8 to t equals infinity, or by merely subtracting actual groundwater discharge from total potential discharge, which gives 422×10^6 ft³.

The second recession has an initial value of about 200 ft³/sec, K_1 and takes about 7.5 months, K_2, to complete a log cycle of discharge. Total potential discharge is calculated

$$Q_{tp} = \frac{K_1 K_2}{2.3}$$

$$= \frac{200 \text{ ft}^3/\text{sec} \times 7.5 \text{ months} \times 30 \text{ days/month} \times 1{,}440 \text{ min/day} \times 60 \text{ sec/min}}{2.3}$$

or about $1{,}400 \times 10^6$ ft^3. The recharge that takes place between recessions is the difference between this value and remaining groundwater potential of the previous recession, or 978×10^6 ft^3.

Other than providing information on basin replenishment, recession characteristics have been usefully applied to investigate the relation between geology and streamflow. Hely and Olmstead (1963) have demonstrated that low flows and baseflows are more directly related to rock type and structure than to physiography. Studies of a similar nature by Cross (1949) suggest that outwash deposits have a higher baseflow component than unglaciated areas or areas underlain by till. Parameters based on baseflow recession curves have the useful property of being independent of areal variations in precipitation, which allows direct examination of relations between low streamflow and geology.

The decline of water levels in wells in the absence of recharge may also be examined in terms of exponential decay. In an attempt to correlate precipitation and water levels in wells on Long Island, New York, Jacob (1943) demonstrated that the decline of water levels on a peninsula in response to an absence of precipitation can be approximated by the equation

$$h = h_0 \exp \frac{-\Pi^2 T t}{4a^2 S} \qquad (2.12)$$

where h_0 is the original height of the water table, h is the height of the water table after a given time t, a is half the width of the peninsula, T is the transmissivity (see Section 4.1), and S is the storativity (see Section 5.2). A plot of h on a logarithmic scale versus time on an arithmetic scale defines a straight line. This useful property is not restricted to peninsulas and can be extended to water-level measurements collected for groundwater basins of any shape, as well as to declining spring discharge. Thus, forecasting the behavior of the response variable of interest under prolonged drought conditions is a relatively simple matter after the first few measurements are obtained. It follows that Eqs. (2.8) through (2.11) may be applied to any hydrologic recession, where the recession represents decline of storage in the absence of replenishment.

Example 2.2 The spring discharge hydrograph in Fig. 2.5 is characterized by two distinct recessions during two periods of drought. If a minimum of 10 ft^3/sec is required at all times, what length of drought period would be required to reduce the spring discharge to this critical minimum?

By extrapolating the recession part of the curve, a period of about 6 months without replenishment would reduce spring discharge to its minimum critical value.

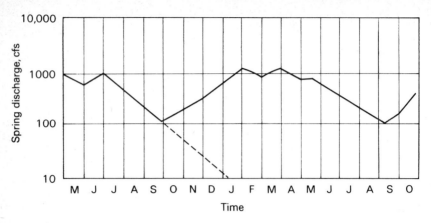

Fig 2.5 Graph of spring discharge versus time.

HYDROGRAPH SEPARATION BY CHEMICAL MASS BALANCE

During periods of low flow, the total flow of a stream is often taken as the baseflow component. During high water or floods, the separation of total runoff into direct runoff and a groundwater component requires some graphical technique by which the groundwater recession curve is extrapolated beneath the flood peak. The actual shape of the recession curve beneath the peak, however, is not known. A method based on chemical mass balance has been proposed which presumably determines this shape.

To obtain the groundwater discharge component of total runoff beneath a storm peak, a chemical mass balance equation of the following form is solved (Pinder and Jones, 1969):

$$C_{tr}Q_{tr} = C_{gw}Q_{gw} + C_{dr}Q_{dr} \tag{2.13}$$

where Q refers to runoff in cubic feet per second (total, groundwater and direct), and C refers to total dissolved solids concentration in parts per million (total, groundwater and direct), and

$$Q_{tr} = Q_{gw} + Q_{dr} \tag{2.14}$$

The value of C_{gw} is taken as the value of C_{tr} at the lowest recorded flow and is assumed constant over the period of storm hydrograph. Figure 2.6 shows the groundwater component calculated for several ions; the upper graph shows the total runoff, and the composite curve based on the mean of discharge for several ions provides the groundwater discharge hydrograph. Hence, this black-box method precludes direct observation of the groundwater discharge component, this factor presumably being reflected in the chemical character of stream discharge. Equation (2.13) is similar in principle to the hydrologic equation in that both express the conservation of some quantity.

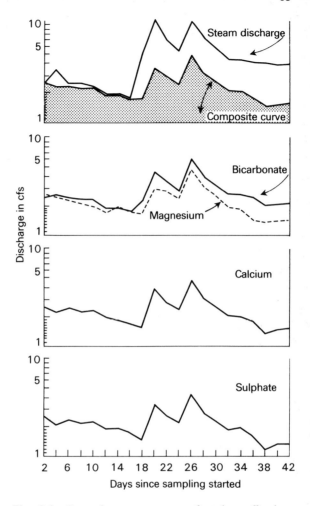

Fig. 2.6 Groundwater component of total runoff, calculated by using selected ions and a composite curve estimated from these ions. (*After Pinder and Jones*, 1969.)

2.2 BASIN ASSEMBLAGES AND REGRESSION ANALYSIS

Several of the examples given in this chapter are similar in either of two ways: (1) They demonstrate the value of some hydrologic variable measured at a given point for several values of time, or (2) they demonstrate the average value of some hydrologic variable measured at several points in one basin for several values of time. Suppose now that a hypothetical study is extended to include an assemblage of small basins in a given geologic province in order to make a set of observations in each individual basin. Usually, individual observations are taken at equal intervals over a time span common to all observations (say, for example, a recession curve of one sort or another for each basin for a common period of time). The ultimate purpose of such an investigation is to identify the geologic controls, if any, on the hydrologic variables.

At one extreme, a prima facie case of geologic control may easily be established if the province as a whole is characterized by several different rock types of which only one forms the surface configuration of each basin. This has been demonstrated by Miller (1961), who examined the solutes in small streams draining single rock types in the Sangre de Cristo Range, New Mexico. Despite large differences in slope, altitude, vegetation, and size of drainage area, waters draining a specific rock type are remarkably uniform in composition (Fig. 2.7). The conditions for homogeneity, however, are seldom realized in most basins, and conclusions drawn from comparative studies are often compromised by our inability to isolate and measure exactly the geologic factors in complex causal relations.

In spite of nature's unwillingness to comply with complete homogeneity, it is not altogether unexpected that spatial averages of certain hydrologic observations taken at specified times in a basin of complex geology may typically reflect, or characterize, the organized complexity of that basin. Similar measurements in adjacent basins of a different but still organized complexity provide the means by which comparative studies may be made. The methods and results of flow variability studies presented by Cross (1949), Hely and Olmstead (1963), and Farvolden (1964) testify to the validity of this approach. These studies demonstrate the influence of geology on certain streamflow parameters because the investigators include, in a small

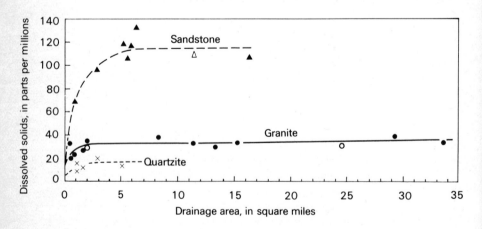

Fig. 2.7 Relation of dissolved solids to drainage area, showing approximate uniformity of waters derived from rock of each type. (*After Miller*, 1961.)

but significant way, elements of the distributed-parameter approach. That is, an assemblage of basins making up a single geologic province can be compared with an assemblage of lumped elements making up a distributed-parameter system in that the spatial variations of the hydrologic variables within each basin (lumped element) are ignored (see Section 6.1). This analogy is illustrative rather than real because the geologic characteristics of any given basin are studied more or less independently of the remaining assemblage of basins.

The basic problem in flow variability studies is to choose those factors or independent variables which may be causally related to the dependent variable under consideration, to express the factors quantitatively, and to apply statistical methods to determine the degree that each independent variable is responsible for variations in the dependent variable. In theory, the independent variables should be independent of each other. In practice, this is seldom possible. For example, variables such as average slope and soil cover might be related to various rock types, and to rainfall as well. Further dependency might prevail between average relief and rock types. Hence, degrees of interdependency are suspected, not only between the topographic variables, but between certain topographic and meteorological variables. In the final analysis, those factors with the least interdependency are chosen.

Flow variability studies may be approached from the point of view of total flow or baseflow or both. Benson (1962), for example, concentrated on the relation between flood peaks and hydrologic controls for 164 drainage areas in New England. Multiple regression equations were developed that relate peak discharge (dependent variable) with basin area, slope, flood storage areas, rainfall intensity, temperature, and an orographic factor (independent variables). Geologic variability, as such, was not included in the investigation. Other studies relating total streamflow to various physiographic and meteorological factors have been conducted by Lane and Lei (1950), Schneider (1957), and Benson (1964).

Walton (1965) investigated the relation between basin characteristics and baseflow for 109 drainage basins in Illinois. A statistical approach provided a quantitative relation between groundwater runoff during years of normal, below normal, and above normal precipitation and basin characteristics, such as geologic variability, topography, and land use. Values of groundwater runoff were plotted against percent of basins on logarithmic probability paper, as illustrated in Fig. 2.8. Groundwater runoff in similar basins is considerably greater when the bedrock is permeable than when it is impermeable. As the amount of surface sand and gravel increases, groundwater runoff increases. This study did not entail regression techniques to account for the flow variability attributed to each basin characteristic.

The most extensive study dealing with geologic influence on total- and

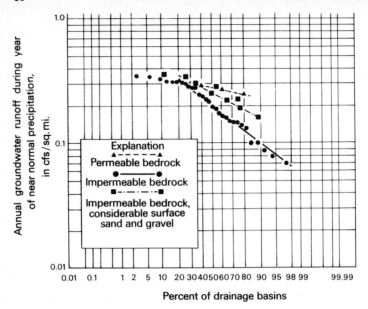

Fig. 2.8 Relation between annual groundwater runoff and character of bedrock. (*After Walton*, 1965.)

low-flow variability has been conducted in the valley and ridge province by Parizek and his colleagues (1969). Fourteen basins underlain by varying amounts of faulted and folded rocks were selected for study. Eighteen variables were selected for analysis with the idea that each could be isolated and measured consistently within the province. In mathematical terms, Y represents either total or low flow, and X_1, X_2, X_3, ..., X_{18} are 18 factors measured in each of 14 basins that are suspected of being important regulators of Y. A simple linear model is proposed,

$$Y = A_o + \sum_{i=1}^{18} A_i X_i \tag{2.15}$$

Three-termed regression equations were computed to explain a certain percentage of the variation in the dependent variable. A summary of the results is given in Table 2.1.

As might be expected, both total-flow and low-flow variability increase as the percent of the land area in slope, X_2, and in shale, X_8, increases. Further, total- and low-flow variability decreases as the percentage of carbonate rocks increases. This suggests that carbonate rocks are important regulators of flow in some basins. The persistence of mean annual temperature as a significant independent variable reflects the influence of evapotranspiration on the flow variable.

HYDROLOGIC MODELS

Table 2.1 Summary of significant bivariant correlations, independent variables included in the best three-termed regression equations, and percent of variation in the flow variable accounted for. *(Reproduced with permission from R. Parizek.)*

Dependent variable	Significant bivariant correlations	Three-termed regression equations	Percent of variation in Y explained
Y_{19}	X_{18}	X_5, X_{12}, X_{18}	68
Y_{20}	X_7, X_{18}, X_2	X_7, X_8, X_{18}	89
Y_{21}	X_7, X_{18}, X_2	X_4, X_7, X_{18}	91.6
Y_{22}	X_{18}	X_{10}, X_{11}, X_{18}	54
Y_{23}	X_7, X_{18}, X_5, X_2	X_7, X_8, X_{18}	85.3
Y_{24}	X_7, X_{18}	X_7, X_8, X_{18}	85.3

Independent variables X
- X_1 Total basin area
- X_2 Percent slope
- X_3 Percent upland
- X_4 Percent valley bottoms
- X_5 Percent forests
- X_6 Percent urbanization
- X_7 Percent carbonates
- X_8 Percent shales
- X_9 Drainage density
- X_{10} Average length, 1st-order streams
- X_{11} Average length, 2d-order streams
- X_{12} Mean elevation
- X_{13} Standard deviation of elevation distribution
- X_{14} Skewness of elevation distribution
- X_{15} Kurtosis of elevation distribution
- X_{16} Mean annual precipitation, total record
- X_{17} Mean annual precipitation, 1 year
- X_{18} Mean annual temperature

Dependent variables Y
- Y_{19} Total flow equaled or exceeded 50 percent of the time
- Y_{20} The average of the total flow equaled or exceeded 25 percent of the time minus the flow equaled or exceeded 75 percent of the time
- Y_{21} The average of the total flow equaled or exceeded 10 percent of the time minus the flow equaled or exceeded 90 percent of the time
- Y_{22} The baseflow equaled or exceeded 50 percent of the time
- Y_{23} The average of the baseflow equaled or exceeded 25 percent of the time minus the baseflow equaled or exceeded 75 percent of the time
- Y_{24} The average of the baseflow equaled or exceeded 10 percent of the time minus the baseflow equaled or exceeded 90 percent of the time

2.3 INFORMATION THEORY AND HYDROLOGIC VARIABLES

In this section, the idea of organized complexity will be advanced from the point of view of information theory. The theory holds promise of application to hydrologic problems at three different levels. At the first and most basic level, a framework of ideas developed in modern communication engineering may be utilized to obtain the number of bits of information required to describe the organization or level of complexity of a hydrologic system. As hydrologists communicate or receive information about nature by measurement, it is not unreasonable to ask how many bits of information are required to describe the fracture pattern of a carbonate terrain, the water-yielding capability of an aquifer, or the runoff pattern of a watershed. The quantity which uniquely allows these calculations is "information" as defined in communication engineering. Such a quantity may provide a "calculus" for the quantification of certain aspects of groundwater hydrology that currently are mostly descriptive.

At the second level, information theory may prove useful as a language or statistic rather than as a unique measure of information. For example, one may wish to compare the variability in the properties of two or more aquifers or correlate the macroscopic structure of a geologic formation with its hydrologic properties. In this context, information theory is not essential to the comparison or correlation in that other statistical approaches may be equally or more useful.

The third level of application deals with information theory and control systems. Although this aspect will not be discussed at length in this section, it is introduced here in order to bring out the concept of redundancy. Redundancy is a property of systems wherein the systems contain more components and connections than would be absolutely required to execute a given process. This ensures a "factor of safety" against malfunction. For example, each of several components of the hydrologic cycle has its own specific function, but if one gets out of order owing to the work of nature or to the exploitation by man, some of the others may work in the direction of error correction. Some examples may be in order:

1. Under natural conditions, the transpiration of groundwater may be the primary way in which a basin discharges groundwater. If the plants are destroyed or otherwise made ineffective, the water table will rise because of the retardation in natural discharge. Once the water table approaches land surface, direct evaporation will take place and the water level may stabilize. In this case, evaporation and transpiration perform the same function and work in the direction of storage stabilization.

2. A groundwater basin may discharge water by transpiration and direct discharge into streams or springs. When the basin is excessively developed by man, the storage may be reduced below the level required to maintain the natural discharge. This will result in the termination of transpiration and spring flow and the induced infiltration of stream flow into the underground basin. In this case, the decrease in transpiration and spring flow and the induced infiltration of streamflow all perform the same function, that is, all work in the direction of augmenting groundwater storage.

3. The low flow of streams is supplied largely by depletion of groundwater storage. However, during flood peaks, the water level in the stream is higher than the water level in the underground reservoir, and there is a net movement of water from the stream into the underground basin. This not only replenishes the basin for later depletion purposes, but is in the direction of flood-flow retardation.

HYDROLOGIC MODELS

In essence, nature has engineered the hydrologic cycle with ample excess capacity so that numerous alternative routes of communication for recharge and discharge of groundwater may be used. The loss of one merely requires the use of another. The concept of redundancy deals with this aspect.

The purpose of this section is to explore the potential usefulness of information theory in hydrological research. In order to focus on the principles concerned, the situations and examples discussed are overly simplified in a numerical sense, but the ideas can be extended to realistic large-scale problems of a similar type.

OVERVIEW OF INFORMATION, ENTROPY, AND UNCERTAINTY

The mathematical models proposed by the communication engineer Shannon (Shannon and Weaver, 1949) operate by computing properties of data that relate to information content, choice, or entropy, all of which serve as a measure of uncertainty. Although it may appear strange that these terms are used in an interchangeable sense, later discussions will support this position. In the simplest case, when dealing with n independent events or measurements, the following formula holds

$$H = -[P_1 \log_2 P_1 + P_2 \log_2 P_2 + \cdots P_n \log_2 P_n] \tag{2.16}$$

or

$$H = -\sum_{i=1}^{n} P_i \log_2 P_i \tag{2.17}$$

where P indicates probability of occurrence, and H is the entropy, information, or uncertainty of some variable X or Y.

Prior to discussing the application of Eq. (2.17), it is worthwhile to examine some of its properties. If all the P_i are zero except one and that one is unity, H equals zero. This follows from the fact that log 1 equals zero. Intuitively, this situation corresponds to one of absolute predictability, or certainty. Thus, H vanishes only when we are certain of the outcome. At the other extreme, H takes on its maximum value, when all P_i are equally likely ($P_i = 1/n$) for n events or measurements, and is equal to

$$H_{\max} = \log_2 n \tag{2.18}$$

These points may be demonstrated by examining a situation of two possibilities with probabilities P_1 and P_2, where P_2 equals $1 - P_1$. Equation (2.17) becomes

$$H = -(P_1 \log_2 P_1 + P_2 \log_2 P_2) \tag{2.19}$$

and is plotted in Fig. 2.9. The maximum entropy is attained when P_1 equals

P_2 equals $\frac{1}{2}$; this is clearly the most uncertain situation. As one of the events becomes more probable than the other, the value of H decreases. As the logarithms of numbers less than 1 are negative, H is always positive.

The base 2 of the logarithm is arbitrary, but its inclusion here implies binary information. In this case, H is measured in *bits,* a contraction for binary digits. The term *binary* refers to one of two things, for instance, open or closed, yes or no, greater than or less than. As an example of binary information, consider a game where a subject is asked to select a number between 1 and 60. An interrogator recognizes that the free choice can include any one of 60 possibilities, all of which are equally likely. This is a position of total ignorance on the interrogator's part, that is, one of maximum uncertainty. For this case, n equals 60 and Eq. (2.18) equals 5.9, or about 6. This means that six bits of information are required to describe this situation. More specifically, the interrogator can perform six "experiments" on the subject, each requiring a yes or no, systematically arriving at the number in question. In order to accomplish this task, the interrogator will first ask whether the number in question is greater than 30. If affirmative, he will next ask whether it is greater than 45, reducing in this manner the number of possibilities and arriving at the selected number after six such questions.

The relations between the terms *information, uncertainty,* and *entropy* should now be at least partly clear. The form of H may be recognized as that

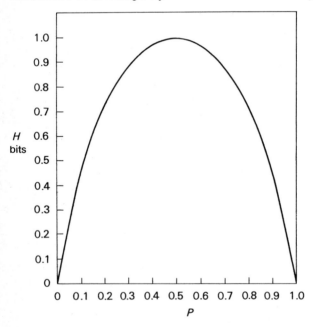

Fig. 2.9 Entropy for two possibilities with probabilities P and $1 - P$.

of entropy in statistical mechanics, where P_i is the probability of a system's being in cell i of its phase space. In statistical mechanics, entropy is a measure of degree of randomness and is a maximum when the system is completely disordered. In information theory, H is a maximum when all probabilities are equally likely. When H in information theory and entropy in statistical mechanics approach their maximum values, the uncertainty of which message was sent in the case of the former and the uncertainty of which phase space the system occupies in the latter case both approach a maximum. In hydrology, the greater H is, the more disordered the system is and the greater is the uncertainty in predicting the value of a given variable or variables. In terms used earlier, the complexity is disorganized. The point of view in each case is different, but the mathematics is the same.

As soon as one situation becomes more probable than the others, the entropy, or uncertainty, of H decreases. A system that usually assumes a given state and only rarely assumes other states is one whose behavior can be readily predicted. One might say that this situation is organized, or not characterized by a large degree of randomness or choice. Attention here may be focused on problems involving homogeneity or heterogeneity on a local or regional scale.

Every time a system is known to occupy a given state, it gives information on itself. This is termed *self-information* and is expressed by Eq. (2.17). The idea of mutual information for combined systems will be introduced later.

APPLICATIONS IN THE BIOLOGICAL AND PHYSICAL SCIENCES

Biological and physical scientists have long recognized the importance of information theory and Eq. (2.17) in particular. MacArthur (1955) recognized that entropy in information theory may be used as a measure of stability in biological communities. Stability means that both number of species and number of individuals per species do not fluctuate greatly over time. In general, the greater the number of species, n, in a community, the more stable the community. If a large number of species is present in equal proportions, H takes on its maximum value, which indicates maximum stability. As some species become more dominant than others, community stability tends to decrease, as does the value of Eq. (2.17). For example, a tropical rain forest is characterized by a large number of species with relatively few individuals per species, and it is stable. The arctic region, on the other hand, is characterized by relatively few species with numerous individuals for some of the species, which results in wide fluctuations in populations. Information theory is used here on the second level, that is, as a mathematical language.

From another point of view, Pelto (1954) recognized that since entropy is a measure of disorder, Eq. (2.17) is useful in mapping multicomponent systems in geology. Uncertainty in terms of geologic classification is maximum when all components are present in equal proportions, and it is minimum or zero when only one component is present. Information theory is useful here also on the second level, this time as a statistic.

From yet other points of view, Leopold and Langbein (1962) applied the concept of entropy in describing the "most probable state of nature" in landscape development, and Matalas and Langbein (1962) drew freely from the ideas of information theory to ascertain the information content of the mean.

SELF–INFORMATION

Example 1.1 may be used to illustrate a problem of self-information. It may be recalled that three events or states were described with probabilities of occurrence as follows:

State I: $\quad \frac{11}{18} = 0.611$

State II: $\quad \frac{4}{18} = 0.222$

State III: $\quad \frac{3}{18} = 0.166$

The entropy or uncertainty characterizing this situation is calculated with Eq. (2.17),

$$H = -(0.611 \log_2 0.611 + 0.222 \log_2 0.222 + 0.166 \log_2 0.166)$$

A short table of logarithms to the base 2 is given in Table 2.2. With data in this table, $\log_2 0.611$ is easily calculated,

$$\log_2 0.611 = \log_2 11 - \log_2 18 = 3.459 - 4.169 - 0.710$$

After making all necessary calculations, H is found to equal 1.3655. This value may be compared with the maximum value of H, which is obtained when all events are equally likely. In this case, each state has a probability of occurrence of 0.333, and Eq. (2.18) becomes

$$H_{\max} = \log_2 3 = 1.5849$$

Relative entropy is the ratio of the actual to the maximum entropy and is easily calculated,

$$\frac{H}{H_{\max}} = \frac{1.3655}{1.5849} = 0.86$$

This means that this simple three-state system, in its selection of any one of

HYDROLOGIC MODELS

Table 2.2 Logarithms to the base 2 of numbers from 1 to 200

n	$\log_2 n$	n	$\log_2 n$	n	$\log_2 n$	n	$\log_2 n$
1	0.00000	51	5.67242	101	6.65821	151	7.23840
2	1.00000	52	5.70044	102	6.67242	152	7.24793
3	1.58496	53	5.72792	103	6.68650	153	7.25739
4	2.00000	54	5.75489	104	6.70044	154	7.26679
5	2.32193	55	5.78136	105	6.71425	155	7.27613
6	2.58496	56	5.80735	106	6.72792	156	7.28540
7	2.80735	57	5.83289	107	6.74147	157	7.29462
8	3.00000	58	5.85798	108	6.75489	158	7.30378
9	3.16993	59	5.88264	109	6.76818	159	7.31288
10	3.32193	60	5.90689	110	6.78136	160	7.32192
11	3.45943	61	5.93074	111	6.79442	161	7.33092
12	3.58496	62	5.95420	112	6.80735	162	7.33985
13	3.70044	63	5.97728	113	6.82018	163	7.34873
14	3.80735	64	6.00000	114	6.83289	164	7.35755
15	3.90689	65	6.02237	115	6.84549	165	7.36632
16	4.00000	66	6.04439	116	6.85798	166	7.37504
17	4.08746	67	6.06609	117	6.87036	167	7.38370
18	4.16993	68	6.08746	118	5.88264	168	7.39232
19	4.24793	69	6.10852	119	6.89482	169	7.40088
20	4.32193	70	6.12928	120	6.90689	170	7.40939
21	4.39232	71	6.14975	121	6.91886	171	7.41785
22	4.45943	72	6.16992	122	6.93074	172	7.42626
23	4.52356	73	6.18982	123	6.94252	173	7.43463
24	4.58496	74	6.20945	124	6.95420	174	7.44294
25	4.64386	75	6.22882	125	6.96578	175	7.45121
26	4.70044	76	6.24793	126	6.97728	176	7.45943
27	4.75489	77	6.26679	127	6.98869	177	7.46760
28	4.80735	78	6.28540	128	7.00000	178	7.47573
29	4.85798	79	6.30378	129	7.01123	179	7.48382
30	4.90689	80	6.32193	130	7.02237	180	7.49185
31	4.95420	81	6.33985	131	7.03342	181	7.49985
32	5.00000	82	6.35755	132	7.04439	182	7.50779
33	5.04439	83	6.37504	133	7.05528	183	7.51570
34	5.08746	84	6.39232	134	7.06609	184	7.52356
35	5.12928	85	6.40939	135	7.07682	185	7.53138
36	5.16993	86	6.42626	136	7.08746	186	7.53916
37	5.20945	87	6.44294	137	7.09803	187	7.54690
38	5.24793	88	6.45943	138	7.10852	188	7.55459
39	5.28540	89	6.47573	139	7.11894	189	7.56224
40	5.32193	90	6.49185	140	7.12928	190	7.56986
41	5.35755	91	6.50779	141	7.13955	191	7.57743
42	5.39232	92	6.52356	142	7.14975	192	7.58496
43	5.42626	93	6.53916	143	7.15987	193	7.59246
44	5.45943	94	6.55459	144	7.16992	194	7.59991
45	5.49185	95	6.56986	145	7.17991	195	7.60733
46	5.52356	96	6.58496	146	7.18982	196	7.61471
47	5.55459	97	6.59991	147	7.19967	197	7.62205
48	5.58496	98	6.61471	148	7.20945	198	7.62936
49	5.61471	99	6.62936	149	7.21917	199	7.63662
50	5.64386	100	6.64386	150	7.22882	200	7.64386

three events, is about 86 percent as free as it possibly could be. Hence, the situation is characterized by a sizeable degree of randomness or choice; that is, the entropy is high. If the calculated H is equal to H_{max}, the system is 100 percent free in its selection of any one of three flows, which characterizes maximum uncertainty.

The percentage of the message that is not governed by free choice, but by the statistical structure of the system itself, is calculated with the formula for redundancy,

$$R = 1 - \frac{H}{H_{max}} \tag{2.20}$$

or, in this case, 14 percent. In the absence of redundancy, all situations would be characterized by maximum disorder. Hence, when H equals H_{max}, redundancy is zero. At the other extreme, redundancy equals 100 percent when one state is favored to the exclusion of all others.

Now, what is the meaning of these calculations? First, from an informational point of view, it takes fewer bits of information to describe the complexity of this system than would be required if the system were as complex as it possibly could be. In terms of order, one might say that the system exhibits some order in that the entropy is somewhat less than the maximum value it could attain. In terms of uncertainty, the uncertainty in predicting what event will occur is likewise somewhat less than it could be if the system were completely disordered. If none of the states are favored and none unfavored so that the redundancy equals zero, no pattern or design emerges. It is within this context that the terms *degree of randomness, choice, uncertainty,* and *entropy* are used interchangeably.

It has been mentioned that statistical interpretation depends as much upon the character of the model adopted as upon the quality of the data. This point may be examined by considering these same data in terms of one-step markovian dependencies. From Example 1.1,

$$\begin{array}{c} & \text{I} \quad \text{II} \quad \text{III} \\ \begin{array}{c} \text{I} \\ \text{II} \\ \text{III} \end{array} & \begin{bmatrix} 0.6 & 0.1 & 0.3 \\ 1.0 & 0 & 0 \\ 0 & 1.0 & 0 \end{bmatrix} \end{array}$$

In this case, each of the three rows is characterized by its own entropy H_i. Carrying out the calculations,

$$H_1 = -(0.6 \log_2 0.6 + 0.1 \log_2 0.1 + 0.3 \log_2 0.3) = 1.25949$$
$$H_2 = -1 \log_2 1 = 0$$
$$H_3 = -1 \log_2 1 = 0$$

HYDROLOGIC MODELS

Hence, whenever the system occupies state II or state III, there is absolute certainty of which state the system will next occupy.

The entropy of the markovian system is defined as the average of all H_i weighted in accordance with the probability of occurrence of the individual states (Shannon and Weaver, 1949)

$$H = \sum_i p_i H_i \qquad (2.21)$$

For this example,

$$H = 0.611 \times 1.25949 + 0 + 0 = 0.7895$$

If this simple problem is viewed from the point of view of uncertainty, the random model is characterized by greater uncertainty than the Markov model. A measure of the uncertainty removed by treating the data as markovian rather than random is easily calculated by

$$\sum P_i \log P_i - \sum P_i H_i \qquad (2.22)$$

which can range from zero, when the data are in fact completely random, to H under conditions of perfect markovian dependence. Equation (2.22) may be used to define the amount of organization introduced into the system by the selection of one model over another. In this sense, organization is a measure of how much information is introduced when one considers interdependency between the values of a given variable.

Certain types of geological problems may be viewed in the same manner. Krumbein (1967) has set up a Markov chain to describe the Chesterian rocks of southern Illinois. From a series of measured sections, he noted whether the rock type at 8-ft intervals was sandstone, shale, or limestone. The various rock types are associated with the following probabilities:

Sandstone	0.255
Limestone	0.289
Shale	0.456

which gives an entropy, or H, value of 1.5368, as opposed to an H_{max} value of 1.5849. Calculations based on the transition matrix,

	Sandstone	Shale	Limestone
Sandstone	0.74	0.23	0.03
Shale	0.10	0.61	0.29
Limestone	0.05	0.38	0.57

give a weighted H of 1.1676. If the beds were part of a cyclothymic sequence, it might be expected that the Markov model introduces more organization into the system than it did in this particular instance.

These ideas may be extended to other spatial distributions of data where the sampling points coincide with the intersections of a uniform grid superposed upon an area. Maximum entropy is taken as the standard for the completely disordered case. At the other extreme, zero entropy is representative of absolute order. For these end members or any intermediate state, the concept of redundancy may be employed to obtain a measure of relative order, or predictability, as measured on a number line ranging from zero to 1. The value $H/H_{\max}$ is relative entropy, or the ratio of the entropy H to the maximum value $H_{\max}$ that it could have while still restricted to the same symbols. A system in maximum disorder is characterized by a collection of measurements, or observations, none of which are particularly favored with others unfavored, so that no pattern or design emerges. In this case, H equals $H_{\max}$ and R equals zero [Eq. (2.20)], which corresponds to a complete lack of predictability, or maximum disorder. At the other extreme, H equals zero when only one possible situation exists and R equals 1, which corresponds to maximum order. As used here, order has a relative connotation rather than an absolute one; that is, order is relative to the maximum disorder that the elements may display.

A criterion developed from the rate of change of redundancy with respect to some variable permits identification of systems that become more ordered as some related variable changes in value. The variable in question might be time or space, but could conceivably be some causative factor, such as temperature, velocity, or some hydrologic property. Differentiating Eq. (2.20) with respect to this variable (von Foerster, 1960),

$$dR = \frac{H \, dH_{\max} - H_{\max} \, dH}{H_{\max}^2} \tag{2.23}$$

If the entropy is decreasing, the system is approaching a more ordered configuration; if the entropy is increasing, the system is becoming more disordered. For the former case, $dR > 0$; and for the latter case, $dR < 0$. Since $H_{\max}^2 > 0$ for all cases except those that are always perfectly ordered, the condition for an increase in order is determined by

$$H \, dH_{\max} > H_{\max} \, dH \tag{2.24}$$

To demonstrate that statement (2.24) describes decreasing entropy, consider the special case where $H_{\max}$ remains constant. This can occur when the number of states constituting a system remain unchanged. As a simple example, if streamflow over time can always take on one of fifty values, $H_{\max}$ is constant, but H changes as more information becomes available. For this and all similar cases, the derivative of $H_{\max}$ vanishes, and statement (2.24) reduces to

$$H_{\max} \, dH < 0 \tag{2.25}$$

which indicates a decrease in entropy.

COMBINED SYSTEMS

It is rare indeed when a single set of probabilities represents a real problem in hydrology. Instead, it is useful to think of two related systems X and Y. The system X may assume any state x_i, and Y can assume any state y_j. The system Y can be observed, but X cannot. How much uncertainty about X can be removed by observing Y?

This is perhaps one of the most fundamental questions in hydrology, if not in all observational and experimental science. An observable system Y is presumed capable of transmitting certain information about X. How well it does so depends on how closely the systems are related. At one extreme, if X and Y are independent, it is not possible to remove any uncertainty at all. At the other extreme, given a one-to-one correspondence between X and Y, all uncertainty vanishes from the problem. That is, we are no longer interested in speculation about X's behavior in that observations of Y tell us all we wish to know about X. In the general case, where there is some interdependency, some of the uncertainty about X can be removed by observing Y. This means that the uncertainty about X when Y is known is always less than the uncertainty calculated from observations of X alone, that is, is less than the self-information of X.

Figure 2.10 illustrates an example in which certain irrigation decisions are arrived at annually on the basis of a "transmission of information" on probable annual water supply. The strategy of the irrigators is easily revealed by the diagram. Alfalfa, a perennial crop, is maintained regardless of the flow of the river, whereas acreage of other crops is adjusted in accordance with flow. It is important to note here that planting decisions for crops other than alfalfa must be arrived at and executed at least 1 month in advance of the runoff season. This is typical in snowmelt areas in western United States, where the accumulated winter snowpack represents about 90 percent of the annual flow and is generally depleted during the months of April, May, June, and part of July.

As the ranchers appear to be making decisions from information received, one may make inquiries about the nature of this information. In this case, a federal agency conducts snowpack measurements, which are converted to runoff, and makes these estimates known at various times previous to the runoff season. The information system is described by Fig. 2.11. The variable Y (snowpack) is observed, but X (runoff) is not directly accessible until after the fact, that is, until all planting decisions are executed. Based on information obtained from Y, a decision is made concerning the state variable, in this case acreage of crops grown other than the perennial stand of alfalfa. If the system is "noiseless," there is a one-to-one relation between X and Y. That is, there is no discrepancy between predicted and actual flows, and the problem is characterized by a complete lack of uncertainty. However, "noise" may enter for several reasons. An unseasonal

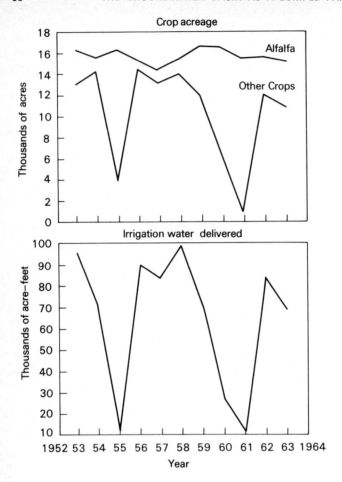

Fig. 2.10 Relation between acres irrigated, crops grown, and surface-water deliveries, Humboldt River, Nevada.

rainfall, for example, may remove much of the snowpack prior to the irrigation season. For this reason, measurements and reports are conducted at various times throughout the winter season. Another source of noise is the conversion of some part of the snowpack into evaporation and groundwater inflow, which may vary from year to year.

A simplified version of this problem is given on the next page, where the numbers 1, 2, 3 represent estimated states or flows from two snowpack surveys Y_1 and Y_2 and actual recorded flows X that were available for irrigation use.

HYDROLOGIC MODELS

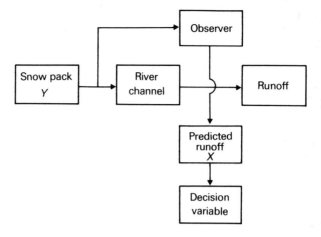

Fig. 2.11 Communication channel for the snowpack-runoff problem.

Y_1	Y_2	X
1	2	1
3	3	3
1	2	2
2	3	2
2	3	3
3	3	3
1	1	1
1	1	1
2	2	2
3	3	3
2	2	2
3	2	2

Several entropies may be calculated here. $H(Y_1)$, $H(Y_2)$, and $H(X)$ may be calculated independently with Eq. (2.17). This gives

$H(Y_1) = 1.58$
$H(Y_2) = 1.55$
$H(X) = 1.55$

This approach, however, does not take full advantage of the interdependency between X and Y. For this case, it is more useful to compute the entropy of X, given knowledge of Y, or

$$H\left(\frac{X}{Y}\right) = \sum_i \sum_j P(x_i, y_j) \log_2 P\left(\frac{x_i}{y_j}\right) \tag{2.26}$$

where $H(X/Y)$ is the uncertainty of X, given knowledge of Y (or the uncertainty in the received signal when the message sent is known), $P(x_i/y_j)$ is the probability of X's being in state x_i when it is known that Y is in state y_j, and $P(x_i, y_j)$ is the joint probability of x_i and y_j. Given a one-to-one correspondence between states x_i and y_j, $P(x_i/y_j) = 1$ and $H(X/Y) = 0$. That is, there is no uncertainty about X's behavior once Y is observed. At the other extreme, where the amount of noise in the system is large so that X and Y are rendered "independent," $H(X/Y) = H(X)$. That is, observation of Y in no way alters the original uncertainty of X.

In summary, then, the uncertainty about X is originally $H(X)$. After observing Y, this uncertainty is reduced to $H(X/Y)$, where $H(X/Y) \leq H(X)$. The decrease in uncertainty is easily calculated by

$$H(X \to Y) = H(X) - H\left(\frac{X}{Y}\right) \tag{2.27}$$

where $H(X \to Y)$ is the uncertainty about X which is removed by observing Y. For a one-to-one correspondence, the total original uncertainty $H(X)$ is fully removed. When X and Y are independent, zero uncertainty is removed. Equation (2.27) defines the amount of information introduced into the system by consideration of the interdependency between X and Y. Table 2.3 shows the calculations for the system Y_1 and X.

These ideas may be extended to spatial distributions of data. For example, there are many possible arrangements of the sediments or fractures which make up a groundwater basin. Hence, given information about subsurface features, such as unconformities or structural deformations in carbonate regions, how much uncertainty can be removed in groundwater exploration? This question can be extended to the nature of information transmitted by geologic maps in general: Given direct information on some geologic aspect, how much uncertainty can be removed in predicting variations in water quality, well yield, or perhaps baseflow components of hydrographs? If petroleum rather than groundwater is of interest, uncertainty in the petroleum prospect is likely subject to reduction by direct information on structural deformation.

In the general case, the problem is to relate some hydrologic or geologic variable in a heterogeneous region to a specific feature within that region. For example, the ability of certain rocks to yield water may differ considerably throughout a region, which gives a disordered pattern when

HYDROLOGIC MODELS

Table 2.3 Calculations for the snowmelt system Y_1 and X

Calculation of probabilities

$P\left(\dfrac{x_i}{y_j}\right)$ for $y_j = 1$ $P(x_i,y_j)$ for $y_j = 1$

$P(\tfrac{1}{1}) = \tfrac{3}{4}$ $P(1,1) = \tfrac{1}{4}$
$P(\tfrac{2}{1}) = \tfrac{1}{4}$ $P(2,1) = \tfrac{1}{12}$
$P(\tfrac{3}{1}) = 0$ $P(3,1) = 0$

$P\left(\dfrac{x_i}{y_j}\right)$ for $y_j = 2$ $P(x_i,y_j)$ for $y_j = 2$

$P(\tfrac{1}{2}) = 0$ $P(1,2) = 0$
$P(\tfrac{2}{2}) = \tfrac{3}{4}$ $P(2,2) = \tfrac{1}{4}$
$P(\tfrac{3}{2}) = \tfrac{1}{4}$ $P(3,2) = \tfrac{1}{12}$

$P\left(\dfrac{x_i}{y_j}\right)$ for $y_j = 3$ $P(x_i,y_j)$ for $y_j = 3$

$P(\tfrac{1}{3}) = 0$ $P(1,3) = 0$
$P(\tfrac{2}{3}) = \tfrac{1}{4}$ $P(2,3) = \tfrac{1}{12}$
$P(\tfrac{3}{3}) = \tfrac{3}{4}$ $P(3,3) = \tfrac{1}{4}$

Calculation of $H(X/Y)$

$y_j = 1$

$$H\left(\frac{X}{1}\right) = \tfrac{1}{4} \log_2 \tfrac{3}{4} + \tfrac{1}{12} \log_2 \tfrac{1}{4} = 0.272$$

$y_j = 2$

$$H\left(\frac{X}{2}\right) = \tfrac{1}{4} \log_2 \tfrac{3}{4} + \tfrac{1}{12} \log_2 \tfrac{1}{4} = 0.272$$

$y_j = 3$

$$H\left(\frac{X}{3}\right) = \tfrac{1}{12} \log_2 \tfrac{1}{4} + \tfrac{1}{4} \log_2 \tfrac{3}{4} = 0.272$$

$$H\left(\frac{X}{Y}\right) = 0.816$$

$$H(X \to Y) = H(X) - H\left(\frac{X}{Y}\right) = 1.55 - 0.816 = 0.734$$

viewed on a map. The disordered pattern may vanish, however, when the uncertainty in sets of values characterizing water-yielding ability X is examined in relation to formations penetrated, formation thickness, structure, or some other geologic variable. The classic problem of this sort deals with carbonate rocks. A well drilled in such an area could be dry or have a very high yield, depending on its location. Hence, the spatial distribution of fractures and fracture density is of particular interest to the hydrologist. With the mathematical formulation provided above, the problem is defined as follows: Take a geological map of a structurally deformed carbonate terrain, and divide the map into arbitrary areas using a suitable grid system. Conduct a detailed field study of the fracture distribution X and several other variables Y that may be deduced with ease from the map. The variables Y may include areas where the rate of change of dip is greatest or areas of unconformities, of weathered zones, or of changes in formation thickness. What variable, or variables, Y removes the most uncertainty in predictions of the fracture distribution X? The question remains whether it is possible to use these variables to make fracture-distribution predictions in other regions by merely examining a geological map.

It is also possible to consider combining two variables X and Y into a joint system. Examples in groundwater hydrology include combinations of the occurrence of master fractures in carbonate or volcanic rocks with high-yield wells; ground- or surface-water quality with rock types, or mineralogy; duration or slope of baseflow recession or quantity of discharge in various basins with dominant rock type. In oil exploration, a logical joint system might be the occurrence of geologic structures of various types and the occurrence of oil. For these and similar cases,

$$H(X,Y) = \sum_i \sum_j P(x_i,y_j) \log_2 P(x_i,y_j) \tag{2.28}$$

where $H(X,Y)$ is the entropy of the joint system, and $P(x_i,y_j)$ is the joint probability of variables x_i and y_j.

The entropy of the joint system may also be expressed

$$H(X,Y) = H(X) + H\left(\frac{Y}{X}\right) \tag{2.29}$$

or

$$H(X,Y) = H(Y) + H\left(\frac{X}{Y}\right) \tag{2.30}$$

If X and Y are independent, $H(X/Y) = H(X)$ and $H(Y/X) = H(Y)$, so that

$$H(X,Y) = H(X) + H(Y) \tag{2.31}$$

That is, when there is no communication or interaction between X and Y, the entropy of the joint system is equal to the sum of the individual entropies. If there is a one-to-one correspondence between variables X and Y [$H(X/Y)$ or $H(Y/X)$ equals zero], the entropy of the joint system is equal to the entropy of the single system X or Y.

2.4 CONCLUDING STATEMENT

This chapter has attempted to show a wide variety of applications of the so-called black-box approach in groundwater hydrology. Some specific conclusions emerge:

1. Some type of model is employed, although the form and specifics of the model can range from the simple deterministic hydrologic equation to models constructed around ideas of uncertainty.
2. The models employed in inventory, yield, and studying recession characteristics are reasonably straightforward and merely require the collection of certain data and some simple calculations.
3. The models employed in studying basin assemblages and information content are quite general and require imagination, judgment, and experience for proper questions to be formulated. Regression analysis, however, is likewise reasonably straightforward once a problem is clearly defined, in that it does one thing well, namely, it explains the variation in a dependent variable as regulated by several independent variables. Information theory, on the other hand, has a broader theoretical construct, and may do several things very well and others not well at all. Both regression analysis and information theory are useful in attempting to identify the environmental controls on a hydrologic variable.

PROBLEMS AND DISCUSSION QUESTIONS

2.1 From the published discussions presented with Conkling's (1946) paper, briefly cite the arguments for and against a safe-yield concept.

2.2 a. Briefly discuss the following inventory in terms of interaction between open subsystems of the hydrologic cycle. Be explicit on what is occurring and when.
  b. What form does the hydrological equation assume to account for this inventory?
  c. What comments can be made about safe yield as a unique and constant value under these circumstances?
  d. From these data comment on the statement, "Safe yield cannot be fully developed until a basin is overdeveloped."
  e. Hypothesize what will happen if pumpage is reduced to 29,000 acre-ft/year.

Time yr	Natural input, Precip. acre-ft	Net natural input, contrib. from streamflow acre-ft	Pumping output acre-ft	Natural output, evapotranspiration acre-ft	Change in groundwater storage acre-ft
1	25,000	0	0	25,000	0
2	25,000	500	10,000	25,000	9,500
⋮	⋮	⋮	⋮	⋮	⋮
7	25,000	1,000	20,000	24,000	18,000
8	26,000	2,000	24,000	22,000	18,000
9	27,000	3,000	30,000	20,000	20,000
10	25,000	4,000	35,000	19,000	25,000
11	26,000	5,000	42,000	18,000	29,000
12	25,000	5,000	50,000	16,000	36,000

Time yr	Natural input, Precip. acre-ft	Net natural input, contrib. from stream-flow acre-ft	Pumping output acre-ft	Natural output, evapotrans-piration acre-ft	Change in groundwater storage acre-ft
13	24,000	5,000	55,000	15,000	41,000
14	25,000	5,000	62,000	18,000	45,000
15	24,000	5,000	72,000	10,000	53,000
16	26,000	5,000	82,000	6,000	57,000
17	25,000	5,000	92,000	3,000	65,000
18	25,000	5,000	100,000	1,000	71,000
19	25,000	5,000	100,000	1,000	71,000

2.3 Assume that the graph of spring discharge versus time in Example 2.2 is actually a graph of flow versus time for a flowing well in a basin.

    a. Calculate the volume of groundwater discharged over the first recession. (Answer: $3,040 \times 10^6$ ft^3.)

    b. Calculate the recharge that takes place between recessions. (Answer: For $K_1 = 2,000$ ft^3/sec and $K_2 = 5$ months, for the second recession recharge = $10,470 \times 10^6$ ft^3.)

2.4 Input-output relations can often be examined quantitatively by considering differential and integral equations. For hydrologic systems, the equations of interest are

$$\frac{dS}{dt} = I(t) - O(t) \tag{1}$$

and

$$S = \int [I(t) - O(t)] \, dt \tag{2}$$

The first of these equations gives the change in storage over time. The second equation, when solved, gives the amount of water stored over a given time period. A passive storage system is assumed to discharge water in proportion to the amount stored. Hence, when inflow is equal to some constant I, and outflow $O(t)$ is taken as proportional to storage, $O(t) = KS(t)$, where K is a constant. This gives

$$\frac{dS}{dt} = I - KS(t) \tag{3}$$

and

$$S(t) = \frac{I}{K} + \left(S(O) - \frac{I}{K}\right)e^{-Kt} \tag{4}$$

where $S(O)$ is original storage.

If inflow is zero, Eq. (4) reduces to

$$S(t) = S(0)e^{-Kt} \tag{5}$$

If original storage is zero, Eq. (4) reduces to

$$S(t) = \frac{I}{K}(1 - e^{-Kt}) \tag{6}$$

    a. What is the value of Eqs. (4) through (6) at time equal to zero? At time equal to infinity?

    b. When inflow equals zero, demonstrate that $S(0)(1 - e^{-Kt})$ equals the volume of groundwater discharge measured under the stream hydrograph. What kind of information does this equation yield that the recession Eqs. (2.7) through (2.11) do not?

HYDROLOGIC MODELS

    c. The quantity $S(t)$ in Eqs. (4) and (6) approaches I/K as time approaches infinity. When original storage equals I/K in Eq. (4), then $S(t)$ equals I/K for all times. Hence, the value I/K may have some equilibrium connotation. Explain.

    d. As K gets very large (approaches unity) in association with very large values of time, $S(t)$ of Eqs. (4) and (6) approaches I. What does this indicate about the residence time of water in storage?

2.5 From Fig. 2.7, what can you infer about the relation between rock mineralogy and the degree of water mineralization? Would you suspect any relation between degree of water mineralization and sediment load?

2.6 Streams flowing in one region are characterized by steep baseflow recession curves, whereas streams in a different region are characterized by flat recession curves.

    a. What can you deduce about the differences in infiltration and groundwater storage capacity of these regions?

    b. If one of the regions in question is granitic rock and the other limestone, which is characterized by the flat recession?

2.7 Discuss the effect of the geology of a watershed both on low flows experienced during periods of prolonged drought and on flood flows. Why is it important to understand the geology (especially rock type) in any precipitation-augmentation program?

2.8 From purely theoretical considerations, would you expect that the slopes of several baseflow recession curves obtained in a given basin are reasonably constant for that basin?

2.9 In a certain region, 25 percent of all wells are in excess of 500 ft in depth. All these wells yield at least 1,000 gpm. Also, all wells drilled to a depth of less than 500 ft have yields of less than 1,000 gpm.

    Setup: $P(x) = 0.25$

                  $P(y) = 0.25$

                  $P\left(\dfrac{y}{x}\right) = 1.00$

                  $P(x,y) = P(x)P\left(\dfrac{y}{x}\right) = 0.25$

                  $P\left(\dfrac{x}{y}\right) = \dfrac{P(x,y)}{P(y)} = 1.00$

    a. If you know that a well is in excess of 500 ft in depth, do you get any further information by being informed of its yield? Explain.

    b. Demonstrate that the uncertainty of the joint system consisting of deep wells and yields of 1,000 gpm or more $[H(x,y)]$ is equal to the uncertainty of the single system x or y.

2.10 In a certain region, 25 percent of all wells are in excess of 500 ft in depth. Of these wells, 75 percent yield at least 1,000 gpm. Also, 50 percent of all wells in the region yield at least 1,000 gpm.

    Setup: $P(x) = 0.25$

                  $P(y) = 0.50$

                $P\left(\dfrac{y}{x}\right) = 0.75$

$$P(x,y) = P(x)P\left(\frac{y}{x}\right) = 0.1875$$

$$P\left(\frac{x}{y}\right) = \frac{P(x,y)}{P(y)} = 0.375$$

    a. If you know that a well is in excess of 500 ft in depth do you obtain any additional information by being informed of its yield? Explain.

    b. Demonstrate that the uncertainty of the joint system consisting of deep wells and yields of 1,000 gpm or more [$H(x,y)$] is less than the sum of the individual entropies $H(x)$ and $H(y)$.

2.11 Select a problem in your area of interest, and discuss how information theory might be applied to that problem.

REFERENCES

Banks, H. O.: Utilization of underground storage reservoirs, *Trans. Amer. Soc. Civil Engrs.*, vol. 118, pp. 220–234, 1953.

Barnes, B. S.: The structure of discharge recession curves, *Trans. Amer. Geophys. Union*, pt. 4, pp. 721–725, 1939.

Benson, M. A.: Factors influencing the occurrence of floods in a humid region of diverse terrain, *U.S. Geol. Surv., Water Supply Papers*, 1580-B, 1962.

———: Factors influencing the occurrence of floods in the Southwest, *U.S. Geol. Surv., Water Supply Papers*, 1580-D, 1964.

Butler, S. S.: "Engineering Hydrology," Prentice Hall, Inc., Englewood Cliffs, N.J., 1957.

Chapman, T. G.: Effects of groundwater storage and flow on the water balance, in "Water Resources Use and Management," Melbourne University Press, Australia, pp. 290–301, 1964.

Conkling, H.: Utilization of groundwater storage in stream system development, *Trans. Amer. Soc. Civil Engrs.*, vol. 111, pp. 275–305, 1946.

Cross, W. P.: The relation of geology to dry weather streamflow in Ohio, *Trans. Amer. Geophys. Union*, vol. 30, pp. 563–566, 1949.

Farvolden, R. N.: Geologic controls on groundwater storage and baseflow, *J. Hydrol.* no. 3, pp. 219–250, 1964.

Fischel, V. C.: Long term trends of groundwater levels in the United States, *Trans. Amer. Geophys. Union*, vol. 37, pp. 429–435, 1956.

Gleason, G. B.: Changes in groundwater elevations of the South Coastal Basin during the past quarter century in comparison to long term mean precipitation and runoff, *Trans. Amer. Geophys. Union*, vol. 23, pp. 108–124, 1942.

Hantush, M. S.: Preliminary quantitative study of the Roswell groundwater reservoir, New Mexico, *New Mexico Inst. Mining and Technol.*, 1957.

Hely, A. G., and F. H. Olmstead: Some relations between streamflow characteristics and the environment in the Delaware River region, *U.S. Geol. Surv., Profess. Papers*, 417-B, 1963.

Jacob, C. E.: Correlation of groundwater levels and precipitation on Long Island, New York, *Trans. Amer. Geophys. Union*, vol. 24, pp. 564–573, 1943.

———: Correlation of groundwater levels and precipitation on Long Island, New York, *Trans. Amer. Geophys. Union*, vol. 25, pp. 928–939, 1944.

Kazmann, R. G.: "Safe yield" in groundwater development, reality or illusion? *Proc. Amer. Soc. Civil Engrs.*, vol. 82, no. IR 3, 1956.

Krumbein, W. C.: Fortran IV computer programs for Markov chain experiments in geology, Computer Contributions 13, *State Geol. Surv., Kansas, Univ. Kansas Publ.*, Lawrence, 1967.

Lane, E. W., and K. Lei: Streamflow variability: *Trans. Amer. Soc. Civil Engrs.*, vol. 150, pp. 1084–1134, 1950.

Lee, C. H.: The determination of safe yield of underground reservoirs of the closed basin type, *Trans. Amer. Soc. Civil Engrs.*, vol. 78, pp. 148–151, 1915.

Leggette, R. M.: Water levels and artesian pressure in observations wells in the United States in 1940, Part I: Northeastern states, section on Long Island, *U.S. Geol. Surv., Water Supply Papers*, 906, pp. 110–115, 1942.

Leopold, L. B., and W. B. Langbein: The concept of entropy in landscape evolution, *U.S. Geol. Surv., Profess. Papers*, 500-A, 1962.

MacArthur, R.: Fluctuations of animal populations and a measure of community stability, *Ecology*, vol. 26, pp. 533–536, 1955.

Malmberg, G. T.: Available water supply of the Las Vegas groundwater basin, Nevada, *U.S. Geol. Surv., Water Supply Papers*, 1780, 1965.

Matalas, N. C., and W. B. Langbein: Information content of the mean, *J. Geophys. Res.*, vol. 67, no. 9, pp. 3441–3448, 1962.

Meinzer, O. E.: Outline of groundwater hydrology, with definitions, *U.S. Geol. Surv., Water Supply Papers*, 494, 1923.

———: Outline of method for estimating groundwater supplies, *U.S. Geol. Surv., Water Supply Papers*, 638-C, pp. 94–144, 1932.

——— and N. D. Stearns: A study of groundwater in the Pomperaug basin, Conn., with special reference to intake and discharge, *U.S. Geol. Surv., Water Supply Papers*, 597, pp. 73–146, 1928.

Meyboom, P.: Estimating groundwater recharge from stream hydrographs, *J. Geophys. Res.*, vol. 66, no. 4, pp. 1203–1214, 1961.

Meyer, O. H.: Analysis of runoff characteristics: *Trans. Amer. Soc. Civil Engrs.*, vol. 105, pp. 83–89, 1940.

Miller, J. P.: Solutes in small streams draining single rock types, Sangre de Cristo Range, New Mexico, *U.S. Geol. Surv., Water Supply Papers*, 1535-F, 1961.

Parizek, R. R., L. J. Drew, and J. W. Bauer: Factors influencing streamflow variability of rivers draining folded carbonate terranes, Paper presented at the Geol. Soc. Amer. Ann. Meeting, Hydrogeol. Div., Atlantic City, N.J., Nov. 10–12, 1969.

Pelto, C. R.: Mapping of multicomponent systems, *J. Geol.*, vol. 62, pp. 501–511, 1954.

Pinder, G. F., and J. F. Jones: Determination of the groundwater component of peak discharge from the chemistry of total runoff, *Water Resources Res.*, vol. 5, no. 2, pp. 438–445, 1969.

Schneider, W. J.: Relation of geology to streamflow in the Upper Little Miami Basin, *Ohio J. Sci.*, vol. 57, pp. 11–14, 1957.

Shannon, C. E., and W. Weaver: "The Mathematical Theory of Communication," The University of Illinois Press, Urbana, 1949.

Thomas, H. E.: "The Conservation of Groundwater," McGraw-Hill Book Company, New York, 1951.

Todd, D. K.: "Groundwater Hydrology," John Wiley & Son, Inc., New York, 1959.

von Foerster, H.: On self-organizing systems and their environments, in "International Tracts in Computer Science and Technology and Their Application, Vol. 2: Self Organizing Systems," Pergamon Press, New York, pp. 31–50, 1960.

Walton, W. C.: Groundwater recharge and runoff in Illinois, *Illinois State Water Surv., Rept. Invest.*, 48, 1965.

Wenzel, L. K., Several methods for studying fluctuations of groundwater levels, *Trans. Amer. Geophys. Union*, vol. 17, pp. 400–405, 1936.

3
Optimization Models

The task of operating groundwater resource systems in a manner which is optimal in some sense is a complex operation. This problem was introduced from a general point of view in Chapter 1 and will be pursued here in more detail. The scope of the problem is restricted to those systems where groundwater is either the only source of water or the major component of a regional water supply.

Optimization is the problem of finding a best course of action from a set of alternatives. From a philosophical point of view, the problem appears quite simple: (1) define the objectives and the manner in which they might be achieved; (2) analyze the various courses of action, and select the one that best meets the objectives. In practice, it is easy to define or agree on those few objectives which reflect parallel interests, but numerous others are often conflicting. Further, it is generally not possible to define all alternative courses of action, nor is it possible to realize all far-reaching consequences of a given decision.

From a mathematical point of view, the problem is not so poorly defined. The methods employed vary considerably in degree of accuracy, complexity, and sophistication. At one extreme, simple graphical or arithmetic schemes may be used and may often give satisfactory results. At the other extreme, it is far easier to formulate more problems than can be solved by the most advanced technique. As in the previous chapter, some of the methods can be used in a reasonably straightforward manner, whereas others are quite general and require judgment and experience in problem formulation.

3.1 OVERVIEW OF OBJECTIVES AND CONCEPTS IN GROUNDWATER MANAGEMENT

This section deals with a brief overview of prevailing objectives and concepts in groundwater management. Material selected for review provides a background for sections to follow and is divided into four categories: (1) conventional concepts of safe and alternative yield; (2) socioeconomic objectives, including the broad theme of welfare economics and its influence on water law; (3) the criterion of optimal yield; and (4) suboptimization.

CONVENTIONAL CONCEPTS OF SAFE AND ALTERNATIVE YIELD

In an effort to remove some of the ambiguity in meaning of the term *safe yield* (Section 2.1), the Committee on Groundwater of the American Society of Civil Engineers introduced four concepts of yield, defined as follows (American Society of Civil Engineers, 1961):

1. Maximum sustained yield is the maximum rate at which water can be withdrawn perennially from a particular source.
2. Permissive sustained yield is the maximum rate at which water can economically and legally be withdrawn perennially from a particular source for beneficial purposes without bringing about some undesired result.
3. Maximum mining yield is the total volume of water in storage that can be extracted and utilized.
4. Permissive mining yield is the maximum volume of water in storage that can economically and legally be extracted and used for beneficial purposes, without bringing about some undesired result.

Sustained yield is usually expressed as a volume per unit time, and it can be maintained perennially. Maximum sustained yield is a use rate that is determined by, and limited to, average natural replenishment. Permissive sustained yield is invariably less than natural recharge owing to physical or man-made limitations. A case in point may be a withdrawal rate less than annual replenishment designed to prevent sea-water encroachment, land subsidence, interference with existing rights, or some other undesirable result.

Mining yield is a volume of extractable, nonrenewable water in a groundwater basin. It is an exhaustible resource of fixed supply, somewhat analogous to a mineral or petroleum deposit. It may be mined slowly, or rapidly, but the duration of extraction is definitely limited. Maximum mining yield is a volume of nonrenewable water that can be economically exploited, whereas permissive mining yield is that part of maximum mining yield that can be exploited without bringing about some undesired result. Development of groundwater resources in some parts of New Mexico is reported to be accomplished on a maximum-mining-yield basis (American Society of Civil Engineers, 1961).

The term *overdraft,* or *overdevelopment,* is generally reserved for the condition where withdrawals exceed sustained yield. This is occurring in several places in the United States. Five types of overdraft have been recognized (Snyder, 1955):

1. *Developmental overdraft,* a necessary first stage in groundwater development in that withdrawals cause a lowering of the water table in areas of natural recharge and discharge. This permits full utilization of the interaction between components of the hydrologic cycle.
2, 3. *Seasonal* (annual) or *cyclical* (periodic) *overdraft,* both characterized by a zero net change in water levels over a specified interval of time. Seasonal overdraft occurs when water levels at the beginning of the pumping season remain the same from year to year, but are in a continual state of decline during pumping seasons. Cyclical overdraft exists when water levels decline over two or more seasons, but eventually return to their original level. These types of overdraft are relatively unimportant and depend a great deal on the seasonal or annual demand for water.
4. *Long-run overdraft* is perennial pumping in excess of natural replenishment, which may lead, ultimately, to depletion.
5. *Critical overdraft* occurs when pumping leads to some undesirable physical result, restoration from which is technologically or economically impossible.

The question whether groundwater should be managed on a sustained- or mining-yield basis is not yet fully resolved and is controlled more by local conditions and demands than by policy decisions in advance of their absolute necessity. This is understandable in that there is likely to be little public sympathy for an announced depletion policy, whereas one of sustained use lends a ring of permanency. Whatever the merits of sustained- and mining-yield concepts, they are definitely ingrained in groundwater management.

SOCIOECONOMIC CONSIDERATIONS

From the viewpoint of social welfare, water has value only by virtue of use, some uses or rates of exploitation often generating a whole suite of interesting social problems. These problems are the spillover effects, external diseconomies, or extrasocial costs, as they are commonly referred to, which are sometimes inflicted on members of society as a feedback of public or private decisions. In particular, questions arise concerning the wisdom of use rates that lead, ultimately, to depletion because certain economic considerations made this the most profitable course. Hence, contrary to conventional mining-yield policies where only entrepreneurial costs and revenues come into consideration, a socially oriented policy on groundwater development would generally consider all costs and returns associated with a particular use rate.

Numerous investigators have questioned the role that water-resource development should play in achieving the objectives of social welfare. In addressing this question, Hufschmidt (1965) considered objectives of water-resource design in terms of human welfare, including income redistribution, increases in national income, economic growth, maintenance of a satisfying level of employment, and maximum productivity. Krutilla and Eckstein (1958) advocated the concept of economic efficiency as an objective. This is defined as

> ...a situation in which productive resources are so allocated among alternative uses that any reshuffling from the pattern cannot improve any individual's position and still leave all others as well off as before.

Hartman (1965) envisioned an economically efficient situation as one in which the incremental addition to income from the last input unit of the resource is equal in all uses. If Z is the quantity of a limited resource and $u_1, u_2, \ldots, u_n$ are valued outputs from uses of Z, then an allocation of Z to produce the highest-valued output will occur when

$$\frac{\Delta u_1}{\Delta Z} = \frac{\Delta u_2}{\Delta Z} = \cdots = \frac{\Delta u_n}{\Delta Z} \tag{3.1}$$

The idea of economic efficiency as described above underlies the theme of welfare economics, a concept which decision makers in the water-resource field are not yet fully prepared to accept. These ideals incorporate goals of income maximization for society in general and profit-maximizing behavior for consumers of resource outputs as acceptable guidelines in resource problems. Usually, private costs and revenues are assumed to be, respectively, social costs and benefits. Thus, the private producer who maximizes the difference between costs and revenues maximizes benefits that society receives from use of its natural resources. Income maximization for consumers of natural resources and an overall increase in national wealth mean also an increase in economic welfare. As pointed out by Scott (1955), optimum benefits cannot accrue to society when the social cost of resource use is in excess of the private cost of producing it. As a guideline, a rule has been formulated which states that any social, legal, economic, or institutional change is desirable which results in (1) everyone's being better off or (2) someone's being better off and no one's being worse off than before the change (Buchanan, 1959). This is equivalent to Krutilla and Eckstein's definition of economic efficiency and is one of the cardinal points of welfare economics.

A graphical example may clear up these ideas. The axes of Fig. 3.1 are presumed to measure the utility of two groups x and y. The concept of utility is used as a measure of "satisfaction with the status quo." It is assumed that point e is common to the utility of groups x and y. If a change can be made to increase the utility of x to point d, the y group is unaffected and presumably indifferent to the change. In accordance with the second half of the rule cited above, the change is desirable. A change bringing about an increase in the utility of y to point b is likewise not contrary to x's interests, which again satisfies the second part of the rule. A change bringing about an increase of utility to point c is beneficial to both groups, which satisfies the first part of the rule. On the other hand, an increase in y's utility to point a decreases x's utility by fb. Two questions then arise:

1. How much is y willing to pay to go to a?
2. How much is x willing to pay to stop y?

If the amount cited in 1 exceeds that of 2, group y can compensate x, which again satisfies the second part of the rule.

The part of the rule requiring that no one be worse off as a result of a change is equivalent to the part of the definition of economic efficiency to "leave all others as well off as before." In simple terms, this is the equalizer required to offset the undesirable aspects associated with unregulated exploitation of a groundwater basin. As policies that make everybody better off are not generally possible in water-resource developments, and for

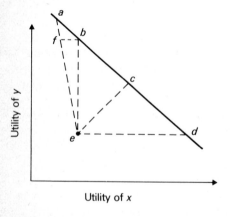

Fig. 3.1 Welfare economics criteria.

groundwater in particular, means for eradication of the undesirable effects have received considerable attention in recent years. The nature of these effects and some proposed solutions are best discussed within the context of a common-pool resource.

COMMON-POOL PROBLEMS AND PROPOSED SOLUTIONS

A common pool is defined as a "fugitive" supply of some commodity that is no one's personal property until it is physically reduced to possession. In that groundwater may be so described, pumpers overlying a basin are said to have *common* ownership in the pool regardless of man's attempt to assign private property rights of pumpage. As the most substantial characteristic of a common pool is withdrawal interdependency, individual withdrawals may reduce the quantity of water available to other users at a given cost level. Hence, with unregulated exploitation, it is seldom in the interest of a single user to conserve water, as what is not withdrawn this year is free to be withdrawn by others and therefore not available at the same cost level at some future date. Financial incentives and uncertainties associated with this type of exploitation require competitive pumpers to maximize short-run values of their withdrawals and thereby consciously or unconsciously to reduce future values of water left in storage to zero (Renshaw, 1963).

The consensus on common-pool problems is that some type of intervention is desirable when the onus of exploitation cost is nonspecifically distributed among members of the pumping community, or society as a whole. Examples in areas other than water resources are common. Federal intervention in the petroleum industry prior to and following World War I to regulate or influence production practices is classic in this respect. Regulatory control has since been relinquished to the Interstate Oil Compact Commission, which serves in an advisory capacity to states on such matters as quota and production practices that are in the best interests of the industry; suggested policies are generally accepted and enforced under an authority tantamount to police power. Another example is provided by the legal institutions for the wildlife and fisheries industry, where constraints on season, size, species, quota, territory, and equipment efficiency, among other things, are collectively interpreted as "management." For a detailed account, the reader is referred to Gordon (1954) for what is probably the best and most adequate contemporary treatment of the economic theory of a common-property resource. Common usage of waterways for waste discharge is another serious problem where exploitation costs are seldom specific to any individual polluter and large-scale intervention is imminent.

Scott (1955) has discussed policies designed to short-circuit unregulated exploitation of a common pool. Some of these policies in one form or another are already incorporated in groundwater laws of several states. The main theme of Scott's thesis is to reduce a nonspecific resource into one possessing "specificity" through simulation of a sole-ownership concept. The term *specificity* is used to classify resources in terms of incidence of cost of output. If all costs arising from an individual's pumping are borne by that individual, the resource is specific to its owner; the resource is nonspecific if the cost of an individual's pumping is shared by the pumping community, the industry as a whole, or society in general.

Advantages of sole ownership are obvious in the petroleum and fisheries industry, and may be equally advantageous in the case of groundwater development. For example, the rational sole producer will reduce user costs whenever possible, which effectively amounts to conservation. More explicitly, it is in the best interests of the sole owner to achieve optimal well spacing, to demand only a reasonable offtake from each well or well field, to prevent waste, and if water levels have to be lowered, to achieve lowering somewhat uniformly. It is also in the sole owner's interest to take steps to arrest salt-water intrusion or land subsidence, and to recharge the groundwater supply artificially if the practice is economically and physically feasible. Indeed, an unsatisfactorily resolved problem with artificial recharge in large basins is the unequal distribution of benefits among those who pay, or, as may be the case, accumulation of benefits to those who do not pay. Sole ownership eliminates this problem. In short, sole ownership makes user costs and benefits specific to one manager.

For reasons cited above, it is not likely that competitive pumping will be established within the structure of a single enterprise, with the result that many unacceptable characteristics of a common pool vanish. Scott (1955) proposed three methods for achieving this end: (1) prorating, or assignment of quotas; (2) unitization, or collectively treating divided interests as undivided interests; and (3) sole ownership itself. In discussing these aspects as they apply to water resources, Hirshleifer and his colleagues (1960) combined sole ownership and unitization under the heading "centralized decision making" and suggested a use tax as another means of accomplishing these ends.

Centralized decision making is most easily managed through a public water district or other entity charged with the responsibility of delivering water to a community. This management scheme allows the district to operate as a sole owner of the resource, thereby becoming the sole recipient of all costs of groundwater development and production. This form of management has been proposed for Las Vegas, Nevada (Leeds, Hill, and Jewett, Inc., 1961), and Smith Valley, Nevada (Domenico, 1967*a*). Assignment of quotas, on the other hand, is an unsatisfactory solution where private property concepts prevail, unless pumping quotas do not exceed the sustained yield of the resource.

Proponents devoted to equalizing social and private costs of development under the broad theme of welfare economics suggest "compensation in kind" as the classic solution. Water-law statutes for the state of Utah contain such a clause (Hutchins, 1965). Section 409 of the Model Water Use Act may also be interpreted as a compensation-in-kind clause (Trelease, and others, 1965). This section contains doctrine that presumably can be used to replace existing statutes pertaining to water-level decline in the states of Nevada, Utah, Washington, Wyoming, and Kansas in the interests of achieving uniformity in water law:

> Where application is made for permit and there is sufficient water available, but the use under the permit would interfere substantially and materially with a domestic use previously initiated, or with the water supply, water diversion facilities, or water power of a preserved use or a use made under a permit previously initiated, the commission may issue a permit subject to the condition that the permit holder furnish to the person whose use is interfered with a quantity of water or power equal to that lost by reason of interference.

The unit of measure of cost inflicted on the hypothetical injured party in the citation above is the deprived benefit, in this case water or increased power requirements as a result of water-level decline.

IMPACT ON WATER LAW

Groundwater resources are appropriated in various manners according to laws and institutions in individual states. In general, prevailing doctrines ignore the use-tax and sole-ownership method and rely on the assignment of quotas within a larger framework of unitization, or centralized state control. The appropriation doctrine is applied almost unequivocally in the most arid states where water is relatively scarce and most uses consumptive (Fig. 3.2). Quotas, or rights to use, are based on priority in time of beneficial use. A quota in this sense implies that each user has a specific share. Centralized, or state, control comes into play in curtailing groundwater development once all assigned quotas effectively exhaust safe yield, as in Nevada and Utah, in dictating minimum well-spacing restrictions in areas of overdraft or in the vicinity of surface-water rights, in revoking rights in the event the criterion of beneficial use is not satisfied, in establishing priorities of use in the event water is scarce and additional water required, or in authorizing withdrawals that limit water-level decline to a specified amount per year.

The riparian, or land-ownership, doctrine is a common policy in eastern and midwestern United States, where water was at least thought to be abundant and most uses nonconsumptive. A quota in the broader sense of the term is a reasonable use with respect to requirements of other riparians. Such a loosely defined quantity constitutes a poor definition of a property right and is only realistic if water is indeed abundant with respect to demands for its use. Under this policy, new rights may be established as long as new

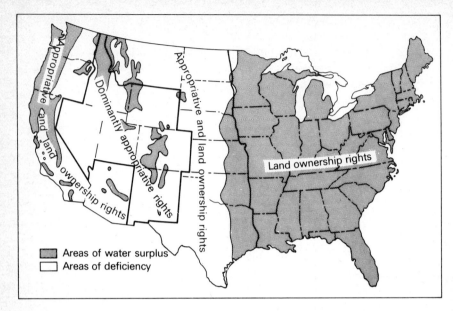

Fig. 3.2 Water-rights doctrines by states, and areas of water surplus and deficiency. (*After Thomas*, 1955.)

use is reasonable, priorities are not assigned on the basis of time, and centralized control is often reduced to a judicial matter in ascertaining the reasonableness of a particular use.

Several versions of this doctrine are applied in practice. The correlative rights doctrine of California, for example, recognizes all rights as equal, and appropriate shares or quotas are assigned on the basis of historic use in the event of shortages. The absolute ownership version, on the other hand, places no limitations on withdrawals or their effect on neighboring riparians. This doctrine provides a classic example of common ownership of a natural resource and exploitation under competitive pumping. Absolute ownership and reasonable-use doctrines do not generally differentiate between a renewable and nonrenewable supply.

Detailed accounts of water law have been presented by Milliman (1959), Bagley (1961), and Trelease (1965). It is not intended here to investigate which of the prevailing doctrines results in the closest agreement between social and private costs and benefits, and whether more emphasis should be placed on a private-property concept or on greater state control. Milliman (1959) stated that state ownership of water resources has always been common in western United States. Current eastern trends also point in this direction. It is sufficient to recognize groundwater law as a control that attempts to

OPTIMIZATION MODELS

minimize extrasocial costs by reducing a nonspecific resource to a quasi-specific state through private-property concepts.

THE CRITERION OF OPTIMAL YIELD

A third approach to resource management entails optimization of an integrated groundwater-surface-water system, or of an aquifer as a separate unit independent of a regional water supply. This approach requires operating rules such that the system is operated in an optimal manner. These are generally determined through some economic or social objective associated with uses to which the water is put. Accordingly, the criterion of safe yield is abandoned in favor of the yield required to meet this objective.

The optimal-planning problem has been described in detail by Bear and Levin (1967). Figure 3.3 shows the relation between inputs, or resources

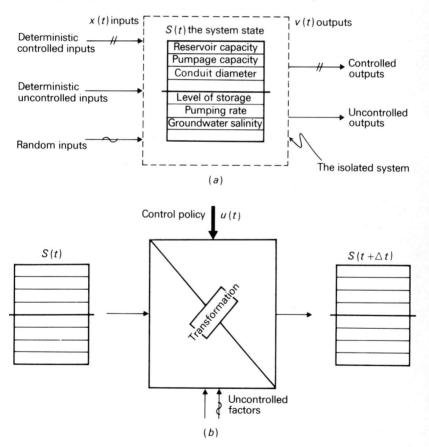

Fig. 3.3 Relation between inputs and outputs in a water-resource system with groundwater as a major component. (*After Bear and Levin*, 1967.)

needed for a production process, and outputs. The inputs are identified in terms of replenishment, both natural and artificial. Artificial replenishment is often deterministic and subject to control, whereas natural replenishment is not. Outputs include pumpage, a controllable variable, and outflows, such as spring discharge, which is generally not directly controllable.

If an operating rule is to be formulated in terms of the state of the system, the state must be known at all times. State variables are separated into those which describe the physical components of the system, such as pumping or conduit capacity, and those which describe the instantaneous level of operation, such as storage volume. Any transformation from state $S(t)$ to state $S(t + \Delta t)$ one time period later will depend on uncontrollable parameters, such as natural recharge; parameters which are determined by the system's state itself, such as a reduction or increase in outflow or spring discharge; and parameters subject to control, such as pumping and artificial recharge rates. For continuous time, these variables are related to each other by the continuity equation

$$\frac{dS}{dt} = R_N(t) + R_A(t) - X(t) - Q(S) \tag{3.2}$$

where t is time, S is storage in the aquifer, R_N is natural replenishment, R_A is artificial replenishment, X is pumpage, and Q is outflow or spring discharge, here considered a function of storage. In this formulation, S is a state variable, R_A and X are decision variables, Q is an uncontrollable (directly) output, and R_N is an uncontrollable input. For discrete time, the transformation from state S_t to state $S_{t+\Delta t}$ one time period later (Fig. 3.3b) depends on the control policy (pumping and artificial-recharge rates) to be executed over this period, as well as uncontrollable factors. Equation (3.2) becomes

$$S_{t+\Delta t} - S_t = \Delta t [R_{N_t} + R_{A_t} - X_t - Q(S_t, S_{t+\Delta t})] \tag{3.3}$$

where Δt is the length of a season, and the subscript t corresponds to the number of the season in a sequence of seasons (Bear and Levin, 1967).

The parameters subject to control in the above-cited problem constitute the decision variables. As the only justification for controlling certain variables is to achieve objectives, the objectives must be identified in terms of value, or values, to be maximized. The task of the rational decision maker then involves (Marshall, 1965):

1. The listing of alternative courses of action
2. The determination of the consequences that follow from each of the alternatives
3. The comparative evaluation of these sets of consequences in terms of the value, or values, to be maximized

OPTIMIZATION MODELS

The value, or values, to be maximized is termed an *objective function* and is taken as synonymous with the concept of benefit. The objective function is a variable since if it were constant with only one possible value, there would be no alternatives from which to choose and therefore no decision problem. The objective function is further defined as a dependent variable, since its value depends on the controllable and uncontrollable factors affecting the state of the system.

The presentation given above applies to a specific plan of operation requiring underground storage of surface waters. Buras (1966) presents good insight into the general problem. According to Buras, water is usually available at times, in places, or of a quality different from those which characterize the demand for it. This statement almost always refers to a surface-water resource. Further, the amounts of water available may be at variance with the quantities required for certain economic activities. This statement may apply to an isolated groundwater basin in a state of partial depletion, or to a surface-water system of inadequate quantity or quality. Considering this situation, Buras recognizes three important questions:

1. What system has to be built in order to minimize the discrepancy (in time, space, and quality) between the natural supply of water and the demand for it?
2. To what extent should the water-resource system be developed, and how extensive should be the region serviced?
3. How should the system be operated so as to achieve a given set of objectives in the best possible way?

Although much has been written about the integrated use of ground- and surface-water resources in general, these three questions drive right to the heart of the problem. They may be viewed as problems in design, development, and operation, respectively. Interdependency prevails, however, in that design of a water-resource system often determines operation.

SUBOPTIMIZATION

The point of view taken thus far is that the objectives of resource management may be classified in three groups, with considerable overlapping. Closer investigation reveals a complex of parallel and conflicting interests between group objectives. Consider, for example, the physical objective of maximizing safe yield. From the expression of continuity

$$S_{t+\Delta t} - S_t = \Delta t [R_{N_t} - X_t - Q(S_t, S_{t+\Delta t})] \tag{3.4}$$

we formulate the relation

$$\max R_N = \max[(Q_t - Q_{t+\Delta t})\chi] \tag{3.5}$$

where R_N is *captured,* or effective, natural discharge from time t to $t + \Delta t$; Q_t is natural discharge at time t; $Q_{t+\Delta t}$ is natural discharge one time period later; and $Q_t - Q_{t+\Delta t}$ is the difference in natural discharge over the time period as a function of the control vector χ of variables χ_i, with $i = 1, 2, \ldots, n$. The simple message of Eq. (3.5) is that the maximum possible value for recharge over a given time period is achieved by maximizing the difference in natural discharge over this period. As a minimum change in storage is desirable from the viewpoint of stability of production, the control vector χ, which maximizes R_N, has a minimizing effect on ΔS.

The vector χ is adjusted in accordance with the objectives of Eq. (3.5). This is generally accomplished in groundwater basins by adjusting the rate of exploitation and its spatial distribution. The components of the control vector were revealed several years ago by Theis (1940):

1. The pumps should be placed as close as possible to areas of rejected recharge or natural discharge where water either cannot enter the ground or is being lost by evapotranspiration.
2. The pumps should be placed as uniformly as possible in areas remote from areas of natural discharge or rejected recharge.
3. The amount of pumping in any one locality should be limited.

Clearly, the objectives cited above and the manner in which they may be achieved are not in any apparent conflict with the broad objectives of social welfare. However, when reduction in natural discharge affects prior rights established on surface-water resources or on spring flow, some conflicting factors arise.

Other conflicts may arise when two groups desire to use a common resource in different ways. This is exemplified by the inevitable conflict between holders of rights to pumpage and operators of a basin for storage and reservoir purposes. Thomas (1957) suggested a compensation policy for removal of this conflict. An additional conflict may occur when one group considers a certain management practice advantageous, whereas another group considers the same practice harmful. An example is the granting of additional rights to pump in a basin that is already overappropriated.

A conflict of interest, then, is the rule rather than the exception, and may arise between group objectives as well as between an individual's objective and the objectives of the group. As demonstrated in Fig. 3.1, when the action of one group or one individual to achieve an objective is independent of the other group, the action is justified for economic efficiency. However, as also shown by this figure, mutually dependent objectives may increase the utility or satisfaction of one group while reducing that of another. By and large, because of an inherent interdependency of withdrawals, this latter condition persists. The ideas presented on sole ownership and compensation are well suited (at least theoretically) to coping with this situation,

OPTIMIZATION MODELS

whereas much of the philosophy of western water law is aimed, rather unsuccessfully, at preventing the condition from happening at all.

The optimization of one objective that results in a lower degree of attainment of certain other objectives is termed *suboptimization*. In that any single perspective, whether physical, economic, or legal, is not adequate in itself, optimization of some objectives will result in a loss of future opportunities owing to suboptimization of others. The key problem concerns the nature of the resource itself and the constraints imposed on its development. However, criteria for optimum performance must be established. This requires a means for measuring whether or not the goals have been achieved. In that costs and benefits associated with various levels of performance are measurable, they are logical choices of the value or values to be minimized or maximized, respectively.

3.2 BASIC ECONOMIC AND HYDROLOGIC CONCEPTS

In private planning, the optimum state of resource development is generally taken as synonymous with a time distribution of use rates that maximizes the present value of future net revenues. This concept has been discussed by Ciriacy-Wantrup (1952) and Davis (1960). Formulated in terms suitable for groundwater problems (Ciriacy-Wantrup, 1942),

$$V = \int_0^T [B(X,t) - C(X,t)]e^{-\gamma t}\, dt \tag{3.6}$$

where V is present worth of future net revenues, T is the length of a planning horizon, B is the net revenue function, exclusive of pumping costs, as a function of rate of pumping X and time t, $C(X,t)$ is the pumping cost function, and $e^{-\gamma t}$ is the present worth factor. Time t enters in this case since benefits and costs are subject to variations related to the state of the economy and therefore beyond control. To simplify the discussion, such variations will be ignored.

As decision making in groundwater basins commonly entails choice between alternative courses of action, the value of alternative decisions may be compared on the basis of profit (or loss) that each will yield in the future. The present-worth calculation serves the purpose of reducing future net revenues in relation to their distance in time from the present. According to Grant (1950), present worth of a prospective series of future money receipts is the present investment that the future receipts would just repay with interest. This may be expressed

$$V = \frac{s_1}{1+\gamma} + \frac{s_2}{(1+\gamma)^2} + \frac{s_3}{(1+\gamma)^3} + \cdots + \frac{s_t}{(1+\gamma)^t} \tag{3.7}$$

where s is the value of the net benefit per time period t, and γ is the rate of interest. The ratio $1:(1 + \gamma)^t$ is termed the *present-worth factor for annual compounding* and is equal to the present value of a unit of net benefit to be obtained at time t. If the number of times the interest is compounded annually approaches infinity, the present-worth factor is expressed as $e^{-\gamma t}$ [Eq. (3.6)]. In a similar fashion, present worth of future disbursements may be thought of as the sum which, if invested now at interest, would provide exactly the funds needed to make these disbursements.

From the formulation of present worth, benefits accruing in future years are not as valuable as those of the same amount accruing closer to the present. Figure 3.4 shows how present worth of a constant annual return is affected both by the discount (interest) rate and the number of years over which the return is expected. This figure is merely a graphical representation of Eq. (3.7). The height of the surface above the point of intersection of the discount rate and the time horizon gives the present worth of the income stream. For these highly idealized conditions, high discount rates favor high rates of groundwater production in the early production periods. Stated in another way, too slow a rate of production may postpone benefits further into the future than warranted by the rate of interest. It is then clear why the length of the planning horizon and the rate of production should be controllable variables in any decision problem where maximization of the income stream is the objective. In reality, the present-worth system is warped and twisted in the direction of time because of technological innovation or advance, a change in prices and costs, and, especially in the case of an uncertain future water supply, a change in the discount factor.

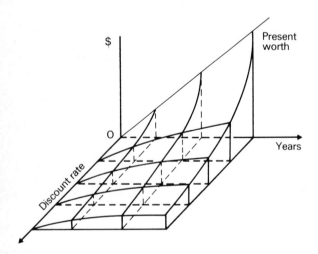

Fig. 3.4 Present-worth surface as affected by time and interest rate. (*After Grubb*, 1966.)

OPTIMIZATION MODELS

As stated, Eq. (3.6) contains no provisions for parameters dealing with natural replenishment or with the cumulative effects of continued use. A decision to pump a certain amount of water in excess of natural replenishment will transform the state of the system from some storage $S(t)$ to some other storage $S(t + \Delta t)$, which will increase pumping costs. Such a change may be functionally described by the hydrologic equation

$$\frac{dS}{dt} = R - X(t) \tag{3.8}$$

where $X(t)$ is the pumping rate as a function of time, and replenishment R is assumed constant. By integration,

$$S(t) = Rt - \int_0^t X(\tau)\, d\tau + B \tag{3.9}$$

where B is a constant of integration to be determined from the initial conditions. For S equals S_0 at t equals t_0, B equals S_0 and

$$S(t) = S_0 + Rt - \int_0^t X(\tau)\, d\tau \tag{3.10}$$

where S_0 is original storage.

It follows that the effects of recharge and changes in storage may be incorporated into Eq. (3.6) by recognizing that pumping costs are functionally related to both. This gives

$$V = \int_0^T \{B(X) - C[S(t)]\} e^{-\gamma t}\, dt \tag{3.11}$$

where the integrand is the objectives function to be maximized, the pumping rate is the controllable variable, and the pumping-cost function is dependent on the level of groundwater storage.

The important feature here is that the pumping-response surface is assumed to be completely described by the simple hydrologic equation. This means that the effect of a given extraction is averaged over the entire basin. In this sense, the response is independent not only of the extraction pattern, but of the hydrologic properties which control the flow of groundwater to wells. The pumping-response surface is thus taken as a featureless surface that is everywhere the same, a gross idealization characteristic of nearly all optimization studies involving groundwater withdrawals.

3.3 STRATEGIES FOR ISOLATED GROUNDWATER BASINS

An isolated groundwater basin is one which interacts with other components of the hydrologic cycle, but is void of sufficient surface-water supplies (Fig. 3.5). For economic or other reasons, it is further stipulated that such a

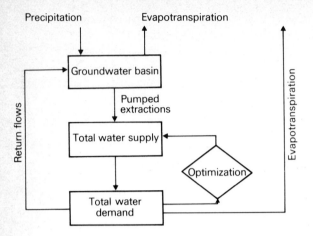

Fig. 3.5 Schematic representation of an isolated groundwater basin.

basin is likely to remain closed to interbasin transfers of water for some time in the foreseeable future. This type of system is common in western United States, and the problems associated with its development are typified by the fact that more than one-fourth of all groundwater withdrawn is mined. Optimization, or the problem of finding a best course of action from a set of alternatives, must then be considered in a somewhat restricted manner in that the alternatives are somewhat restricted. At one extreme, excessive unregulated mining inescapably tends toward exhaustion of the storage reserve during a finite time period. At the other extreme, infinite preservation of storage associated with a safe-yield policy may be excessively conservative. Clearly, the "best" course, if one exists, lies intermediate between these extremes. This suggests that any optimal scheme is one that somehow incorporates consideration of both the present and the future, or, more specifically, one that considers the value of present and future benefits and costs.

THE VALUE OF PRESENT USES AND FUTURE COSTS: TWO EXAMPLES

An example or two taken from the literature may demonstrate the concept of value of present uses and future costs. Renshaw (1963) recognized that in an area of active groundwater mining, such as the Southwest's high plains, withdrawals may eventually be reduced to safe yield, with no further lowering of the water table. Hence, water left in the aquifer during the period of mining has value to the extent that it will reduce pumping costs once basin draft is limited to natural replenishment. The problem then deals with the comparison of present values associated with present uses and future costs

OPTIMIZATION MODELS

under a sustained-yield policy. This comparative evaluation is made in rather simple terms by considering an ideal pumping-response surface. Specific yield (see Section 6.2) is defined as the volume of water derived from storage in response to a change in water level over a given area of aquifer, and it may be expressed

$$\text{SY} = \frac{\Delta V}{\Delta h\, A} \tag{3.12}$$

where SY is specific yield, ΔV is the volume of water drained from the aquifer in response to the change in water level, Δh, which is effective over the area A. The release of 1 acre-ft of water, ΔV, from an aquifer underlying an area of 1 acre, A, will result in a water-level change of

$$\Delta h = \frac{1}{\text{SY}} \tag{3.13}$$

Introducing the marginal cost of pumping, mc, as the cost required to lift 1 acre-ft of water 1 ft (consisting largely of power charges, but properly including some charges on equipment), annual savings resulting from leaving water in storage may be expressed

$$\frac{\text{mc}\, R}{\text{SY}}$$

where R is recharge to be pumped under a safe-yield policy, in acre-feet per acre per year. In that this savings is available in perpetuity, the appropriate interest rate, or capitalization factor, is $1/\gamma$, giving

$$\frac{\text{mc}\, R}{\text{SY}\, \gamma}$$

If the marginal cost of pumping is taken as $0.05 per acre-foot per foot of lift, and specific yield as 0.2, the removal of 1 acre-ft of water from storage will lower the water level by 5 ft, which will permanently increase the future cost of pumping 1 acre-ft of water by $0.25. In that this is a permanent cost, its capitalized value at an interest rate of 5 percent is $5.00. If the recharge to be pumped from this area under a future safe-yield policy is 2 acre-ft/acre, the capitalized value of leaving 1 acre-ft of water in storage is $10.00. This can be compared with the value of 1 acre-ft of water in current use.

Another example is provided by Kelso (1961), who examined a question of alternatives in an area of central Arizona where 808,000 acres are supplied irrigation water from groundwater pumping. Natural replenishment is sufficient for irrigating 150,000 acres, the remaining 658,000 acres being irrigated by mining from groundwater storage. Under the present (1961) practice of applying $5\frac{1}{4}$ acre-ft/(acre)(year), the net return per acre before paying for water is about $100. By assuming that the marginal cost of pumping is $0.04 per acre-foot per foot of lift, the break-even point (that is, the point where water costs equal net returns) is about 455 ft. By reducing the per acre application of water to $3\frac{2}{3}$ acre-ft, net revenue will drop to $75 per acre, and the break-even depth increases to 511 ft. Calculations indicate that at a depth of 342 ft, net returns will be the same ($25 per acre-foot) whether $5\frac{1}{4}$ or $3\frac{2}{3}$ acre-ft are used. At the current rate of exploitation, this depth will be reached in 24 years beyond 1960.

The alternatives addressed by Kelso are:

1. Continue using $5\frac{1}{4}$ acre-ft/acre for 24 years, and then reduce this application to $3\frac{2}{3}$ acre-ft.
2. Immediately reduce applications to $3\frac{2}{3}$ acre-ft.

The results are shown in Fig. 3.6. The ordinate is the net return minus pumping costs. Line A shows the effect of continued withdrawals at the present rate, line AB shows the effect of reduction to $3\frac{2}{3}$ acre-ft 24 years hence, and line C shows the effect of immediate reduction. Note the life span of the resource associated with each alternative.

It is clear from this figure that the revenue gained by exercising option 2 ($WXYZ$) is larger than the revenue sacrificed (UVW). The problem reduces to ascertaining the present value of revenues gained and foregone. At an 8 percent rate, the present worth of the foregone revenue, which is closest to the present, is more than twice as great as the present worth of future gains. Clearly, the gains do not offset the losses. At a rate of $4\frac{1}{2}$ percent, the gains are just balanced by the losses.

The examples cited above were the forerunners of more sophisticated optimization studies for isolated groundwater basins. Although the techniques currently employed are considerably more sophisticated, the

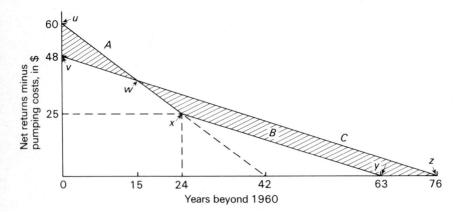

Fig. 3.6 Net returns minus pumping costs per acre for groundwater, with different amounts of water applied per acre. (*After Kelso*, 1961.)

OPTIMIZATION MODELS

fundamental idea of discounted future benefits and costs and the ideal lumped-parameter response surface remain essentially unchanged.

SAFE–YIELD VERSUS ALTERNATIVE–YIELD POLICIES

Safe-yield and alternative-yield policies dominate the real-life decision process in that they are consistent with the constraints imposed by man and nature. Hence, it is not unreasonable to ask which of these policies is best for achieving some balance between present and future costs and benefits. With the assumption that each of the four policies described in Section 3.1 is feasible within the constraints that are operative, they are considered the only admissible inputs to the problem. Implementation of each policy yields an output, in this case the value of one objective function common to all policies. In principle, a satisfactory method would rank the input-output pairs in accordance with the above stated criterion, selecting the one that is most satisfying in this regard (Duckstein and Kisiel, 1968).

In order to formulate the problem so that it can be examined in one sweep instead of in pieces, it is necessary to incorporate the economic ramifications of each policy in one mathematical statement. This can be accomplished in the following manner: It is assumed first that water is applied in some productive process, with $p(t)$ designating net profit per unit pumped as a function of time. Profit per unit time is expressed

$$p(t)X(t) = p(t)[R(t) + q(t)] \tag{3.14}$$

where rate of pumping $X(t)$ is the sum of recoverable replenishment $R(t)$ and any given rate of storage withdrawal $q(t)$. By designating the time when such pumping is no longer economical as *time of exhaustion, T*, present worth of a maximum-mining-yield policy may be formulated in the same general form as Eq. (3.6),

$$V_{mm} = \int_0^T [p(t)q(t) + p(t)R(t)]e^{-\gamma t}\,dt \tag{3.15}$$

where V_{mm} is present worth.

On the other hand, if withdrawals are to be reduced to recoverable replenishment at some time $t' < T$ in accordance with a permissive-mining yield policy, present worth must be reformulated to incorporate the value of perennial use of recharge for all $t > t'$. Further, if natural replenishment is assumed to be recoverable only as long as water levels in the reservoir are being lowered, $R(t)$ is a variable for all $t < t'$ and equals some constant $R(t')$ thereafter. Viewed in this light, the problem can be formulated as the sum of three definite integrals ranging over a time period from the present to infinity (Domenico and others, 1968).

$$V(t') = \int_0^{t'} p(t)q(t)e^{-\gamma t}\, dt$$
$$+ \int_0^{t'} p(t)R(t)e^{-\gamma t}\, dt$$
$$+ \int_{t'}^{\infty} p(t)R(t')e^{-\gamma t}\, dt \tag{3.16}$$

The three terms of Eq. (3.16) give, respectively, present worth of net receipts forthcoming from use of groundwater in storage, use of recoverable replenishment over the mining period, and use of recoverable replenishment after mining ceases. The third term is an evaluation of the remaining worth of the basin after it has been partially depleted. As t' approaches T, the value of this term approaches zero, and the formulation reduces to that of maximum-mining yield [Eq. (3.15)]. On the other hand, under a sustained-yield policy, the first and last terms are zero, and the upper limit of the integral of the second term approaches infinity. Hence, all policies are incorporated in this one mathematical statement. The best policy may be deduced by merely determining the maximum value of Eq. (3.16) with respect to the time mining should stop. Designating T' as the optimal time for reducing withdrawals to recoverable replenishment,

$$\frac{dV(T')}{d(T')} = p(T')q(T') + \frac{p(T')}{\gamma}\frac{dR(T')}{dT'} + \frac{R(T')}{\gamma}\frac{dp(T')}{dT'} = 0 \tag{3.17}$$

The first term of Eq. (3.17) is the value of the groundwater mined during the last year T' of mining. The second term is the present value of natural replenishment recovered during the last year of mining. Once captured, this incremental volume is available in perpetuity; thus, its worth takes the form of a perpetual annuity starting at T'. The third term, which is negative as long as profit per unit pumped diminishes with water-level decline, gives the present worth of the loss in value of recharge to be pumped after mining ceases that is due to the increased lift incurred during the last year's mining. This is a cost in perpetuity. The condition expressed by Eq. (3.17) takes account of the simple fact that groundwater mining in agricultural areas:

1. Generates funds from use of the one-time reserve of storage that become available on a year-to-year basis.
2. May increase the annual natural replenishment to the basin and thereby increase the rewards associated with its use in perpetuity.
3. Reduces in perpetuity the value of the resource as a future supply of water in proportion to the annual increase in pumping lifts.

OPTIMIZATION MODELS

In that recharge is finite, this happy situation cannot go on indefinitely. A literal interpretation of Eq. (3.17) is that it is profitable to continue groundwater mining until the net annual gain from use of groundwater in storage plus the capitalized annual net gain of increased natural replenishment equals the capitalized annual net loss in value of the resource as a future supply of water. This decision rule provides the basis by which various sustained- and mining-yield policies may be compared.

The fundamental ideas of this analysis can best be examined by assuming that natural replenishment is constant and independent of the state of storage development, so that the middle term of Eq. (3.17) vanishes. This gives

$$p(T')q(T') + \frac{R}{\gamma}\frac{dp(T')}{dT'} = 0 \tag{3.18}$$

The decision rule now states that groundwater mining is profitable as long as annual net revenue generated from mining exceeds the capitalized annual loss in value of recharge. Figure 3.7 shows the application of this modified rule for some hypothetical data. The height of curve 1 for any year represents the net receipts from mining for that year. The height of curve 2 for any year represents the capitalized annual loss in value of the resource as a future supply of water. The net for any given year is the vertical distance between the curves, and it is positive for all values for $t < T'$, zero for t equals T', and negative for $t > T'$. Viewed another way, curve two represents a part of the annual gains from mining which if invested at γ at their time of receipt, would repay for all time the reduction in value of the supply attributable to mining operations for that year. For any year's operation, this sum represents a hypothetical return to the mine to make all future operating costs identical with what they were previous to mining. Incremental returns from mining beyond T' are not sufficient to pay for incremental losses by the amount designated by the vertical distance between the curves. This demonstrates why $V(T')$ is a maximum if mining stops at T', clearly showing the role that present and future costs and benefits play in the analysis.

For the assumptions used in arriving at Eq. (3.18), Eq. (3.16) reduces to

$$V(T') = \int_0^{T'} p(t)q(t)e^{-\gamma t}\,dt + \int_0^{T'} p(t)Re^{-\gamma t}\,dt + \frac{Rp(T')e^{-\gamma T'}}{\gamma} \tag{3.19}$$

A maximum $V(T')$ corresponding to this statement is clearly the discounted net between the two curves of Fig. 3.7, plus the value of

$$V = \int_0^{\infty} \varepsilon Re^{-\gamma t}\,dt \tag{3.20}$$

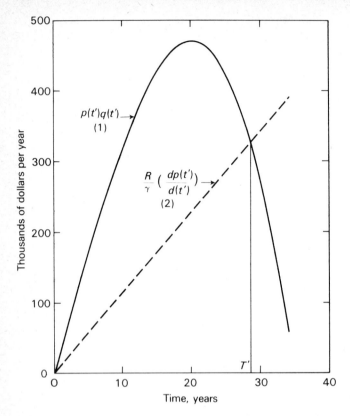

Fig. 3.7 Plot of net annual gains from use of groundwater storage, and capitalized annual loss in value of the resource as a future supply of water under conditions of constant recharge. (*After Domenico and others*, 1968.)

or $\varepsilon R/\gamma$, where ε is a constant profit per unit pumped under a safe-yield policy. Equation (3.20) gives present worth of future net receipts associated with perennial use of natural replenishment and is derived from Eq. (3.16) when withdrawals never exceed safe yield. If the mining period extends beyond T', the discounted net is reduced by the area between the curves beyond T'

Figure 3.8 demonstrates an example where recoverable replenishment is dependent on the state of storage development. Interpretation is carried out in the same manner as in Fig. 3.7.

To arrive at less abstract decision rules than those expressed by Eqs. (3.17) and (3.18), it is necessary to introduce some simply structured profit function $p(t)$ and to relate it to an equally simple function describing water-

OPTIMIZATION MODELS

level decline in response to storage withdrawals. The profit function is selected by

$$p(t) = \varepsilon - mc\, h(t) \qquad (3.21)$$

so that net profit per unit pumped can vary linearly with cumulative water-level decline $h(t)$. The constant ε in this equation is the net unit value of water at time t equals zero. The analytical relation between water-level decline and storage withdrawals must, of mathematical necessity, exclude all considerations of space and is of the form

$$h(t) = \int_0^t \frac{dh}{dS} q(\tau)\, d\tau \qquad (3.22)$$

where S is the volume of groundwater removed from storage, and dh/dS is

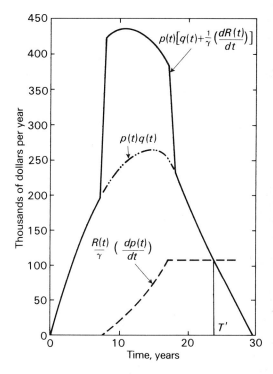

Fig. 3.8 Plot of net annual gains from use of groundwater storage and capitalized annual loss in value of the resource as a future supply of water when recharge is a function of storage. (*After Domenico, 1967.*)

assumed constant for the entire basin. This allows Eq. (3.21) to be expressed as

$$p(t) = \varepsilon - mc\,\bar{K} \int_0^t q(\tau)\,d\tau \qquad (3.23)$$

where $\bar{K}$ is dh/dS, or water-level decline per acre-foot of storage withdrawal. By recognizing further that

$$\frac{dS(t)}{dt} = q(t) \qquad (3.24)$$

substitution of Eqs. (3.21) and (3.24) in Eq. (3.18) yields, eventually (Domenico and others, 1968),

$$a_m = \frac{\varepsilon}{mc\,\bar{K}} - \frac{R}{\gamma} \qquad (3.25)$$

In this equation, a_m is designated *optimal-mining yield*, or the volume of groundwater storage that may be withdrawn in accordance with the condition stipulated by the decision rule of Eq. (3.18).

It may be noted that the quantity a_m gets larger with increasing interest rates, with increasing valued uses to which the water is put, decreasing values of the idealized lumped parameter $\bar{K}$, decreasing values of natural recharge, and decreasing marginal costs of pumping. The first part of this equation, $\varepsilon/mc\,\bar{K}$, gives the total volume of minable water above the economic limit of pumping. This is easily illustrated by setting the profit function of Eq. (3.23) equal to zero and rearranging terms. This corresponds to maximum-mining yield. The second part of this equation, R/γ, represents the capitalized annual volume of natural replenishment. Hence, the mining volume required to satisfy the decision rule of Eq. (3.18) is merely the difference between maximum-mining yield and the capitalized annual volume of natural replenishment. At one extreme, as the recharge rate becomes small, or the interest rate becomes large, the optimal volume approaches maximum-mining yield. At the other extreme, as recharge becomes large, or the interest rate becomes small, the minable volume approaches zero, which indicates that a safe-yield policy is most desirable. Hence, the optimal-mining volume is a variable that may take on values ranging from zero (safe yield) to the maximum amount of usable water in storage (maximum-mining yield), depending on the value of pertinent hydrologic and economic variables.

A TEMPORAL ALLOCATION

Inasmuch as the volume of minable storage of the previous discussion was arrived at independently of the rate of production that would exhaust it, the total supply is rendered fixed from the point of view of current technology,

OPTIMIZATION MODELS

enterprise selection, natural recharge, and the rate of interest. Discussions in Section 3.2, however, suggest that a most profitable rate of production is the factor sought in most optimizing schemes. Although such a rate cannot be deduced from what has gone before, a few facts about it are known. First, in accordance with our criterion of best as it applies to both present and future values, whatever the optimal withdrawal rate, it must eventually converge on natural replenishment and thereafter remain steady. Further, according to Fig. 3.4, too slow a rate of convergence may postpone benefits further into the future than warranted by the rate of interest. In the absence of the latter consideration, the minable storage volume determined previously must be considered optimal only for the conventional alternative-yield policies regarded as admissible, and the possibility of greater economic advantages should be considered by admitting policies formulated as use rates.

The question of an optimal temporal allocation has been addressed in a number of papers by Burt (1964a, 1964b, 1966, 1967a, 1967b). We shall examine Burt's (1964a, 1967a) approximate decision rule for isolated groundwater basins in order to promote an understanding of the economic forces involved. In its simplest form, the approximate decision rule is expressed

$$\frac{\partial G(X,S)}{\partial X} = \frac{1}{\gamma} \frac{\partial G(X,S)}{\partial S} \tag{3.26}$$

where $G(X,S)$ is the net benefit derived from use of water as a function of the withdrawal rate X and the amount of groundwater in storage S, and γ is the rate of interest. Assuming that groundwater is used in agricultural production, the net benefit function can be expressed

$$G(X,S) = B(X) - C(S)X \tag{3.27}$$

where $B(X)$ is the net annual benefit before paying for water, and $C(S)$ is the cost per unit pumped as a function of the amount of water in storage. Substituting Eq. (3.27) in Eq. (3.26) and carrying out the differentiation,

$$\frac{B'(X) - C(S)}{X} = -\frac{1}{\gamma} C'(S) \tag{3.28}$$

Interpreted literally, pumpage from the basin is expanded to the point where marginal net benefit per unit pumped [left-hand side of Eq. (3.28)] equals the negative of capitalized marginal pumping costs with respect to water in storage (Burt, 1964a). This rule applies to seasonal pumping quantities and emphasizes that water not used in current production has value to the extent that it reduces future pumping costs, the savings being in the form of a perpetual annuity. Hence, the concept of optimization applied to the isolated groundwater basin is once more identified with the value of present and future benefits and costs.

In order to put this program into action, it is again necessary to consider a differentiable benefit function and its relation to some simple function describing water-level response to pumping. If the benefit function is assumed to be quadratic in X (Fig. 3.9),

$$B(X) = a_1 X - b_1 X^2 \tag{3.29}$$

where $B(X)$ is in thousands of dollars per year, and X is in thousands of acre-feet per year. Cost per unit pumped is expressed

$$C(S) = a_2 - b_2 S \tag{3.30}$$

where a_2 is the unit maximum-pumping cost, and b_2 is defined as

$$b_2 = mc \frac{dh}{dS} \tag{3.31}$$

where mc is the marginal cost of pumping, and dh/dS is water-level decline with respect to storage withdrawals. As in previous discussions, spatial considerations of the pumping-response surface are ignored, and dh/dS is treated as a lumped average for the entire basin. This gives

$$C(S) = mc\, H - mc\, \bar{K} S \tag{3.32}$$

where H designates the thickness of the saturated and unsaturated portions

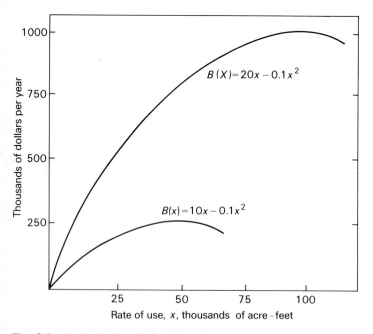

Fig. 3.9 Quadratic benefit functions.

of the basin, and $\bar{K}$ is a constant water-level decline per acre-foot of storage withdrawals. Note that mc H represents the cost per unit pumped at maximum-pumping depth, and mc $\bar{K}S$ is the unit cost for various levels of storage S, with $\bar{K}S$ representing the pumping lift in feet. The product $\bar{K}S$ equals H for full saturation.

With appropriate substitution, the net benefit function of Eq. (3.27) becomes

$$G(X,S) = a_1 X - b_1 X^2 - \text{mc } HX + \text{mc } \bar{K}SX \tag{3.33}$$

and, in accordance with the decision rule of Eq. (3.26),

$$\frac{\partial G(X,S)}{\partial X} = a_1 - 2b_1 X - \text{mc } H + \text{mc } \bar{K}S \tag{3.34}$$

and

$$\frac{\partial G(X,S)}{\partial S} = \text{mc } \bar{K}X \tag{3.35}$$

Substituting Eqs. (3.34) and (3.35) in Eq. (3.26) and solving for the use rate X,

$$X = \frac{a_1 - \text{mc }(H - \bar{K}S)}{(1/\gamma) \text{ mc } \bar{K} + 2b_1} \tag{3.36}$$

The units of X are easily verified

$$\frac{(\$/\text{acre-ft}) - (\$/\text{acre-ft}^2)(\text{ft} - \text{ft})}{\text{yr }(\$/(\text{acre-ft}^2)(\text{ft}/\text{acre-ft}) + [(\$ \text{yr})/(\text{acre-ft})^2]} = \frac{\text{acre-ft}}{\text{yr}}$$

For a given problem, the optimal-withdrawal rate X is solved as a function of storage, and the other parameters are assigned constant values. For Eq. (3.36), the decision rule is linear (Fig. 3.10). Hence, given a particular value of storage, the amount to be withdrawn for current use is determined from Eq. (3.36), or its graphical solution. Since one value of storage will correspond to a use rate X_R equal to natural replenishment, equilibrium storage $\bar{S}$ is found by solving Eq. (3.36) for S in terms of the use rate X_R equal to natural replenishment,

$$\bar{S} = \frac{[(1/\gamma) \text{ mc } \bar{K} + 2b_1]X_R + \text{mc } H - a_1}{\text{mc } \bar{K}} \tag{3.37}$$

where X_R is a use rate equal to natural replenishment. This value of storage is given on the curve of Fig. 3.10 where rate of use equals rate of replenishment. The optimal policy then specifies use rates above natural replenishment when storage is above its equilibrium value, and use rates below natural replenishment when storage is below its equilibrium value, the

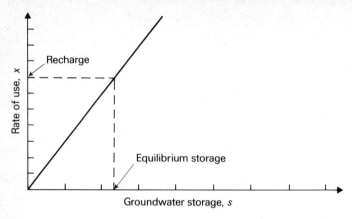

Fig. 3.10 Graphical presentation of linear decision rule for a temporal allocation of groundwater in an isolated basin.

tendency being to move toward equilibrium storage from either above or below.

In examining the optimal policy for quadratic benefit functions, the following conclusions are offered (Burt, 1967a):

1. Low values of recharge tend to decrease equilibrium storage.
2. An increase in the interest rate increases rate of use at a given storage level and decreases equilibrium storage.
3. An increase in net price received for goods that are dependent on the water supply increases the rate of use at a given storage level and decreases equilibrium storage.
4. An improvement in pumping technology (reflected through mc) increases rate of use for a given storage level and decreases equilibrium storage.

These conclusions may be verified by examining Eqs. (3.36) and (3.37). Note that the economic and hydrologic forces tending to either increase or decrease the terminal position of equilibrium storage correspond to those tending to increase or decrease optimal-mining yield (a decrease in equilibrium storage is synonymous with an increase in optimal-mining yield). It should be noted that the rule just given is only approximate and is valid only if storage is large relative to the difference between rate of use and natural replenishment.

It is now clear that both the length of a planning horizon and the rate of production should be controllable variables in any decision problem where maximization of an income stream is the objective. Hence, there are definite economic advantages in formulating a policy of annual withdrawals as a function of the state (storage) of the system, rather than as a definite (safe-yield) or unregulated (mining-yield) quantity.

3.4 RELATED PROBLEMS IN VALUATION

The ideas presented in the preceding section are difficult to apply in most real world situations. In the first place, the benefit function is not easily arrived at when water is put to domestic, municipal, or other "intangible" uses not evaluated through the market place. When water is employed as a factor in production, as in agriculture or some other industries, approximating functions may be found. Even in these instances, it is seldom possible to devise planning periods and rates of use in conjunction with known revenue and cost functions such that a maximum present worth is a planned objective. Experience indicates that groundwater is developed, the enterprise or community dependent on the supply expands, and a greater use rate is demanded. For those numerous cases where groundwater mining has already become a way of life, intervention schemes that require, eventually, a use rate no greater than safe yield in the interests of optimization are clearly contrary to short-term private interests. Generally, they are unenforceable. In other instances, when the importation of surface water might be a strong possibility at some point in the future if an appropriate economic base is established, any intervention scheme that attempts to drive use rates toward safe yield is a gross case of suboptimization. For these as well as other reasons, the more expedient approach has been to devise intervention schemes that mitigate private exploitation only when continuity of resource use is threatened. Some areas require these safeguards sooner, and some later, but the end, historically, has been the same for all. A few such schemes will be discussed below.

THE COST DEPLETION DEDUCTION

With much of the recent effort in water resources dealing with use tax and other monetary transfers aimed specifically at reducing the problems of commonality and, in general, tending toward conservation, it is in keeping with differing philosophies in development of water resources to find forces of equal or greater magnitude working in the opposite direction, that is, toward depletion, and apparently with some success. Reference is made specifically to the decision by the United States Court of Appeals in 1965 allowing a tax deduction for depletion of groundwater in the test case of the United States v. Shurbet (*The Cross Section,* June, 1965). The trial court's findings not only covered the taxpayer concerned, but those of the entire southern High Plains and the Ogallala groundwater reservoir within it. Although the Court stated specifically that its decision was not intended to furnish a precedent except under the peculiar conditions of the southern High Plains, it is rather difficult to conceive of exhaustion of capital investment in groundwater being peculiar to any one area in the United States. This is due to the common practice of capitalizing the value of a pseudofree water supply into land value.

Perhaps the one point of this decision peculiar to the High Plains is the formula devised to measure the allowable deduction. The taxpayer, among other things, must prove purchase price of land and allocate part of this price to water. In the High Plains, where the direction has been from dryland farming to irrigated land, the allocation of price to water in the simplest case is the price differential between irrigated land and otherwise comparable dry land. If, for example, the price differential is $400 per acre and if there was 200 ft of water under the land when purchased, water is valued at $2 per foot. Given an annual decline of 4 ft/year, the depletion deduction is $8 per acre.

It is not intended here to comment on the merits of this decision, but merely to investigate its influence on continued resource use. A depletion allowance acts to reduce tax payments, which introduces credits on the cost side of the ledger. In the models introduced earlier, this is rectified by increasing the net benefit function by the amount of the allowed deduction times an appropriate tax-rate schedule.

On the assumption that the allowance is discontinued after exhaustion of capital investment, its effect on continued resource use depends on whether the assigned saturated thickness terminates above or below what can be considered an economical pumping lift for that area. If pumping is still profitable once the assigned thickness is depleted, the true state of affairs of the allowance is a long-term transfer of government funds to private pumpers, in essence, a temporary increase in the unit value of water through subsidy. The allowance increases present worth, but adds nothing to the life of the resource for the particular use upon which its value was predicated. In fact, the more rapid the depletion, the higher the present worth, which favors early depletion by pumpers.

If, on the other hand, termination of the assigned thickness is well below the level from which water can economically be developed, the life of the resource is theoretically extended. It follows, of course, that capital investment may not be fully recoverable or may be recoverable farther in the future than warranted by present investments to increase efficiency of extraction.

It appears, therefore, that the tax deduction for depletion of groundwater is merely a temporary subsidy which, like the resource itself, provides funds subject to reduction in value in relation to their time distance from the present. The more rapid the depletion, the more profitable the return. The economic life of the resource is certainly not extended by such an allowance, but, on the contrary, may be shortened.

TIMING OF SURFACE-WATER IMPORTATION

During the course of continued overdevelopment, many groundwater basins deteriorate gradually, which leads to lower efficiency and higher extraction costs. History teaches us that numerous such isolated systems evolve, sooner or later, into integrated systems where the groundwater component is the major unit. This has already occurred in many parts of California, in Las Vegas Valley, Nevada, in parts of Arizona, and in parts of Israel. Much of the current effort in the High Plains area after delaying tactics, such as the depletion allowance, is aimed specifically at integrating that system with a surface-water supply. Indeed, a fundamental question arises whether or not the best use of groundwater in a given area is one of rapid exploitation, contributing to the development of an economic base which can support a more expensive and more efficient integrated system.

Given the situation of gradual deterioration, it is suggested that the data on deterioration can help determine the conditions which justify the termination of storage depletion and partial replacement with a more dependable supply. The effect of pumping costs in determining this condition should not be underestimated. For example, if groundwater storage is to be reduced to the extent of perennially increasing pumping costs as much as $0.25 per acre-foot, 1 acre-ft of water pumped each year would—at present value calculated at 5 percent interest—increase the unit cost by $5. If the amount to be pumped with a supplemental surface-water supply is on the order of 100,000 acre-ft, the additional cost is $500,000. Hence, surface water used directly is worth at least as much as the marginal value of water in storage, plus pumping costs, corrected for importation costs.

If it is recognized that costs as well as benefits incurred at different points in time have different values when viewed from the present, an immediate inference from Eq. (3.16) is a companion equation for discounted future disbursements. Provided that the surface supplies will be utilized to sustain all use rates in excess of basin replenishment, the companion equation can be formulated for an anticipated transition from groundwater storage to an imported supply at some time t',

$$c_{\text{PW}}(t') = \int_0^{t'} [q(t) + R(t)]c(t)e^{-\gamma t}\, dt + \int_{t'}^{\infty} R(t')c(t')e^{-\gamma t}\, dt$$
$$+ Ce^{-\gamma t'} + \int_{t'}^{\infty} s(t)c_1(s)e^{-\gamma t}\, dt \quad (3.38)$$

where the subscript $_{\text{PW}}$ denotes present worth of future disbursements, $c(t)$ is unit pumping cost, C represents the capital investment in importation facilities, $s(t)$ is the importation rate, $c_1(s)$ is the power and associated operation charges on delivery of imported water, and all other terms are as previously defined. By designating T' as the optimal time for a transition from groundwater storage to imported supplies, the decision rule becomes

$$\frac{dc(T')}{dT'} = q(T')c(T')e^{-\gamma T'} + \int_{T'}^{\infty} \frac{d}{d(T')}[R(T')c(T')]e^{-\gamma t}\, dt$$
$$- \gamma Ce^{-\gamma T'} - s(T')c_1(s)e^{-\gamma T'} = 0 \quad (3.39)$$

If recharge is assumed constant and cost per unit pumped is constant for all $t > T'$,

$$q(T')c(T')e^{-\gamma T'} + \frac{R(T')}{\gamma}\frac{dc(T')}{dT'}e^{-\gamma T'} = \gamma C e^{-\gamma T'} + s(T')c_1(s)e^{-\gamma T'} \quad (3.40)$$

Interpreted literally, to make a transition from groundwater storage to imported supplies is economic when the cost of mining (left-hand side) equals the cost of importation (right-hand side).

All costs in the developments cited above are considered in terms of present worth. The cost of mining includes not only current pumping charges, but the capitalized value of all future pumping charges associated with pumping sustained yield. The cost of importation includes the initial investment as well as operating costs.

In the general case, the interest rate for the right-hand side of Eq. (3.40) will not be the same as that for the left-hand side. For example, there is often inadequate incentive on the part of pumpers to make a substantial investment in expensive surface-water facilities. This has promoted considerable governmental activity in this area, most often associated with project allocations with a low, or "social," rate of interest. To examine the potential influence of this activity, it is assumed that virtually no difference exists between a private and a social rate of interest, which gives

$$q(T')c(T') + \frac{R(T')}{\gamma}\frac{dc(T')}{dT'} = \gamma C + s(T')c_1(s) \quad (3.41)$$

The interpretation here is basically the same as for Eq. (3.40). Note now, however, that a low rate of interest operating on the left-hand side of Eq. (3.41) tends to place greater emphasis on the future costs of pumping. A low rate of interest on the right-hand side tends to favor lower cost importation. The net result tends to act against excessive depletion of groundwater storage, while at the same time it promotes water importation and is in the direction of conservation without impairing economic growth.

As with all previous models in this chapter, Eq. (3.41) is a quantitative arrangement of economic and physical facts which serve as a substitute for rule of thumb judgment. The models employed have merit to the extent that they provide a conceptual guide to resource development. As a predicting tool, a forecaster, or a handbook approach to management problems, they are severely limited by the assumptions required to obtain simply structured decision rules. Such assumptions include continuous and continuously differentiable cost functions, an "averaged" basin-response surface, and static economic conditions, among others.

OPTIMIZATION MODELS

CONTROL OF GROUNDWATER STORAGE BY PUMPING TAX

At this point, it may be worthwhile to review the chapter. The divergence between the interests of private pumpers and the state when public policy is expressed through the conservative concept of safe yield is recognized as an important initiating conflict. By ignoring the relatively large amounts of water in storage, this idealistic precaution establishes annual replenishment as the limiting factor in the potential growth of an area. In that groundwater mining does not necessarily imply catastrophic effects and is often profitable from a private pumper's point of view, the situation often evolves into one of overdraft. Several reasons other than economic incentives may be cited for this:

1. The relatively large quantity of groundwater in storage in contrast to mean annual supply
2. The difficulties encountered in estimating recharge
3. The fact that in many arid areas, safe yield cannot be fully developed until the basin is overdeveloped

Perhaps the most important factor contributing to overdraft is the inability to recognize it when it is occurring. At any rate, overdraft constitutes the current state of affairs in numerous areas and may at least be examined from an optimization point of view. Invariably, by treating the basin as an isolated system, an inescapable proposition is that use rates must somehow be made to converge on safe yield.

The efforts of action-oriented groups to supplement an overdeveloped basin with imported water are instigated by the facts that annual replenishment is unacceptable as a factor limiting economic growth and that any enterprise built on the hopeful premise of permanency of groundwater storage must ultimately decline and break down. Given this situation, it is important that the importation be logically and effectively planned and constitute a rational first step in the orderly achievement of total resource management. This aspect may also be examined in terms of optimal importation schemes; inevitably, however, the timing of importation will be influenced more by political and institutional constraints than by the rate and technology of groundwater exploitation.

Given that these conflicts are resolved by some form of suboptimization and water becomes available for importation, a third in a series of conflicts becomes evident. This conflict is between the selected best physical and economical plan of operation and the pumpers. A case in point may be the reluctance on the part of certain pumpers to purchase surface water because the groundwater supply is comparatively inexpensive. Hence, by administratively treating groundwater storage as a "free good," there is a normal tendency against payment for its replacement. It follows, of course, that storage is not free, but possesses value at least to the extent of reducing future pumping costs.

The ideas presented on sole ownership and the public water district in particular are well suited to coping with these problems (Section 3.1). This is not to say that all these ideas apply to all cases. In general, however, the

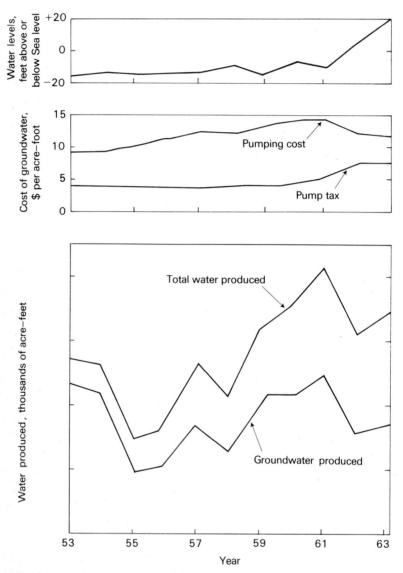

Fig. 3.11 Water production, costs, and water levels, Orange County Irrigation District, California, 1953 to 1963. (*From data given by Weschler*, 1968.)

OPTIMIZATION MODELS

concepts are broad enough to incorporate many of the conflict areas between pumper and plan. The oft-cited case of the Orange County Water District in southern California as a "management model" exemplifies the resolution of some of these conflicts.

At the present time, continued water use in the Orange County Water District is dependent on imported supplies utilized in conjunction with a partially depleted groundwater reservoir threatened by sea-water intrusion. Among the responsibilities of the district as a management agency are included:

1. Protection of the water rights of inhabitants of the basin
2. Obtaining maximum importation of outside water for direct use, for basin replenishment, and for retardation of sea-water intrusion
3. Encouragement of maximum beneficial use consistent with stability of supply

Of these activities, 3 exemplifies the conflict between the plan of operation and the pumpers.

A dual tax structure provides the financial base for payment of the imported water and its recharge by artificial means to the basin. The first of these, the ad valorem, or property tax, affects all inhabitants of the area. The second, termed a *replenishment assessment,* or *pump tax*, extracts money from each pumper to overcome overdraft to the extent that he contributes to it. This is the interesting tax in that it constitutes a powerful tool in achieving maximum beneficial use of the present supply. In effect, the pump tax raises the cost of groundwater to individual pumpers in the interests of protecting the groundwater supply, which forces the use of more expensive surface water (Weschler, 1968). This management scheme has resulted in an increase in water levels via an increase in relative amounts of surface water used (Fig. 3.11).

3.5 MATHEMATICAL PROGRAMMING

Emphasis thus far has focused on simple models demonstrating efficiency in development of a renewable groundwater supply. The term *efficiency* is interpreted as profit maximization consistent with stability of supply. This allowed the derivation of mathematical decision rules which present telling insight into the efficient use of groundwater resources. As demonstrated, points of maxima or minima for problems formulated with one or a few continuous and differentiable functions are easily determined by standard methods of calculus. Given now more difficult situations dealing with inequality constraints, or discrete changes from one operating level to another, the calculus is extremely limited or altogether inadequate, and simply structured decision rules are almost impossible to derive.

Because the more complicated situations are considerably closer to real world occurrences, it is of some importance to develop a set of mathematical techniques aimed specifically at the problem of how to develop and use limited resources to the best advantage. The time-consuming search for, or development of, a particular technique to solve a particular problem is eliminated, it is hoped, by consideration of a small set of techniques which are applicable to whole classes of problems. This is the idea behind mathematical programming. The term *programming* refers to the orderly process by which a given class of problems is solved, and *mathematical programming* is the name given to the set of techniques.

The method to be pursued in this section is more heuristic than rigorous, and will be developed along two lines. First, certain problem types will be cited in order that the features they share in common might be revealed. As some types of solutions are more desirable than others among the set that is feasible, the more desirable aspects of a given solution will be discussed with the hope of determining what exactly is to be expected from a mathematical technique. Second, the techniques themselves are presented in the light of the structuring of the original problems. Although it is not generally possible to solve these problems without the aid of a digital computer, it is possible to study the methodology and logic of the techniques without reference to computing facilities.

PROBLEM STRUCTURING AND CLASSIFICATION

1. The first problem relates to the one-time allocation of limited resources to meet some desired objective. Consider, for example, the problem of allocating certain crops of different values to irrigable acreage of different quality, so that net profit is maximized. Usually land, water, and labor are available in limited quantities, so that production scheduling is accomplished within given constraints. Similarly structured problems might include the allocation of funds to various phases of exploration, the allocation of storage space in a multipurpose reservoir to its various purposes, the allocation of reservoirs to geographical areas, or the allocation of wells to localities within a given area. A cost minimization problem might consist of allocating several sources of water to a given municipal, industrial, and agricultural demand in a highly urbanized area. Constraints include the amount of water required by each entity and water availability from each source. Several alternative allocations may be employed, only one of which minimizes cost to some centralized water-management agency. Whatever scarce resource is being allocated, the problem of maximization of benefits or minimization of costs, subject to certain constraints, is essentially the same provided that the objective can be stated mathematically, the resources are known quantitatively and are in limited supply, several alternative courses of

action are open, and the allocation is done at one point in time. This is termed a *single-stage process*.

2. The allocation problem cited above is not realistic for long planning horizons, where costs, benefits, and the dependability of resources are subject to change. For example, if competitive pumping exists in the crop allocation problem discussed above, the cost of water will increase with increasing pumping lift, and lower net returns will be realized for a set allocation. It is logical, therefore, to think in terms of a succession of choices at various stages of time. It follows that a maximization of benefits for each of a long series of irrigation seasons under competitive pumping will entail seasonal adjustments in crop selection, as well as in resource use. Whatever process is used to optimize the single-stage problem cited in 1, it should be extendable to couple N such stages, each of which is optimized successively.

3. Successive optimization of the type described in 2 does not necessarily lead to a long-term optimum. For example, a decision to pump at a given point in time will transform the state (storage), but the future effect of this transformation is not taken into account in deciding how much to pump. This is demonstrated in Fig. 3.12, where the decisions indicate water pumpage during the first period as a function of initial aquifer storage. These decisions lead to maximum returns over the number of years considered. In stage 1, only a 1-year planning horizon is considered. The large withdrawals typify disregard for future operations. This is analogous to the successive optimization described above. Longer planning horizons result in conservative first-year water allocations.

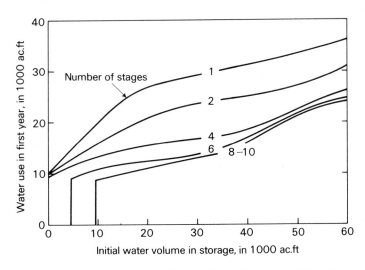

Fig. 3.12 Water allocation policy varying with stages. (*After Aron*, 1969.)

It follows that successive optimization of the type described in 2 is only justifiable under competitive pumping, as what is not pumped during a given season may not be available in future seasons. Given now a centralized management agency, the role of time and storage control may be considered in a more realistic manner. What is required here is a mathematical description of the problem wherein the state of the system is transformed by a certain decision, which yields a financial reward, and N sequential decisions must be made so that the cumulative reward is as large as possible. This is the multistage sequential decision problem discussed in Section 3.1, where the state variable is groundwater storage and the decision variable is groundwater pumping. It follows that the dynamics remain essentially unchanged for a whole class of similarly structured problems, including those with two or more state variables, a stochastic state variable, or two or more decision variables. The ideas developed above may thus be extended to the integrated use of ground- and surface-water supplies.

It is now possible to examine more closely the cited problem types in order to focus on their similarities and differences. First, it is clear that in a single-stage process, all decisions are made simultaneously. In a multi-stage process, the decisions are made sequentially. Hence, one of the dimensions of the latter type of problem must be time. However, whether or not the time element comes into consideration, the three problems share many features in common. In each case, for example, there is a dependent variable that is a function of a few or more independent variables, not all of which are subject to direct control. This may be expressed

$$Y = f(x_1, x_2, x_3, x_4, \ldots, x_n) \tag{3.42}$$

In general terms, Eq. (3.42) represents the mathematical statement to be maximized, or minimized. However, the proposition is not as straightforward as one may suspect. There are, for example, constraints on the solution because of natural, economic, or man-made limitations. Further, these constraints may be in the form of inequalities, which reflect the fact that only minimum or maximum requirements need be met. In the allocation of water to various uses, for example, the amount delivered from any one source cannot exceed the delivery capacity of that source, but will fall within the range $0 \leq Q \leq C$, where Q is the amount delivered, and C is the delivery capacity. Further, the total amount delivered from all sources cannot exceed the total delivery capacity or the total demand. Similar statements can be made if groundwater is one of the supplies and the amount of pumping is limited by pumping capacity or safe-yield considerations. It follows that the inequality expressions dealing with constraints are likely to be a function of many variables.

The decision problem to be solved thus consists of an objective function containing both decision and state variables and of constraints that limit the values which the decision variables may assume. Any solution satisfying the constraints is a feasible solution to the problem. An optimal solution is a feasible solution that either maximizes the return or minimizes the cost, depending on the nature of the problem addressed.

LINEAR PROGRAMMING

Linear programming can be applied to decision problems where the objective function is linear and where the constraints are statements of linear equalities or inequalities. Linearity implies proportionality, which means that if one unit of resource will return or cost X dollars, ten units of the same resource will return or cost $10X$ dollars. Linear programming, along with its logical extension, nonlinear programming, is a single-stage optimization technique and deals with the simultaneous optimization of several decision variables.

The general linear programming problem can be formulated: Maximize

$$Z = c_1 x_1 + c_2 x_2 + c_3 x_3 + \cdots + c_n x_n \tag{3.43}$$

subject to

$$\begin{aligned} x_1 &\geq 0 \\ x_2 &\geq 0 \\ x_3 &\geq 0 \\ &\cdots \\ x_n &\geq 0 \end{aligned} \tag{3.44}$$

and

$$\begin{aligned} a_{11} x_1 + a_{12} x_2 + a_{13} x_3 + \cdots + a_{1n} x_n &\leq b_1 \\ a_{21} x_1 + a_{22} x_2 + a_{23} x_3 + \cdots + a_{2n} x_n &\leq b_2 \\ &\cdots \\ a_{m1} x_1 + a_{m2} x_2 + a_{m3} x_3 + \cdots + a_{mn} x_n &\leq b_m \end{aligned} \tag{3.45}$$

where Eq. (3.43) is the objectives function; $a_{11} \cdots a_{mn}$, $b_1 \cdots b_m$, and $c_1 \cdots c_n$ are sets of constants; and $x_1 \cdots x_n$ is a set of variables. The objectives function is linear in x and may be expressed

$$Z = f(x) \tag{3.46}$$

Although the objectives function is defined for all values of x, the solution is restricted by Eqs. (3.44) and (3.45). Equations (3.44) stipulate that all components should be nonnegative entities. Equations (3.45) stipulate that the components must satisfy m linear statements. The minimization problem is treated by describing Eqs. (3.45) as $\geq$ rather than $\leq$, and by multiplying Eq. (3.43) by -1, which maximizes $-Z$. Any problem which can be described in this fashion is a typical linear programming problem.

A graphic interpretation of a simple linear programming problem is shown for two variables x_1 and x_2 in Fig. 3.13a. The objective might be to maximize profits from the sale of two commodities (say, crops x_1 and x_2), with four fixed resources allocated to the production process. For this problem, the coefficients c_1 and c_2 of the objectives function indicate the fixed unit profit received from sale of each commodity, and the variables x indicate the amount of each produced. The feasible-policy space is bounded by three of the four linear constraints, which gives the polygon $ABCDE$. Each intersection specifies a set of feasible combinations of x_1 and x_2, among which the one corresponding to the maximum of Z is to be found. One of the constraints plots outside the polygon, which constitutes a resource in excess abundance or redundant constraint.

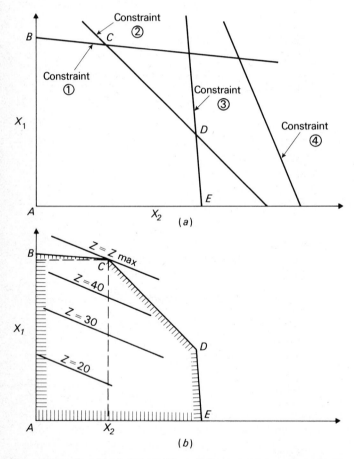

Fig. 3.13 Graphic interpretation of linear programming, showing (a) the position of linear constraints, and (b) the feasible-policy space.

The objectives function is also linear and variable in x, and may be plotted on the two-dimensional graph. It is nearer the origin for small profit and farther removed from the origin as the profit increases (Fig. 3.13b). The linear objectives function attains its maximum value at an extreme point of the feasible-policy space, or point C. This gives the optimal production levels for x_1 and x_2 consistent with resource constraints.

The graphic approach is clearly inadequate when the number of variables exceeds two (or, more correctly, three, in that three dimensions may be portrayed graphically). One method of solution for these more complicated problems is termed the *simplex method*. It is not intended here to discuss the simplex method, but its systematic logic can be briefly examined (Shinners, 1967). First, it is recognized that a linear objectives function will have a maximum only if the constraints establish a finite boundary in the direction of increasing profits. Second, a maximum point will always correspond to a boundary; that is, it will always correspond to one or more of the constraints. Third, provided the objectives function is not parallel to one of the edges of the feasible region, the maximum point will correspond to an intersection of constraints. Hence, a systematic method of solution programmed for an electronic computer would include the following steps (Makower and Williamson, 1967):

1. Locate an extreme point of the feasible region.
2. Examine each boundary edge intersecting at this point to see whether movement along any edge increases the value of the objectives function.
3. Move along the edge in the direction of increasing profit to the adjacent extreme point.
4. Repeat 2 and 3 until further movement along an edge no longer increases the value of the objectives function.

Linear programming has been employed in virtually hundreds of instances, and is adequately described in numerous books and articles on the subject (Sasieni and others, 1959; Hillier and Lieberman, 1967; Levin and Lamone, 1969). A detailed nonmathematical approach to the technique is given by Heady and Candler (1958), Chapter 7 of which treats the resource valuation problem. Typical applications in water resources include a groundwater allocation problem (Castle and Lindeborg, 1961) and a marginal-value analysis of irrigation water in Colorado (Hartman and Whittelsey, 1960). Extensions of the technique include nonlinear programming, where the proportionality assumptions are deleted for nonlinear objectives and constraints functions, which is difficult mathematically; and stochastic programming, where some or all of the parameters of the linear programming model are random rather than deterministic.

The chief advantage of linear programming is the simplicity in program formulation. This has resulted in the development of standard computer programs for problem solution. Chief limitations include the linearity conditions imposed on objectives functions and constraints and the "all-or-nothing" optimal allocation which results from correspondence of maximum points and intersecting constraints. Since this particular single-stage optimization procedure is of limited value in realistic groundwater problems, we shall move on now to more realistic formulations.

LINEAR PROGRAMMING AND MULTISTAGE DECISION PROBLEMS

Another limitation in linear programming arises in attempting to allocate several resources to several uses over a sequence of time periods. In this situation, the number of equations may become overly large and so create computational problems. This can be demonstrated for a problem having only one state and one decision variable, say, groundwater storage and pumping. Let S and X equal the maximum storage and seasonal pumping capacity, respectively. For an N-stage process, where N is the number of pumping seasons,

s_i = storage at the beginning of the ith period

r_i = recharge during the ith period

x_i = pumping allocation during the ith period

q_i = spring outflow during the ith period

The constraints are easily identified in terms of nonnegative entities and the continuity equation, respectively,

$$\begin{align} s_i &\leq S \\ s_i &\geq 0 \\ x_i &\leq X \\ s_i + r_i - x_i - q_i &= s_{i+1} \end{align} \tag{3.47}$$

The objective function is to be maximized and is of the form

$$Z = \sum_{i=1}^{N} (B_i - C_i) x_i \tag{3.48}$$

where B is unit seasonal benefit, and C is unit seasonal cost over the N periods of interest.

The simple problem cited above has $4N$ equations to be dealt with. This number increases considerably if the number of state and decision variables is increased. For example, consider the extension of this problem to two surface-water reservoirs in conjunction with a groundwater supply (three state variables) to meet certain water requirements. Decisions to be implemented include surface-water diversions for direct use (2), groundwater pumping (1), and surface-water diversions for recharge (2), which gives a total of 5. The total number of constraints include at least the following:

$3N$ upper-limit constraints on storage
$3N$ lower-limit constraints on storage
$3N$ expressions of continuity
$5N$ capacity constraints

or a total of $14N$ equalities or inequalities.

Dracup (1966) solved a linear programming problem of the type under discussion for the San Gabriel Valley, California. Five sources of water were to be optimally used to satisfy three water requirements. The objective function called for minimizing the cost of supplying water to the area to meet the demand. The system was operated over 30 one-year periods (1960 to 1990), utilizing three decision rules, but only one state variable (groundwater storage). Twelve constraints were employed, including supply and demand equations, the continuity equation, maximum storage capacity of the basin, the quantity available from the surface source, etc. An additional constraint dealing with the maximum change in storage was made subject to variation in accordance with the decision rule to be employed. Three hundred and thirty-one equations were used.

Three points of special interest in the paper include the decision rules themselves, the parametric approach, and the solution of the dual linear problem. The decision rules put arbitrarily contrived controls on the terminal groundwater storage level. They could be implemented by adherence to a safe-yield policy, a maximum-withdrawal policy, or a pumping and artificial-recharge policy. Analysis of the results was conducted within the framework of each of the decision rules. The parametric analysis included variations of the objective-function cost coefficients and the right-hand-side terms of the constraints. The dual solution is generally normal procedure in linear programming and describes the maximum value associated with one unit of resource. These values indicate the amount that the objective function of the dual could be increased or decreased per unit increase or decrease in each variable, which would allow further cost minimization. The dual concept is important from an economic point of view in that it provides answers concerning the value of alternative courses of action.

Another application deals with acceptance of the concept of competitive pumping in an isolated groundwater basin with concentration on predictions in farmer response to a falling water table. In other words, this model typifies a real world situation by attempting to incorporate the things that competitive pumpers actually do. This attitude suggests that a set of services from use of the resource is available at a given cost only during one period of time. If the resource is not used during that period, it is lost forever at that given cost level. Such an example is provided by Stults

(1966; 1967) and is of some importance in that the ideas employed may be extended for application in simulation studies (Chapter 7).

The linear programming model employed by Stults (1966) is a simple one, with nine activities (crops), one state variable (storage), and six constraints. The constraints include governmental control of cotton allotments and acres and water availability, but do not consider labor and capital limitations. For predicting adjustments, the 1966 net returns and rate of water use were projected for 10 years. Then a new solution with a new rate of decline and higher pumping cost was determined. The higher costs are reflected in lower net returns for the crops in the model. This was repeated at 10-year intervals until the year 2006. Hence, this multistage decision process coupled N single-stage linear programming models, with N equal to 5. Costs and benefits were assumed constant during a given stage (10-year period). However, since the cost of water will increase with additional pumping lift, there will follow certain adjustments in cropping patterns and use of water. The state variable is transformed by the simple continuity equation, and a new set of coefficients characterize the beginning of each of the N stages.

Over the period 1966 to 2006, total acreage decreased by about 50 percent, water use decreased by about 42 percent, but net income decreased by only 20 percent. This proportionally small change in net revenue reflects a decrease in resource allocations to lower-valued crops.

DYNAMIC PROGRAMMING

We shall now take up the processes of sequential optimization known by the name of *dynamic programming*. The word *programming* is again used in a mathematical sense of selecting an optimal allocation of resources, and the word *dynamic* means that decisions are to be made at several distinct stages. In particular, emphasis is on problems whose dynamics may be as follows:

1. The state of the system changes with time because of natural or uncontrollable events.
2. N decisions concerning the system will be made, which will alter the system's state N consecutive times.
3. The system's state at any point in time is determined by its state at a previous time, as well as by previous decisions.

From 3, the following relation is established:

$$y = f(y, u, t) \tag{3.49}$$

which expresses the fact that the state of the system, y, is a function of the state variable y, the decision variable u, and time t. Suppose, for example, the system is in state y_1 and a decision u_1 concerning the system is to be made.

OPTIMIZATION MODELS

This decision will transform the system from state y_1 to state y_2, the new state y_2 being a function both of the state in effect at the time the decision is executed and of the nature of the transformation itself. That is,

$$y_2 = f(y_1, u_1, t) \tag{3.50}$$

As decisions are generally made in order to achieve objectives, the decision u_1 not only transforms the state, but generates some reward or benefit. The reward RW_1 associated with the decision u_1 is a function of that decision and of the state in effect when the decision is executed, or

$$RW_1 = G(y_1, u_1) \tag{3.51}$$

If only one decision is to be made, the problem is merely to choose u_1 in order to maximize RW_1. This is the familiar single-stage decision process. The state y_2 which automatically accompanies decision u_1 would most likely affect the nature of the reward associated with decision u_2, but there is no desire to execute u_2 in a single-stage policy. For the single-stage policy, the maximum return is given by

$$RW_{1\max} = \max_{u_1} G(y_1, u_1) \tag{3.52}$$

which is the objective function to be maximized, with u_1 indicating the decision variable.

For a two-stage policy, a second decision u_2 is required, producing a reward RW_2 and, as well, the state y_3, where

$$y_3 = f(y_2, u_2, t) \tag{3.53}$$

Similarly a three-stage policy produces the state y_4, where

$$y_4 = f(y_3, u_3, t) \tag{3.54}$$

For an N-stage policy, this process continues until the terminal state y_N is achieved by making a decision u_{N-1} on state y_{N-1}, or

$$y_N = f(y_{N-1}, u_{N-1}, t) \tag{3.55}$$

To execute this policy, there are N decisions $u_1, u_2, u_3, \ldots, u_N$, the problem being for a given y_1 to select the sequence of decisions u_i which maximize the objective function

$$RW_N = \max_{u_i} \sum_{i=1}^{N} G(y_i, u_i) \tag{3.56}$$

To solve the optimization problem of (3.56) analytically, it is necessary to obtain the partial derivatives of the N equations, set them equal to zero, and solve them simultaneously. This is clearly too complex a procedure when N is large. On the other hand, a sheer enumeration approach runs into dimensionality problems (Bellman and Dreyfus, 1962). Suppose, for example, that there is 1 variable which can range over 10 discrete values for each of the 10 stages. The 10-stage maximization process will then entail 10^{10} combinations of choices. With 2 variables and 20 stages, 10^{40} different combinations must be enumerated to arrive at a maximum. As pointed out by Hall (1966), a high-speed computer capable of assessing the returns from one such alternative in $\frac{1}{1000}$ sec. will require 3.2×10^{32} years to grind its way through the required combinations.

To reduce the dimensionality of this problem will require two principles:

1. After decision u_i, the returns associated with the remaining decisions depend only on y_{i+1} and the remaining decisions u from u_{i+1} to u_N (Watt, 1968).
2. Any policy which is optimal over the interval $(0,T)$ is necessarily optimal over the interval (t,T), when $0 \le t \le T$. In other words, an optimal policy can only be formed from optimal subpolicies. This is Bellman's (1957) principle of optimality in a slightly modified form.

It follows that if $F_N(y_1)$ represents the maximum return from an N-stage process with initial state y_1,

$$F_N(y_1) = \max_{u_1} [G(y_1,u_1) + F_{N-1}(f(y_1,u_1))] \qquad (3.57)$$

where $G(y_1,u_1)$ is the return from the first decision u_1 applied to the initial state y_1, and state y_2 equals a function of y_1 and u_1, that is, equals $f(y_1,u_1)$. To determine the maximum return $F_N(y_1)$, it is necessary to determine the u_1 that maximizes $G(y_1,u_1)$. Then, for the $f(y_1,u_1)$ produced by decision u_1 that maximizes $G(y_1,u_1)$, seek out the u_2 that maximizes $G(y_2,u_2)$, and so forth. Hence, the total return of an N-stage process is reduced to the problem of solving a sequence of N single-stage decision processes. In this manner, the problem containing 10^{40} combinations of choices is reduced to one containing $20(10^2)$, or 2,000 combinations, by merely rejecting all but the maximum values at each stage.

Unlike linear programming, no standard computer program is available for dynamic programming. However, the basic formulation is flexible enough to describe a whole class of problems. Take, for example, the classic groundwater problem for isolated basins, where the state variable is groundwater storage S, and the decision variable is the amount of water to be pumped per unit time period to maximize some benefit function. The function $f(y_1,u_1)$ is easily accounted for by the continuity equation

$$f(y_1,u_1) = S_2 = S_1 - X_1 + R \qquad (3.58)$$

where X is pumpage and R is recharge. Equation (3.57) then becomes

$$F_N(S_1) = \max_{X_1} [G(S,X) + F_{N-1}(S_1 - X_1 + R)] \qquad (3.59)$$

where the constraint $0 \le X \le S$ is met for magnitudes of withdrawals. The quantity $G(S,X)$ is the net benefit resulting from X units of pumping withdrawal at a given storage level S. In general terms, with a provision for discounting,

$$F_N(S) = \max_{X_N} [G_N(S,X_N) + (1+\gamma)^{-1} F_{N-1}(S - X_N + R_N)] \qquad (3.60)$$

which may be interpreted as the maximization with respect to water use at stage N of the immediate benefit $G_N(S,X_N)$, plus the discounted net benefit in the $N-1$ remaining stages, given that an optimal policy will be employed during the remaining $N-1$ stages. Figure 3.9 is an example of the relation $G(S,X_N)$ for a given level of storage.

Equation (3.60) is a simplified version of a similar maximization problem solved by Burt (1964b), where recharge was associated with probability terms, which resulted in expected rather than deterministic benefits. If inflow can take any value $R_{N,i}$, where i equals $1 \cdots m$ with probability P_i, the dynamic programming formulation for maximization of expected benefits is

$$F_N(y) = \max \sum_{i=1}^{m} P_i [G(y,u) + B F_{N-1}(f(y,u)_i)] \qquad (3.61)$$

where B is the discount factor.

The problems discussed above are characterized by one state variable and one decision variable. This formulation can be readily extended to include two decision variables related to the same state variable

$$F_N(S) = \substack{\max \\ \text{or} \\ \min} [A_N(y;u_a,u_b) + F_{N-1}(f(y;u_a,u_b))] \qquad (3.62)$$

where the problem may be one of maximization, or minimization, depending on whether $A_N(y;u_a,u_b)$ represents benefits or costs. Groundwater problems characterized by the above formulation have been thought of almost exclusively in terms of pumping, artificial recharge, and groundwater storage. This is demonstrated in Fig. 3.14. In the case of coastal aquifers (Fig. 3.14a), it is recognized that excessive pumpage lowers the water table, which decreases the loss of freshwater to the sea, but increases seawater intrusion. A limitation is thus imposed on pumping unless imported water is used for artificial recharge. Considering the effect of pumping and artificial recharge on the state variable in coastal aquifers, a worthwhile objective might be to maximize net benefits from operating a coastal aquifer between the levels $S_{\max}$ and $S_{\min}$ (Fig. 3.14a) so that long-term damages from sea-water intrusion are minimized. The optimization problem of Fig. 3.14b encompasses similar

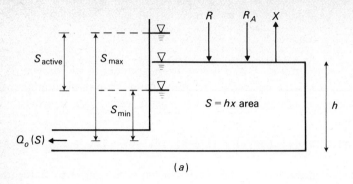

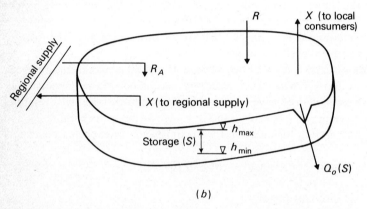

Fig. 3.14 Physical models of management situations characterized by one state variable and two decision variables. (a) An ideal coastal aquifer. (*After Buras and Bear*, 1964.) (b) An ideal integrated use system. (*Modified after Bear and Levin*, 1967.)

objectives in the absence of seawater intrusion. In both cases, natural outflow Q_o occurs and must be empirically related to the state variable

$$Q_o = f(S) \tag{3.63}$$

The coastal aquifer problem has been addressed by Buras and Bear (1964), with the following dynamic programming formulation

$$F_N(S) = \min_{X, R_A} [C_N(S; X, R_A) + F_{N-1}(S + R + R_A - X - Q_o)] \tag{3.64}$$

subject to the normal constraints limiting pumping and artificial recharge to values ranging between zero and maximum capacity. In the above formulation, R_A is artificial replenishment and Q is outflow to the sea, the latter empirically related to storage,

$$Q = f(S) = 0.1S \tag{3.65}$$

The cost function includes cost of pumping, cost of meeting the demand from sources outside the region, benefit realized from exporting water to the outside, and cost of artificial recharge.

The optimization problem associated with Fig. 3.14b has been solved by Bear and Levin (1967) and has already been briefly discussed in terms of its contributions to the idea of optimal yield (Section 3.1).

The optimization problem dealing with pumping and artificial recharge as decision variables and with groundwater storage as the relevant state variable may now be expressed in general terms (Bear and Levin, 1967): Given the initial value S of storage in the aquifer, determine the operation policy in terms of the decision variables related to storage levels at the beginning of each season such that benefits are maximized (or costs are minimized) subject to certain constraints. Buras and Bear (1964) report this policy in the form of tables. Bear and Levin (1967) graphically describe a typical form of the optimal policy (Fig. 3.15). Note that both pumping and artificial-recharge rates are determined by a given level of storage, and an inverse relation exists between storage and the amount to be recharged. Hence, given the economic variables describing the situation, studies of this type examine the logic of artificially supplementing the natural flow.

In some simple cases, one of the decision variables may be only indirectly related to the system state, and the problem can be reduced considerably. Consider, for example, the integrated use of a groundwater supply and an "infinite" reservoir, with the only limitation being the size of the surface-water delivery system. (A reservoir is infinite, for all practical purposes, when the dependable overyear storage plus the expected annual inflows are several times larger than the annual demand on that storage.) If the problem is to determine how much groundwater to pump and how

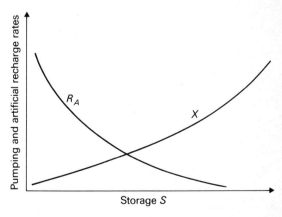

Fig. 3.15 General form of optimal pumping and artificial-recharge decisions for low-valued uses of water.

much surface water to import to meet a deterministic demand over N time periods, a decision to use a given amount of water from one source means that the remainder must come from the alternative source. That is, if α is the percentage of the total water demand D_N to be pumped, subject to capacity constraints, $(1 - \alpha)D_N$ is the amount of surface water to be imported. Hence, there is in reality only one decision variable. If total cost is to be minimized over an N-stage planning horizon,

$$F_N(S) = \min_\alpha [C_1(\alpha D_N) + C_2((1 - \alpha)D_N) + F_{N-1}(S - \alpha D_N + R)] \tag{3.66}$$

where $F_N(S)$ is the total minimum cost, $C_1(\alpha D_N)$ is the cost function for groundwater, and $C_2((1 - \alpha)D_N)$ is the cost function for importation.

This problem has been evaluated by Cochran (1968) for the importation of Lake Mead water to Las Vegas Valley, Nevada, for several alternative schemes. The alternatives dealt basically with internal policy choices of the management water agency and were formulated as problem constraints. A deterministic pumping policy is indeed useful—if not realistic for long planning horizons—in that the results are readily demonstrated (Fig. 3.16).

Given a resource allocation problem within the framework of the system dynamics spoken of earlier, there appears to be no limitation to the usefulness of dynamic programming. If there are two decision and two state variables they are easily accommodated in problem formulation

$$F_N(y_a, y_b) = \genfrac{}{}{0pt}{}{\max\ \text{or}\ \min}{u_a, u_b} [G(y_a, u_a; y_b, u_b) + F_{N-1}(f(y_a, u_a; y_b, u_b))] \tag{3.67}$$

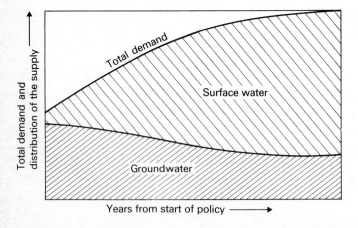

Fig 3.16 Graphic presentation of a deterministic pumping and water-importation policy.

OPTIMIZATION MODELS

Three or more state and decision variables may likewise be easily accounted for. These problems, however, are not so easily solved. A problem with 2 dimensions (2 state variables), each having 10 levels, necessitates 10^2 calculations for each step. A problem with 3 dimensions, each having 10 levels, necessitates 10^3 calculations at each step. Hence, the computational tasks increase exponentially with the number of state variables. For this reason, the number of state variables is defined as the *dimensionality of the problem*. At this date, problems with more than three dimensions are regarded as impractically large for most computing facilities. Hence, although there is practically no limit to the problems that may be formulated by dynamic programming, there is a definite limit to those that are computationally feasible.

The question now arises how many state variables are required in realistic mathematical models of complex water-resource systems in which the aquifer is an integral unit. This question may often be answered by addressing first a series of more fundamental questions: To what degree are the individual subsystems interdependent, which of the individual subsystems may be analyzed separately, and which of the individual subsystems must be treated jointly? The key to this problem lies in the fact that most large-scale systems are composed of several smaller subsystems coupled to each other, and the physical situation will generally indicate which parts of the whole may be decomposed into subsystems. Once this is observed, the decentralization principle may be invoked: A system may be decomposed into subsystems which are optimized individually if all subsystems operate independently from each other (Lasdon and Schoeffler, 1966). In principle, this statement is correct. In practice, interdependency is an inherent property of conjunctive use systems, and it is difficult to detect independent subsystems.

Several realistic problems of integrated use may be recognized as two or three dimensional without the detail of the preceding paragraph. Two-dimensional problems include (1) an aquifer and a surface-water reservoir with decisions focused on pumping and direct releases, (2) two aquifers with associated pumping decisions, and (3) one aquifer and a variable water volume in an artificial-recharge area. If artificial recharge is included in (1), a third state variable may be required. Buras (1963), for example, solved a problem with three decision variables (irrigation releases, releases for recharge, and pumping for irrigation) and three state variables (volumes of water in a surface-water reservoir, in an aquifer, and in the recharge grounds). Aron (1969), on the other hand, examined the conjunctive use of surface water and groundwater by using three state variables and twelve decision variables, the state variables consisting of two aquifers in two different basins and seven surface-water reservoirs lumped into one subunit.

Although a method of solution for the dynamic programming formulation of Eq. (3.57) has been suggested, no clues have been given to how this may be accomplished. In practice, dynamic programming is a procedure that moves backward in time. From the scale diagram of Fig. 3.17, the solution is obtained first for each state of the last time period (stage 1), with successive solutions computed for each state of the remaining stages until the entire solution is obtained for an N-stage policy. This procedure terminates at the initial period looking N stages into the future, with an optimal plan for managing the resource. Since the initial state is known, the initial decision is also known. Subsequent decisions are determined in accordance with the state we find ourselves in in subsequent stages.

The general procedure goes as follows: At any given stage, Eq. (3.57) may be thought of in terms of the present, $G(y,u)$, and the future, $F_{N-1}(f(y,u))$, the latter term representing the value of the resource (water) left over. After completion of an N-stage planning horizon, it is often assumed that any water left over has zero value. Hence, starting at stage 1 (last time period)

$$F_{1-1}(f(y,u)) = F_0 = 0$$

so that only $G(y,u)$ has to be evaluated for feasible states. The assignment of a value greater than zero is, of course, also possible. On moving now to the second stage, two sequential decisions must be made, one for the second stage (present) and one for the remaining stage (future). The future, in this case, is

$$F_{2-1}(f(y,u)) = F_1$$

for which an optimal decision has already been found in stage 1. Hence, whatever the first decision in the second stage may be, it will propel the system in a remaining state for which an optimal decision has been found in the first stage. The method proceeds by recursion, arriving in this backward fashion to the last stage N (first time period). Hence, the dynamic programming formulation is a recursive relation that identifies the optimal policy for each state with N stages remaining, given the optimal policy for each state with $N-1$ stages remaining.

The idea in the above discussion is to select a first decision as a function of a beginning state, so that the combined reward of the two sequential decisions are maximized. According to Bellman (1957),

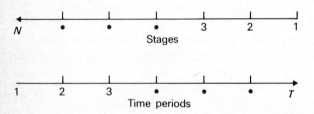

Fig. 3.17 Stages and time periods in dynamic programming.

OPTIMIZATION MODELS

An optimal policy has the property that whatever the initial state and initial decision are, the remaining decisions must constitute an optimal policy with regard to the state resulting from the first decision.

This is the principle of optimality referred to earlier and the basic method by which a multistage decision process may be converted into a series of single-stage problems. For most of the pumping problems discussed in this section, this may be restated (Cochran, 1968):

An optimal groundwater pumping policy has the property that whatever the initial state of the groundwater storage reservoir and the initial decision to pump a given volume of water are, the remaining pumping decisions must constitute an optimal policy with regard to the state of the groundwater storage reservoir resulting from the first pumping decision.

Optimization in space rather than time is also readily accomplished by dynamic programming, most of the applications dealing with resource allocation to various uses (Hall, 1961; Hall and Buras, 1961). Further, there are numerous other optimization techniques that have been employed in problems of one sort or another. These techniques have not received much attention in groundwater problems.

Example 3.1 Problem examples of dynamic programming, however simplified, are somewhat involved and require careful study. This example is presented in detail to illustrate the procedure.

Two sources of water are to be used over a 3-year period to satisfy an annual demand of 50 million gal. One source, a well, will cost $0.50 per million gallons per foot of lift. The second source, a pipeline, will cost $35 per million gallons.

The minimum depth to water in the well is 50 ft, and the well depth is such that it is not possible to extract water below a depth of 90 ft. Hence, the effective saturated thickness is 40 ft. Recharge to the aquifer is 10 million gal/year, and water levels fall at the rate of 1 ft/million gal of storage removal. The objective is to satisfy the water demand at minimum cost. An interest rate of 10 percent will be used. In the event there is no difference in price at any stage, the decision favoring preservation of groundwater storage will be selected.

The recursive relation is

$$F_N(S) = \min [C_1 X_1 + C_2 X_2 + (1 + \gamma)^{-1} F_{N-1}(S - X_1 + R)]$$

where X_1 = volume of water pumped, million gal
X_2 = volume of water taken from pipeline, million gal
L = depth to water, ft
$C_1 L$ = 0.5L, \$/million gal
C_2 = 35, \$/million gal
$(1 + \gamma)^{-1}$ = discount factor
R = rate of recharge, 10 million gal/year

As pumping costs are related to water levels, it is convenient to utilize the average water level as a measure of storage. The relation desired is

$$L - \bar{K}(R - X_1)$$

where $\bar{K}$ is water-level decline per million gallons produced, in this case equal to 1. Hence, if 20 million gal are produced from a pumping level of 50 ft, the new pumping level is easily determined

$$50 - 1(10 - 20) = 60 \text{ ft}$$

Using now the level of storage, L, as the state variable,

$$F_N(L) = \min_{X_1} [C_1 X_1 + C_2 X_2 + (1 + \gamma)^{-1} F_{N-1}(L - \bar{K}(R - X_1))]$$

In the solution that follows in tabular form, the storage levels are incremented in intervals of 10 ft, and the pumping variable is incremented in intervals of 10 million gal for each level of storage. The solution columns mean the following:

(1) Storage levels L (the state variable)
(2) Pumping decisions at a given level
(3) Importation volumes associated with pumping decisions of column 2
(4) Changes in storage resulting from pumping decisions $\bar{K}(R - X_1)$
(5) New level of storage in stage $N - 1$ resulting from decisions in stage N, namely, $L - \bar{K}(R - X_1)$
(6) Average pumping lift used to calculate costs for the decisions in column 2, namely, $L - \Delta L/2$
(7) Cost of pumping
(8) Cost of pipeline water
(9) Total cost, columns 7 + 8
(10) Minimum cost to pump from the level in column 5 (determined from previous state)
(11) Present worth of column 10
(12) Minimum total cost to end of 3-year period from the given stage, columns 9 + 11

$T = 3, N = 1$

(1) L_N	(2) X_1	(3) X_2	(4) ΔL	(5) L_{N-1}	(6) L	(7) $C_1 X_1$	(8) $C_2 X_2$	(9) Total
50	50	0	−40	90	70	1,750	0	1,750
	40	10	−30	80	65	1,300	350	1,650
	30	20	−20	70	60	900	700	1,600
	20	30	−10	60	55	550	1,050	1,600*
	10	40	0	50	50	250	1,400	1,650
	0	50	0	50	50	0	1,750	1,750
60	50	0	−40	Not feasible				
	40	10	−30	90	75	1,500	350	1,850
	30	20	−20	80	70	1,050	700	1,750

OPTIMIZATION MODELS

$T = 3, N = 1$

(1) L_N	(2) X_1	(3) X_2	(4) ΔL	(5) L_{N-1}	(6) L	(7) $C_1 X_1$	(8) $C_2 X_2$	(9) Total
	20	30	−10	70	65	650	1,050	1,700
	10	40	0	60	60	300	1,400	1,700*
	0	50	+10	50	55	0	1,750	1,750
70	50	0	−40	Not feasible				
	40	10	−30	Not feasible				
	30	20	−20	90	80	1,200	700	1,900
	20	30	−10	80	75	750	1,050	1,800
	10	40	0	70	70	350	1,400	1,750
	0	50	+10	60	65	0	1,750	1,750*
80	50	0	−40	Not feasible				
	40	10	−30	Not feasible				
	30	20	−20	Not feasible				
	20	30	−10	90	85	850	1,050	1,900
	10	40	0	80	80	400	1,400	1,800
	0	50	+10	70	75	0	1,750	1,750*
90	50	0	−40	Not feasible				
	40	10	−30	Not feasible				
	30	20	−20	Not feasible				
	20	30	−10	Not feasible				
	10	40	0	90	90	450	1,400	1,850
	0	50	+10	80	85	0	1,750	1,750*

* Indicates optimal policy.

$T = 2, N = 2$

(1) L_N	(2) X_1	(3) X_2	(4) ΔL	(5) L_{N-1}	(9) Total cost	(10) Minimum cost	(11) Present worth of col. 10	(12) Total cost to end
50	50	0	−40	90	1,750	1,750	1,590	3,340
	40	10	−30	80	1,650	1,750	1,590	3,240
	30	20	−20	70	1,600	1,750	1,590	3,190
	20	30	−10	60	1,600	1,700	1,545	3,145
	10	40	−0	50	1,650	1,600	1,454	3,104*
	0	50	0	50	1,750	1,600	1,454	3,204
60	50	0	−40	Not feasible				
	40	10	−30	90	1,850	1,750	1,590	3,440
	30	20	−20	80	1,750	1,750	1,590	3,340
	20	30	−10	70	1,700	1,750	1,590	3,290
	10	40	0	60	1,700	1,700	1,545	3,245
	0	50	+10	50	1,750	1,600	1,454	3,204*

$T = 2, N = 2$

(1)	(2)	(3)	(4)	(5)	(9)	(10)	(11)	(12)
							Present	Total
					Total	Minimum	worth of	cost
L_N	X_1	X_2	ΔL	L_{N-1}	cost	cost	col. 10	to end
70	50	0	−40	Not feasible				
	40	10	−30	Not feasible				
	30	20	−20	90	1,900	1,750	1,590	3,490
	20	30	−10	80	1,800	1,750	1,590	3,390
	10	40	0	70	1,750	1,750	1,590	3,340
	0	50	+10	60	1,750	1,700	1,549	3,299*
80	50	0	−40	Not feasible				
	40	10	−30	Not feasible				
	30	20	−20	Not feasible				
	20	30	−10	90	1,900	1,750	1,590	3,490
	10	40	0	80	1,800	1,750	1,590	3,390
	0	50	+10	70	1,750	1,750	1,590	3,340*
90	50	0	−40	Not feasible				
	40	10	−30	Not feasible				
	30	20	−20	Not feasible				
	20	30	−10	Not feasible				
	10	40	0	90	1,850	1,750	1,590	3,440
	0	50	10	80	1,750	1,750	1,590	3,440*

* Indicates optimal policy.

$T = 1, N = 3$

(1)	(2)	(3)	(4)	(5)	(9)	(10)	(11)	(12)
							Present	Total
					Total	Minimum	worth of	cost
L_N	X_1	X_2	ΔL	L_{N-1}	cost	cost	col. 10	to end
50	50	0	−40	90	1,750	3,340	3,036	4,786
	40	10	−30	80	1,650	3,340	3,036	4,686
	30	20	−20	70	1,600	3,299	3,000	4,600
	20	30	−10	60	1,600	3,204	2,913	4,513
	10	40	0	50	1,650	3,104	2,822	4,472*
	0	50	0	50	1,750	3,104	2,822	4,572
60	50	0	−40	Not feasible				
	40	10	−30	90	1,850	3,340	3,036	4,886
	30	20	−20	80	1,750	3,340	3,036	4,786
	20	30	−10	70	1,700	3,297	3,000	4,700
	10	40	0	60	1,700	3,204	2,913	4,613
	0	50	+10	50	1,750	3,104	2,822	4,572*

OPTIMIZATION MODELS

$T = 1, N = 3$

(1)	(2)	(3)	(4)	(5)	(9)	(10)	(11)	(12)
							Present	Total
					Total	Minimum	worth of	cost
L_N	X_1	X_2	ΔL	L_{N-1}	cost	cost	col. 10	to end
70	50	0	−40	Not feasible				
	40	10	−30	Not feasible				
	30	20	−20	90	1,900	3,340	3,036	4,936
	20	20	−10	80	1,800	3,340	3,036	4,836
	10	40	0	70	1,750	3,299	3,000	4,750
	0	50	+10	60	1,750	3,204	2,913	4,663*
80	50	0	−40	Not feasible				
	40	10	−30	Not feasible				
	30	20	−20	Not feasible				
	20	30	−10	90	1,900	3,340	3,036	4,936
	10	40	0	80	1,800	3,340	3,036	4,836
	0	50	+10	70	1,750	3,299	3,000	4,750*
90	50	0	−40	Not feasible				
	40	10	−30	Not feasible				
	30	20	−20	Not feasible				
	20	30	−10	Not feasible				
	10	40	0	90	1,850	3,340	3,036	4,886
	0	50	+10	80	1,750	3,340	3,036	4,786*

* Indicates optimal policy.

Optimal policies determined by initial states

Starting level	T = 1		T = 2		T = 3		Starting level	PW min. cost
	Pump X_1	Final level	Pump X_1	Final level	Pump X_1	Final level		
50.......	10	50	10	50	20	60	50	4,472
60.......	0	50	10	50	20	60	60	4,572
70.......	0	60	0	50	20	60	70	4,663
80.......	0	70	0	60	10	60	80	4,750
90.......	0	80	0	70	0	60	90	4,786

3.6 CONCLUDING STATEMENT

In most optimization schemes, the groundwater response to extensive pumping is most often viewed as illustrated in Fig. 3.18. The effects of pumping on natural discharge may or may not be included, depending on the natural situation.

It is possible, in many cases, to incorporate differences in yield characteristics by dividing the aquifer into individual cells, or subunits, which are relatively homogeneous, each with its own average response surface. This may be done by using a given value of specific yield for each cell. The idea here is much the same as discussed in Section 2.2, where an assemblage of lumped elements operated independently of each other. A specified value for specific yield within a cell indicates a lack of spatial variations in the response surface within that cell.

The optimal one-time reserve in Fig. 3.18 is most often allocated over time, but may also be considered, in accordance with Western water law, a volume allocation independent of the withdrawal rate. Regardless of which allocation is used, the equilibrium position of the water table indicates that one additional unit of groundwater storage used in current production is of no greater value than the unit of storage left unexploited to reduce future pumping costs. This is the idea behind an "infinite period" of water use.

Groundwater response to pumping in integrated systems may also be viewed as illustrated in Fig. 3.18. In most cases, however, artificial recharge is an additional input which raises the water table uniformly, again in accordance with some averaged basin-response parameter.

With regard to optimization itself, the following statements are offered:

1. Essentially all existing techniques and applications should be regarded as guidelines which serve as a substitute for rule-of-thumb judgment.
2. In general, there is a preoccupation with the sophistication of a technique

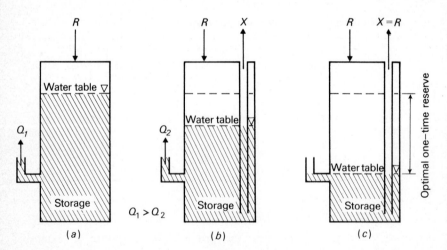

Fig. 3.18 Diagram of planned exploitation of a basin, showing (*a*) conditions prior to exploitation, (*b*) conditions during exploitation, and (*c*) exhaustion of the optimal one-time storage reserve.

and the preclusion of a practical or theoretical appreciation for a meaningful objective function. More thought should be given to this criterion problem, and to its multidimensional and semiquantitative nature. Stated specifically, what is an effective measure of how well a system meets certain specifications, and what are these specifications? Although the systems subject to analysis are becoming more complex in their component parts, the problem of deciding what is relevant insofar as criteria are concerned has been handled, paradoxically, by assuming no problem exists.
3. It appears that, for some time in the future, optimization will be suboptimal in character; that is, the optimization of some subsystems will result in a loss of future opportunities because of the suboptimization of other subsystems. This is largely the result of a lack of clearly defined objectives at the state or national level.
4. Perhaps the most useful aspect of optimization theory is that it can contribute to the breakdown of institutional barriers in water-resource development. In that the analysis is approached from the physical and economic point of view, the cost of letting other criteria dominate the decision process becomes all too clear to decision makers. Sensitivity analysis, or the sensitivity of model solutions to variations in cost or to uncertainty in data, can contribute here. Model solutions may be employed to examine the worst case (pessimism), the best case (optimism), or any intermediate ground.

PROBLEMS AND DISCUSSION QUESTIONS

3.1 Basic to many of the concepts and models put forth in this chapter is the idea that a future value or cost should be measured in present terms. Explain.

3.2 Although it is often true that present groundwater use competes with future groundwater use, it is also possible that present groundwater use is complementary to future total water-resource development. Explain, and give examples.

3.3 According to the experience of the Orange County Water District, is it possible to impose a pump tax in the absence of sole ownership? Does this mean that monopoly makes for conservation?

3.4 The principle of equimarginal value in use is a rule governing the division of a given quantum of a commodity among uses (or among individuals). Whenever marginal values in use are unequal, opportunities for mutually advantageous exchanges exist between uses (or individuals).
   a. Relate this concept to Eq. (3.1).
   b. Does the point of transition from groundwater mining to natural replenishment, as described in the material on optimal mining yield, reflect a marginal-value rule at the time of transition? The division of groundwater storage among what uses is considered at this transition point?
   c. Does the point of transition from groundwater mining to imported supplies [Eq. (3.40)] reflect a marginal-value rule at the time of transition? The division of groundwater storage among what uses is considered at this transition point?

d. Does Eq. (3.28) reflect a marginal-value rule? What is the role of time in this formulation, and how does it differ from the formulations cited above? The division of groundwater storage among what uses is considered?

3.5 Explain how the following concepts are reflected in Eqs. (3.25), (3.28), and (3.40):
  a. The divergence between a private and a so-called social rate of interest is an important factor contributing to depletion of groundwater resources.
  b. Inflationary periods are accompanied by practices that tend toward rapid depletion of groundwater storage.
  c. Recessionary periods are accompanied by practices that favor preservation of groundwater storage.

3.6 How would an optimal time distribution of groundwater use be different for an undeveloped (or developing) nation than for a highly developed nation?

3.7 Discuss the relations demonstrated in Fig. 3.12. How would you account for the effects of present pumping on future pumping costs for an isolated groundwater basin to be operated over an infinite planning horizon?

3.8 A rancher grows two crops, referred to as I and II. One acre of crop I requires 3 acre-ft of water during the growing season. One acre of crop II requires 2 acre-ft of water during the growing season. During a given season, the rancher's well can deliver a maximum of 180 acre-ft of water. Land conditions are such that no more than 40 acres of crop I can be grown, and no more than 60 acres of crop II can be grown. The rancher expects to make \$30 on the sale of 1 acre of crop I and \$50 on the sale of 1 acre of crop II. What is the best combination of crops I and II to maximize profits?

Linear programming formulation Maximize $30x_1 + 50x_2$, where x_1 refers to the acres of crop I and x_2 refers to the acres of crop II subject to

$$x_1 \leq 40$$
$$x_2 \leq 60$$
$$3x_1 + 2x_2 \leq 180$$
$$x_1 \geq 0 \quad x_2 \geq 0$$

  a. How many feasible combinations of x_1 and x_2 are there among which the maximum of Z is to be found? What are the allocations and values of Z at these intersections? (Answer: $x_1 = 0$, $x_2 = 0$, $Z = 0$; $x_1 = 0$, $x_2 = 60$, $Z = 3{,}000$; $x_1 = 20$, $x_2 = 60$, $Z = 3{,}600$; $x_1 = 40$, $x_2 = 30$, $Z = 2{,}700$; $x_1 = 40$, $x_2 = 0$, $Z = 1{,}200$.)
  b. Explain how you would couple N of these single-stage linear programming models into a multistage decision model under conditions of competitive pumping. Given that part *a* represents the result of allocations over the first stage, what additional information is required to proceed to the following stages?

3.9 Examine Example 3.1. Note that whenever a storage level is 70 ft or greater, the minimum cost is associated with decisions to utilize the pipeline to meet the full demand, with no water taken from the well. What particular feature about the 70-ft level makes this so?

3.10 Set up and solve a problem identical with Example 3.1 with the following modifications: The well water will cost \$0.25 per million gallons per foot of lift, and the pipeline water will cost \$20.00 per million gallons. The minimum depth to water is 50 ft, and the well depth is such that it is not possible to extract water below a depth of 100 ft.

3.11 This chapter has been designed to illustrate how some broad hydrological and economic principles operate in a number of optimization schemes for groundwater basins. The reader has been cautioned that all existing techniques and applications should be regarded as guidelines providing insight into given general problems, but do not provide an adequate basis for management. We obviously require more detailed information on the spatial variations in the pumping-response surface (Chapter 7), and on the associated undesirable mechanisms of sea-water intrusion and land subsidence (Chapters 4 and 5). However, optimization studies are often very instructive when we are attempting to gain insight into given problems. Describe in detail any insight you can now achieve regarding the following broad range of problems:

a. How can you determine whether a groundwater basin is economically under- or overexploited? The overexploitation question is of interest because it may represent a current state of affairs in numerous areas, and is certain to become more widespread with continued unregulated mining. The underexploitation problem is significant because total resource development may be impaired. Examine the developments associated with Eqs. (3.18), (3.28), and (3.40) to arrive at your answer.

b. How significant, both operationally and economically, is the situation where recharge is storage dependent as opposed to storage independent? Examine the developments associated with Eqs. (3.17) and (3.39) in arriving at your answer.

c. How does the value of water affect the economics of artificial recharge as opposed to direct use of imported supplies? Examine Fig. 3.15 and associated developments in arriving at your answer.

d. Can the lack of variability in crop or enterprise selection under conditions of competitive pumping result in a premature decline of the agricultural community? In arriving at your answer, examine the allocation problem for long-planning horizons, where costs, benefits, and resource availability are subject to change.

REFERENCES

American Society of Civil Engineers: Groundwater Basin Management, Manual of Engineering Practice, no. 40, 1961.

Aron, G.: Optimization of conjunctively managed surface and groundwater resources by dynamic programming, *Univ. Calif., Water Res., Contribution no.* 129, Davis, Calif., 1969.

Bagley, E. S.: Water rights law and public policies relating to groundwater "mining" in the southwestern states, *J. Law and Econ.,* vol. 4, pp. 144–174, 1961.

Bear, J., and O. Levin: The optimal yield of an aquifer, *Intern. Assoc. Sci. Hydrol. Bull., Symp. Haifa,* Publ. 72, pp. 401–412, 1967.

Bellman, R.: "Dynamic Programming," Princeton University Press, Princeton, N.J., 1957.

——— and S. Dreyfus: "Applied Dynamic Programming," Princeton University Press, Princeton, N.J., 1962.

Buchanan, J. M.: Positive economics, welfare economics, and political economy, *J. Law and Econ.,* vol. 2, pp. 124–138, 1959.

Buras, N.: Conjunctive operation of a surface reservoir and a groundwater aquifer, *Intern. Assoc. Sci. Hydrol.,* Publ. 63, pp. 492–501, 1963.

———: Dynamic programming in water resource development, in V. T. Chow (ed.), "Advances in Hydroscience," Academic Press, Inc., New York, vol. 3, pp. 367–412, 1966.

―――― and J. Bear: Optimal utilization of a coastal aquifer, *Proc. Intern. Congr. Agri. Eng.*, 6th, Lausanne, Switzerland, September, 1964.

Burt, O.: Optimal resource use over time with an application to groundwater, *Management Sci.*, vol. 11, pp. 80–93, 1964a.

――――: The economics of conjunctive use of ground and surface water, *Hilgardia*, Calif. Agri. Expt. Sta. Publ., vol. 36, pp. 36–111, 1964b.

――――: Economic control of groundwater reserves, *J. Farm Econs.*, vol. 48, pp. 632–647, 1966.

――――: Temporal allocation of groundwater, *Water Resources Res.*, vol. 3, no. 1, pp. 45–56, 1967a.

――――: Groundwater management under quadratic criterion functions, *Water Resources Res.*, vol. 3, no. 3, pp. 673–682, 1967b.

Castle, E., and K. Lindeborg: Economics of groundwater allocation, *Oregon State Coll., Agri. Expt. Sta., Misc. Papers*, 108, Corvallis, 1961.

Ciriacy-Wantrup, S. V.: Private enterprise and conservation, *J. Farm Econ.*, vol. 24, pp. 75–96, 1942.

――――: Resource "Conservation, Economics and Policies," University of California Press, Los Angeles, 1952.

Cochran, G.: Optimization of conjunctive use of ground and surface water for urban supply, M.S. thesis in Civil Engineering, University of Nevada, 1968.

Cross Section, Water depletion claim upheld by high court, article in the monthly publication of the High Plains Underground Water Conserv. Distr. No. 1, June, 1965.

Davis, I.: The economic picture, *Colo. Expt. Sta., Bull.* 506-S, Ft. Collins, 1960.

Domenico, P.: Economic aspects of conjunctive use of water, Smith Valley, Nevada, *Intern. Assoc. Sci. Hydrol., Symp. Haifa*, Publ. 72, pp. 474–482, 1967a.

――――: Valuation of a groundwater supply for management and development, *Desert Res. Inst., Publ.*, no. 3, Center for Water Resource Res., Univ. of Nevada, Reno, 1967b.

――――, D. Anderson, and C. Case: Optimal groundwater mining, *Water Resources Res.*, vol. 4, no. 2, pp. 247–255, 1968.

Dracup, J.: The optimum use of a groundwater and surface water system—a parametric linear programming approach, *Univ. of Calif. (Berkeley), Water Res. Contrib.* no. 107, 1966.

Duckstein, L., and C. Kisiel: General systems approach to groundwater problems, *Proc. Nat. Symp. Analysis Water Resource Systems*, Denver, Colo., Amer. Water Resource Assoc., pp. 100–115, 1968.

Gordon, H. S.: The economic theory of a common-property resource, *J. Pol. Econ.*, vol. 44, pp. 124–142, 1954.

Grant, E. L.: "Principles of Engineering Economy," The Ronald Press Company, New York, 1950.

Grubb, H. W.: Optimum utilization of groundwater resources, *Proc. Econ. Water Resources Develop. Western Agri. Econs. Res. Council*, Las Vegas, Nev., Rep. 5, pp. 113–122, December, 1966.

Hall, W. A.: Aqueduct capacity under an optimum benefit policy, *Proc. Amer. Soc. Civil Engrs., Irrigation Div., IR* 3, pp. 1–11, 1961.

――――, Systems Engineering: Modern Engineering Program (short course), Univ. of California, Los Angeles, 1966.

―――― and N. Buras: Optimum irrigation practice under conditions of deficient water supply, *Trans. Amer. Soc. Agri. Engrs.*, vol. 4, pp. 131–134, 1961.

Hartman, L. M.: Economics and groundwater development, *Groundwater,* vol. 2, no. 2, pp. 4–8, 1965.

——— and N. Whittelsey: Marginal values of irrigation water, *Colo. Agri. Expt. Sta., Tech. Bull.* 70, Colo. State Univ., Fort Collins, 1960.

Heady, E., and W. Candler: "Linear Programming Methods," Iowa State Press, Ames, 1958.

Hillier, F., and G. Lieberman: "Introduction To Operations Research," Holden-Day, Inc., San Francisco, 1967.

Hirshleifer, J., J. C. De Haven, and J. W. Milliman: "Water Supply, Economics, Technology and Policy," The University of Chicago Press, 1960.

Hufschmidt, M.: The methodology of water resource system design, in I. Burton and R. Kates (eds.), "Readings in Resource Management and Conservation," University of Chicago Press, pp. 558–570, 1965.

Hutchins, W. A.: The Utah law of water rights, *Bull. State Eng. Utah,* Salt Lake City, 1965.

Kelso, M. M.: The stock resource value of water, *J. Farm Econ.,* vol. 43, pp. 1112–1129, 1961.

Krutilla, J., and O. Eckstein: "Multiple Purpose River Development," The Johns Hopkins Press, Baltimore, 1958.

Lasdon, L., and J. Schoeffler: Decentralized plant control, *Instr. Soc. Amer. Trans.,* vol. 5, pp. 175–183, 1966.

Leeds, Hill, and Jewett, Inc.: Water supply for Las Vegas Valley: *Rept. to Director Conserv. Nat. Resources,* State of Nevada, 1961.

Levin, R., and R. Lamone: "Linear Programming for Management Decisions," Richard D. Irwin, Inc., Homewood, Ill., 1969.

Makower, M. S., and E. Williamson: "Operational Research," The English Universities Press, London, 1967.

Marshall, H.: Rational choices in water resources planning, in I. Burton and R. Kates (eds.), "Readings in Resource Management and Conservation," University of Chicago Press, pp. 529–543, 1965.

Milliman, J. W.: Water law and private decision making—a critique, *J. Law and Econ.,* vol. 2, pp. 41–63, 1959.

Renshaw, E. F.: The management of groundwater reservoirs, *J. Farm Econ.,* vol. 45, pp. 285–295, 1963.

Sasieni, M., A. Yaspan, and L. Friedman: "Operations Research—Methods and Problems," John Wiley & Sons, Inc., New York, 1959.

Scott, A.: "Natural Resources—The Economics of Conservation," University of Toronto Press, Toronto, 1955.

Shinners, S. M.: "Techniques of System Engineering," McGraw-Hill Book Company, 1967.

Snyder, J. H.: Groundwater in California—The experience of Antelope Valley, *Univ. Calif. (Berkeley), Agri. Expt. Sta., Giannini Found. Study,* no. 2, 1955.

Stults, J. M.: Predicting farmer response to a falling water table—an Arizona case study, *Proc. Econs. Water Resource Develop. Western Agri. Econs. Res. Council, Las Vegas, Nev., Rept.* no. 5, pp. 127–141, December, 1966.

———: Predicting farmer response to a falling water table—an Arizona case study, Ph.D. dissertation, University of Arizona, 1967.

Theis, C. V.: The source of water derived from wells—essential factors controlling the response of an aquifer to development, *Civil Eng.,* Publ. of Amer. Soc. Civil Engrs., pp. 277–280, May, 1940.

Thomas, H. E.: Water rights in areas of groundwater mining, *U.S. Geol. Surv., Circ.* 347, 1955.

Thomas, R. O.: Planned utilization, *Trans. Amer. Soc. Civil Engrs.,* vol. 122, pp. 422–433, 1957.

Trelease, F. J.: Policies for water law—property rights, economic forces, and public regulation, *Nat. Res. J.,* vol. 5, pp. 1–48, 1965.

———, H. S. Bloomenthal, and J. R. Gerand: "Natural Resources," American Casebook Series, West Publishing Company, St. Paul, Minn., 1965.

Watt, K. E. F.: "Ecology and Resource Management," McGraw-Hill Book Company, New York, 1968.

Weschler, L. F.: "Water resources management, The Orange County Experience," *Calif. Gov. Series* no. 14, *Univ. Calif.,* Davis, 1968.

part two
Conservation Principles and Applications

4
Energy and Its Transformations

In the lumped-parameter approach to groundwater in the hydrologic cycle, the following assertions are immediately clear: The detailed mechanics of groundwater movement or response to external stimulation, either natural or man-made, have to be ignored when the total system is regarded as located at a single point in space. It follows that any significant geological parameters, even if well known, constitute extraneous information in that they generally cannot be incorporated in the system model. On the other hand, implicit in the idea of a distributed-parameter system is the realization that:

1. The problem must be formulated with reference to a space coordinate system.
2. The system parameters are fundamentally geological and are assumed known and distributed in either a continuous or a discrete manner.
3. Specification of the dependent variable requires specification of the location of the point under consideration within the system.

That is to say, the problem is formulated as a partial differential equation or finite difference equivalent. The space variables x, y, and z and the time variable t (when it appears) are designated as independent variables. This chapter and the next are concerned primarily with the identification and interpretation of system parameters, the dependent variable to be specified, and their formulation in partial differential equations. In addition, as we are dealing with the main theme of energy and mass conservation laws, opportunities arise to examine other hydrological phenomena, measurements, and interpretations whose fundamentals share this same theme.

In hydrological studies, two conservation laws are generally considered:

1. The law of conservation of mass (continuity principle)
2. The law of conservation of energy (first law of thermodynamics)

A third basic concept, the second law of thermodynamics, is of equal importance when one is interested in the nature of the loss in mechanical energy accompanying the flow of groundwater, and when, in chemical thermodynamics, one is investigating the establishment of chemical equilibrium between mineral matter and aqueous solutions.

The continuity equation permits an accounting of the flow entering or leaving any volume chosen on the flow path. Equation (1.1) is a form of continuity for a lumped-parameter system in that it accounts for the difference between inflow and outflow without the benefit of a space coordinate system. For a differential formulation of this principle for distributed-parameter systems, the continuity equation stipulates that the rate of increase of fluid mass in a differential element be equal to the difference between the element's influx and efflux; that is, there can be no loss or gain in fluid mass. The flow is steady if the difference between influx and efflux is zero, and unsteady if the difference is not zero. Another statement for the conservation of mass is the law of mass action, which is important in geochemical studies.

The generalization known as *the first law of thermodynamics* states that any closed system possesses a fixed amount of energy, and that this quantity can neither increase nor decrease in amount. More specifically, the first law states that the total energy of a closed system is constant. The strict letter of the law is applicable to all forms of energy, such as mechanical, chemical, or electrical, but does not predict the manner in which one kind of energy changes in form from place to place. The generalization known as the first law is important both to groundwater flow and to all theoretical aspects of chemical geohydrology.

4.1 MECHANICAL ENERGY AND FLUID FLOW

The movement of water may be examined from at least three points of view, or scales: (1) molecular, (2) microscopic, (3) macroscopic. At the molecular level, one is interested in irregularities in pore space, which suggests that statistical mechanics may be employed. At the microscopic level, the detailed molecular mechanisms are ignored, the fluid in the pores being considered a continuous medium, or continuum. In that it is not possible to observe individual molecular behavior or velocity distributions that one may calculate at either the molecular or microscopic level, a macroscopic approach is ultimately required. The flow can then be treated in terms of a few variables that are easily determined in field practice. Hence, a macroscopic control volume is viewed as large with respect to an individual pore, but small with respect to the space within which significant variations of the macroscopic variables may be anticipated.

Given that a macroscopic, or bulk, description of flow complements the fact that all measurements on porous materials are made at the macroscopic level, one of two approaches to arrive at such a description may be employed. One can follow the example of Hubbert (1956) and demonstrate that a microscopic theory can lead, eventually, to certain macroscopic laws, as in the derivation of Darcy's law from the Navier-Stokes equation. This is often the method in the study of hydrodynamics. On the other hand, one can accept, as given, the macroscopic laws developed by experimentation. The latter approach has deep roots in the study of groundwater hydrology and is the course to be followed here.

The content of this section leans heavily on developments presented in a number of papers by Hubbert (1940; 1953; 1956) and on Darcy's law.

PRELIMINARY CONSIDERATIONS

Flow may be classified in numerous ways, for example, as laminar or turbulent, ideal or real, steady or unsteady. In laminar flow, fluid particles move along smooth paths, with every fluid particle following the same path as the preceding particle. The paths of flow are called *streamlines,* and this type of flow is sometimes referred to as *streamline flow.* As the velocity of movement increases, or as constrictions and obstructions in the porous media cause abrupt deviations of the streamline, laminar flow may become turbulent. In turbulent flow, the fluid particles move in very irregular paths, which causes the irreversible transfer of kinetic energy into thermal energy through the establishment of shearing stresses in the fluid.

Intuition suggests there is a threshold value of velocity at which the effects of turbulence first become perceptible. Experimentation by Reynolds (1833) showed that there is actually a combination of four factors which determine whether flow is laminar or turbulent. The combination is a pure number, its value being the same for any consistent set of units. The number is referred to as *Reynolds number* and is expressed

$$N_r = \frac{\rho_w v d}{\mu} \qquad (4.1)$$

where ρ_w is the density of the fluid, v is the average velocity, d is the characteristic length of the solid matrix, which could be the diameter of passage or the grain diameter, and μ is the viscosity.

The Reynolds number may be viewed as a ratio between shear stresses due to inertia [numerator in Eq. (4.1)] and shear stresses due to viscosity [denominator in Eq. (4.1)]. For small Reynolds numbers, viscous shear is more important than inertial shear, and the velocity of movement is low. Experimentation cited by Todd (1959) shows that the first perceptible sign of turbulence in porous media is difficult to measure and may occur at a Reynolds number of 60 or of 600 to 700. In pipeline studies, it is assumed that the transition from laminar to turbulent flow occurs at a Reynolds number of 2,000 (Streeter, 1966). For almost all natural groundwater movement—except in rock containing open fractures or solution channels or under very steep hydraulic gradients—groundwater flow is generally treated as laminar.

An ideal fluid is considered both frictionless and incompressible. A real fluid possesses viscosity. When the fluid is viscous, flow is accompanied by an irreversible transfer of mechanical to thermal energy through the establishment of frictional forces between the rock and the fluid.

Steady flow exists if conditions at any point in the flow field remain constant with respect to time. And unsteady flow exists if conditions change with respect to time. For example, if the pressure of a fluid at a certain point is a certain amount and it remains that amount indefinitely, the flow is steady and can be expressed $\partial P/\partial t = 0$, where P and t are pressure and time, respectively.

The velocity of movement in porous media is not restricted to one direction, but instead has components in three coordinate directions. Flow may then be classified in accordance with its direction. Groundwater flow in three mutually perpendicular directions is referred to as three-dimensional flow. However, the problems arising from solving a distributed-parameter problem with three coordinate axes are often such that idealization is generally required. These idealizations generally recognize only two and sometimes one direction. If one coordinate direction is ignored, the flow is considered two dimensional. An example of a two-dimensional problem may be flow in a relatively horizontal aquifer confined above and below by low-permeability units. In this case, the boundary streamlines are tangential to the low-permeability units, which results in a series of near-horizontal flow paths. In the cartesian coordinate system, this is referred to as flow *in the x and y directions,* with a zero component of flow in the z (vertical) direction. The use of cylindrical coordinates, which require symmetry about one axis, is referred to as *radial* flow.

Flow restricted to one coordinate direction is referred to as *one-dimensional*. This is perhaps the most limiting assumption, but because this condition approximates many field situations, it has found rather wide application in some problems. In the analysis of "leaky" aquifers, flow in the aquifer is assumed to be two dimensional, while flow through the low-permeability layer is assumed to be one dimensional. Another widely accepted application of one-dimensional flow is the theory dealing with vertical shortening (consolidation) of compressible units subjected to loading beyond that under which they received their initial consolidation.

It is recognized at the outset that all field, or distributed-parameter, problems are three dimensional, but some of them at least approach two- or one-dimensional flow.

DARCY'S EXPERIMENTAL LAW

As early as the first half of the nineteenth century, it was recognized that the velocity of laminar flow of water in porous media is proportional to the slope of the hydraulic gradient. However, the generalization of the fundamental physical law governing such flow is credited to Henri Darcy (Darcy, 1856). Through experiments on the vertical flow of water through filter sands (Fig. 4.1), Darcy concluded that the flow of water was directly related to the slope of the hydraulic gradient by a constant of proportionality which depends on the permeability. Darcy's insistence on recourse to experimental observation as a scientific method did much to advance the science of hydrology, and it is fitting that the compact statement which resulted from his studies bears his name. Darcy's statement of the law is

$$Q = KA \frac{h_1 - h_2}{L} \tag{4.2}$$

or

$$\frac{Q}{A} = q = K \frac{h_1 - h_2}{L} \tag{4.3}$$

where Q is the volume of flow through the filter bed in time t; K is a constant of proportionality; q is the specific discharge, or discharge per unit area; and all other factors are shown in Fig. 4.1. The equation most often cited as Darcy's law is written

$$\frac{Q}{A} = q = Ki \tag{4.4}$$

where i is the hydraulic gradient. As the hydraulic gradient is dimensionless, the proportionality constant has dimensions of a velocity.

It is clear from Fig. 4.1 that Darcy's description of velocity was in average quantities from the gross cross-sectional area A. The specific discharge of Eq. (4.4) is therefore a superficial velocity, taking no account of

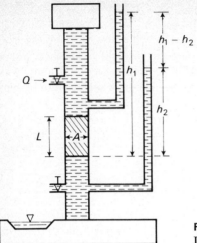

Fig. 4.1 Apparatus for demonstration of Darcy's law.

the fact that flow takes place through the pores. The average velocity can be formulated as follows:

$$v = \frac{Q}{nA} \tag{4.5}$$

where v is average velocity, and n is porosity. From Eq. (4.4)

$$Q = KiA \tag{4.6}$$

Substituting Eq. (4.6) in (4.5) and cancelling terms,

$$v = \frac{Ki}{n} \tag{4.7}$$

The specific discharge-to-head relation expressed in Darcy's law holds only as long as the flow of water through the medium is laminar. With the development of eddies under conditions of turbulent flow, the water particles take more circuitous paths. Hubbert (1956) concluded that the Reynolds number at which turbulence begins is much higher than the one at which Darcy's law fails. The reason cited is that Darcy's law governs flow only when resistive forces predominate (denominator of Reynolds number), and once inertial forces approach the same order of magnitude, Darcy's expression no longer describes the flow regime. The upper limit of Darcy's law is taken as somewhere between $N_r = 1$ and 10.

The experimental evidence upon which Darcy's law is considered valid was originally restricted to one-dimensional flow in which the direction of the flow and the resultant hydraulic gradient coincide. Many studies since then

ENERGY AND ITS TRANSFORMATIONS

have demonstrated that Darcy's formula is applicable for a wide range of hydraulic gradients, effective for nearly all gradients under which groundwater may occur over appreciable areas. Early studies by King (1899) and Slichter (1899, 1905) and their reviews of early European experiments suggested that this was so. Later studies by scientists of the United States Geological Survey demonstrated in a series of experiments that the relation held for gradients ranging from at least 270 ft to 2 or 3 in./mile (Stearns, 1928; Meinzer and Fischel, 1934; Fischel, 1935). Experiments by Hubbert (1940) showed that the relation established by Darcy is invariant with respect to direction of flow.

The validity of Darcy's law as expressed in Eq. (4.2) or (4.4) is an essential assumption in the following theories and methods:

1. Quantitative theory of incompressible flow in aquifers (Laplace equation)
2. Quantitative theory of compressible flow in aquifers (diffusion equation)
3. Theory of consolidation (diffusion equation)

In the application of Darcy's law, the constant of proportionality is often referred to as the *coefficient of permeability* or, in increasingly more cases, as the *hydraulic conductivity*. As will be discussed shortly, the term *permeability* will be used in reference to a property of the medium for conducting any fluid, and *hydraulic conductivity* in reference to a property of both the medium and of the fluid.

A medium is isotropic with respect to permeability if it is equally permeable to flow in all directions. Since Darcy's law expresses flow in one direction, generalization is required to signify the more complicated process of flow in two or three directions in anisotropic media. The generalizations can be expressed in differential form,

$$q_x = -K_x \frac{\partial h}{\partial x}$$

$$q_y = -K_y \frac{\partial h}{\partial y} \qquad (4.8)$$

$$q_z = -K_z \frac{\partial h}{\partial z}$$

where x, y, and z are the principal directions of the anisotropy. The minus sign indicates that the flow is in a direction opposite to the increase in head. Darcy's law may also be expressed as a vectorial equation

$$\mathbf{q} = -K \text{ grad } h \qquad (4.9)$$

where the vector quantity grad h is read *gradient of h*.

PHYSICAL INTERPRETATION OF DARCY'S CONSTANT OF PROPORTIONALITY

The physical significance of Darcy's constant of proportionality has been investigated in a number of ways. Two investigations will be considered here, Hubbert's empirical experiments (1956), and comparison with the Hagen-Poiseuille equation. A reformulation of Darcy's law by using potential theory follows the discussion of groundwater potential.

By experimentally varying fluid density, viscosity, and the geometrical properties of sands, Hubbert (1956) reported that the parameter K varied in the following manner:

$$K \propto \rho_w$$

$$K \propto \frac{1}{\mu}$$

$$K \propto d^2$$

where ρ_w is fluid density, μ is viscosity, and d is the mean grain diameter. He converted to an empirical equation

$$K = \frac{K'\rho_w d^2}{\mu} \qquad (4.10)$$

where K' is a new factor of proportionality containing variables not evaluated. Equation (4.9) then becomes

$$\mathbf{q} = \frac{-K'\rho_w d^2}{\mu} \operatorname{grad} h \qquad (4.11)$$

A further breakdown of the parameter K' is possible by comparing Darcy's equation with the Hagen-Poiseuille equation governing laminar flow through small-diameter passages. The Hagen-Poiseuille equation is written

$$q = \frac{Q}{A} = \frac{N\rho_w g R^2 S^*}{\mu} \qquad (4.12)$$

where N is a dimensionless-shape factor related to the geometry of passage, R is the diameter of passage, g is the acceleration due to gravity, and S^* is the hydraulic gradient. Combining Eqs. (4.4) and (4.12),

$$q = Ki = \frac{N\rho_w g R^2 S^*}{\mu} \qquad (4.13)$$

Recognizing that S^* equals i,

$$K = \frac{N\rho_w g R^2}{\mu} \qquad (4.14)$$

By assuming that the diameter of passage R is essentially equal to the mean grain diameter d, the parameter K is expressed

$$K = \frac{N\rho_w g d^2}{\mu} \tag{4.15}$$

where Nd^2 characterizes the properties of the medium, and $\rho_w g/\mu$ the properties of the fluid. The permeability of the medium is thus taken as

$$k = Nd^2 \tag{4.16}$$

and has the dimensions of a length squared. The hydraulic conductivity is expressed

$$K = \frac{k\rho_w g}{\mu} \tag{4.17}$$

and has the dimensions of a velocity. Hence, only under isothermal conditions (constant density and viscosity) will changes in conductivity necessarily reflect changes in the geologic medium. It follows that Darcy's law may be expressed

$$q = \frac{k\rho_w g}{\mu} i \tag{4.18}$$

which indicates that the proportionality constant is the hydraulic conductivity and is a property of both the medium and the fluid.

As defined above, hydraulic conductivity is a quantity of water flowing in one unit of time through a face of unit area under a driving force of one unit of hydraulic head change per unit length. In Fig. 4.2, it is the volume of water flowing through opening A in one unit of time. Water-resource evaluations in the United States often utilize gallons for quantity, days for time, and square feet for area, which results in an expression for conductivity in gallons per day per square foot. The gallon-day-foot units may be referred to as the practical *American hydrologic system* of units.

A related concept, transmissivity, is a macroscopic property of an aquifer, that of transmitting water through its entire thickness. Transmissivity is defined as the rate of flow of water at the prevailing temperature through a vertical strip of aquifer one unit wide, extending the full saturated thickness of the aquifer, under a unit hydraulic gradient. In Fig. 4.2, it is the quantity of water flowing through opening B, which is one unit wide and equal in length to the saturated thickness. Hence, the transmissivity is equal to the product of the hydraulic conductivity and the saturated thickness, and may be expressed in gallons per day per foot of aquifer width or, in the standard English system, square feet per day (dimensions $L^2 T^{-1}$).

Permeability, as defined earlier, is a macroscopic property of porous

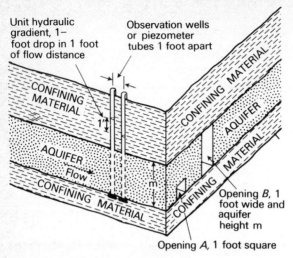

Fig. 4.2 Diagram for hydraulic conductivity and transmissivity. (*After Ferris and others*, 1962.)

material. Since the actual porous structure is random in character, several attempts have been made to relate structure to permeability. Included here is the work of Kozeny (1927) and Carman (1956), leading to what is known as the Kozeny-Carman equation, and the dimensional analysis of Muskat and Botset (1931). An excellent review of the subject is given by Leonards (1962). Further, Darcy's law has been described exclusively in experimental terms as an empirical law. As mentioned previously, this need not be so. Hubbert (1956), for example, derived the law from the fundamental Navier-Stokes equation and thereby established its theoretical basis. However, the empirical law as introduced provides us with the macroscopic law that was sought to develop a description of flow compatible with the level at which field measurements are generally made.

Example 4.1 To obtain an idea of the rate of groundwater movement, assume a head loss of 5 ft for every 1,000 ft of flow ($i = 0.005$), a porosity of 20 percent, and a conductivity of 1×10^2 ft/day. From Eq. (4.7),

$$v = \frac{Ki}{n} = \frac{1 \times 10^2 \text{ ft/day} \times 5 \times 10^{-3}}{20 \times 10^{-2}} = 2.5 \text{ ft/day}$$

For comparative purposes, specific discharge is determined by

$$q = Ki = 1 \times 10^2 \text{ ft/day} \times 5 \times 10^{-3} = 0.5 \text{ ft/day}$$

ENERGY AND ITS TRANSFORMATIONS

Example 4.2 The Darcy-Weisbach equation is used rather extensively to determine head loss in pipe flow and is expressed

$$\Delta h = f \frac{\Delta L}{d} \frac{v^2}{2g}$$

where Δh is the head loss over the length of pipe ΔL of diameter d, v is velocity, and f is a friction factor. For laminar flow, f is determined empirically to be $64/N_r$, where N_r is a Reynolds number. Making this substitution and solving for v,

$$v = \left[\frac{d^2 \rho_w g}{32\mu}\right] \frac{\Delta h}{\Delta L}$$

The quantity in the brackets is analogous to hydraulic conductivity for flow in continuous round pores in that it incorporates properties of the fluid and the medium.

FIELD INTERPRETATION OF HYDRAULIC HEAD AND THE NATURE OF IRREVERSIBLE LOSSES IN GROUNDWATER FLOW

At this point, we wish to investigate the field counterpart of hydraulic head demonstrated in the simple apparatus of Fig. 4.1. The total argument can be presented in terms of the conservation of energy law which, for groundwater flow, states that at all positions in steady, incompressible flow along a streamline in a closed system, the total energy is constant. By initially assuming an ideal fluid (frictionless flow), only mechanical energy terms need be considered, and the conventional Bernoulli equation applies. The assumption of an ideal fluid acting in a closed system will later be relaxed by consideration of real fluids and open systems.

The conventional Bernoulli equation states that at all positions in steady, frictionless, incompressible flow along a streamline in a closed system, the total energy is constant. This may be written as

$$gz + \frac{P}{\rho_w} + \frac{v^2}{2} = \text{constant} \qquad (4.19)$$

where g is the acceleration due to gravity, z is the vertical distance from any arbitrary datum to the point on a streamline, P is the pressure of the fluid, ρ_w is the fluid density, and v is velocity. The dimensions of Eq. (4.19) are L^2/T^2, or energy per unit mass. Dividing through by g,

$$z + \frac{P}{\rho_w g} + \frac{v^2}{2g} = \text{constant} \qquad (4.20)$$

This equation has the units of energy per unit weight, or foot-pounds per pound, or simply feet.

The energy of a body is defined as capacity to do work, and it can be classified in accordance with various forms, such as mechanical, thermal, electrical, or atomic. Mechanical energy, or work done by displacement of

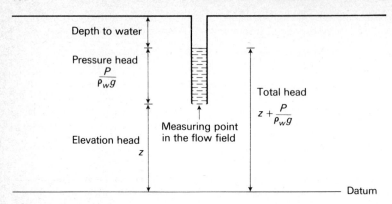

Fig. 4.3 Diagram showing elevation, pressure, and total head for a point in the flow field.

a fluid, can be subdivided into potential energy and kinetic energy. The terms of Bernoulli's equation represent these particular forms of energy. The term z in Eq. (4.20) is the gravitational potential energy per unit weight and represents the work required to increase the elevation of a unit weight of fluid from datum to the height of z (Fig. 4.3). The quantity $P/\rho_w g$ is the flow work or flow energy per unit weight and represents the work that a fluid is capable of doing because of its sustained pressure (Fig. 4.3). The sum of the first two terms is referred to as the *potential energy* of the fluid. The third term, $v^2/2g$, is the *kinetic energy*, or energy due to the fluids motion.

The three terms of Bernoulli's equation, when expressed as energy per unit weight, are referred to as *gravity head, pressure head,* and *velocity head,* respectively. Bernoulli's equation, then, states that the sum of the potential and kinetic energy per unit mass, or weight, along a streamline in a closed system is constant.

The few statements above pertain to ideal fluids, or frictionless flow situations. Groundwater is, of course, a real fluid, possessing the property of viscosity. This provides the means by which mechanical energy is converted to thermal energy. As thermal energy is generally (but not always) not available to do work, there is a net loss involved. Further, the velocity of groundwater flow is so low that the third term of Bernoulli's equation can be ignored. The potential energy per unit mass at an upstream point is then equal to the potential energy at a downstream point plus the loss between the two points

$$gz_1 + \frac{P_1}{\rho_w} = gz_2 + \frac{P_2}{\rho_w} + L_{1-2} \qquad (4.21)$$

ENERGY AND ITS TRANSFORMATIONS

Figure 4.4 diagrammatically shows how the available energy decreases along a streamline. The difference in the level of the water in the piezometers, or measuring devices, demonstrates the loss of potential energy in this system. At point 1, the water rises to a; at point 2, to b; and at point 3, to c. At point 2, the water rises to the same height for piezometers installed at three different depths. Measurements here are taken along a line (vertical, in this case) at which the same amount of potential energy exists. This is termed an *equipotential line*.

Bernoulli's equation is a statement of available energy only. In this sense, it is not a complete statement of the first law of thermodynamics, which includes both available and unavailable energy. Prior to further modification of the equation for application to the flow of viscous fluids, it is profitable now to examine the concept of available energy in thermodynamic terms in order to investigate the nature of reversible and irreversible flow associated with ideal and real fluids. Thermodynamics, as the name implies, is a study of the flow of heat, but it will be seen that the subject treats the more general quantity, energy.

Consider, for example, the steady flow of groundwater between two parallel flow lines, or through a flow tube (Fig. 4.5). Groundwater enters the flow tube with a velocity v_1 and exits with a velocity v_2. If the flow is steady, the unit mass of fluid about to enter the tube corresponds to the unit mass of fluid about to exit from the tube in a unit of time. The energy at the inlet of the tube is the sum of the potential energy, plus any heat that is transferred to the fluid mass, plus the internal energy of the system. From an atomic point of view, the internal energy of the system is the sum of the kinetic k_m and potential P_m energies of the fluid molecules apart from the

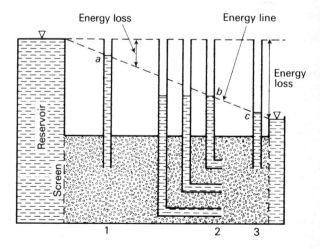

Fig. 4.4 Diagram showing loss in mechanical energy.

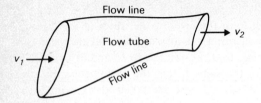

Fig. 4.5 Illustrative flow tube.

kinetic and potential energy of the system as a whole. Designating the internal energy as I and heat added as H_q, the total energy at inlet and outlet for the flow of a real fluid is balanced,

$$gz_1 + \frac{P_1}{\rho_w} + I_1 + H_q = gz_2 + \frac{P_2}{\rho_w} + I_2 \qquad (4.22)$$

In differential form,

$$dH_q = g\,dz + d\!\left(\frac{P}{\rho_w}\right) + dI \qquad (4.23)$$

Bernoulli's equation for the flow of an ideal fluid can also be expressed in differential form,

$$g\,dz + \frac{dP}{\rho_w} = 0 \qquad (4.24)$$

As Eq. (4.23) is a statement for more than one form of energy, and Eq. (4.24) is for potential energy only, the second is subtracted from the first,

$$dH_q = d\!\left(\frac{P}{\rho_w}\right) - \frac{dP}{\rho_w} + dI \qquad (4.25)$$

The first and second terms on the right-hand side of Eq. (4.25) can be further reduced,

$$d\!\left(\frac{P}{\rho_w}\right) - \frac{dP}{\rho_w} = P\,d\!\left(\frac{1}{\rho_w}\right) + \frac{dP}{\rho_w} - \frac{dP}{\rho_w} \qquad (4.26)$$

or

$$d\!\left(\frac{P}{\rho_w}\right) - \frac{dP}{\rho_w} = P\,d\!\left(\frac{1}{\rho_w}\right) \qquad (4.27)$$

Hence, by substitution in Eq. (4.25),

$$dH_q = dI + P\,d\!\left(\frac{1}{\rho_w}\right) \qquad (4.28)$$

For fluids of constant density,

$$d\!\left(\frac{1}{\rho_w}\right) = 0 \qquad (4.29)$$

ENERGY AND ITS TRANSFORMATIONS

Equation (4.28) then reduces to

$$dH_q = dI \qquad (4.30)$$

or any heat added to the fluid mass is compensated for by an increase in the internal energy of the molecules. Interpreted literally, the temperature of the fluid is increased. It follows that if the flow process is not only reversible (no heat of friction), but adiabatic (no heat added or taken from the fluid mass)

$$dI = 0 \qquad (4.31)$$

or the internal energy of the system is constant; that is, there is no change in the temperature of the fluid. This, of course, has already been implied in the definition of an ideal fluid.

Entropy s^* is defined as heat divided by temperature. In differential form,

$$ds^* = \left(\frac{dH_q}{T_e}\right)_{\text{Rev}} \qquad (4.32)$$

where dH_q is the amount of heat added to the system whose temperature is T_e, and the subscript rev refers to a reversible process. The term *entropy* serves as a measure of unavailable energy. As the discussions above describe frictionless flow, Eq. (4.30) can be expressed

$$T_e \, ds^* = dI \qquad (4.33)$$

or entropy (unavailable energy) increases in commensurate amounts with the change in internal energy. For an adiabatic process, $dI = 0$ [Eq. (4.31)], and

$$T_e \, ds^* = 0 \qquad (4.34)$$

or the unavailable energy of the system remains constant. Stated in other terms, the change in entropy is zero when a system undergoes reversible processes (Prigogine, 1967).

Turning attention now to real fluids, the Bernoulli equation with the incorporated loss term [Eq. (4.21)] is expressed in differential form

$$g \, dz + \frac{dP}{\rho_w} + d\,(\text{losses}) = 0 \qquad (4.35)$$

Subtracting this equation from Eq. (4.23),

$$dH_q = d\left(\frac{P}{\rho_w}\right) - \frac{dP}{\rho_w} + dI - d\,(\text{losses}) \qquad (4.36)$$

or

$$d\,(\text{losses}) = P \, d\left(\frac{1}{\rho_w}\right) + dI - dH_q \qquad (4.37)$$

Recognizing again that $d(1/\rho_w) = 0$ for fluids of constant density,
$$d(\text{losses}) = dI - dH_q \tag{4.38}$$
For an adiabatic process, dH_q equals zero, and

$$d(\text{losses}) = dI \tag{4.39}$$

or losses in potential energy are compensated for by an increase in internal energy. Hence, when heat and work are accepted as forms of energy, our experience in noting that the temperature of water is raised by the expenditure of mechanical work is an illustration of the conservation of energy law. Equation (4.38) is valid for groundwater systems that interchange heat energy with the outside environment, as is generally the case, and Eq. (4.39) describes the ideal closed system that exists only as a theoretical model.

Entropy can again be introduced by substituting Eq. (4.33) in Eq. (4.38)

$$d(\text{losses}) = T_e\, ds^* - dH_q \tag{4.40}$$

Rearranging terms,

$$T_e\, ds^* = dH_q + d(\text{losses}) \tag{4.41}$$

If the flow is adiabatic,

$$T_e\, ds^* = d(\text{losses}) \tag{4.42}$$

Hence, for the flow of real fluids, a spontaneous change characterized by a decrease in potential energy is accompanied by an increase in unavailable energy; that is, entropy always increases with irreversibilities (Prigogine, 1967). Stated in terms of irreversible viscous flow, friction merely means a degradation of mechanical energy, not a loss of energy. This is a direct consequence of the second law of thermodynamics.

Example 4.3 Friction is known to produce heat. Calculate the heat of friction associated with a 1,000-ft drop in potential energy.

$$\Delta\Phi = 10^3\ \text{ft-lb/lb} = 10^3\ \text{ft} \times \frac{1}{453.6}\ \text{lb/g} \times 0.3239\ \text{cal/ft-lb}$$

which gives $\Delta\Phi$ in calories per gram.

$$\Delta T_e = \frac{\Delta\Phi\ \text{cal/g}}{Q\ \text{cal/(g)(°C)}}$$

where Q is specific heat. This gives

$$\Delta T_e = \frac{10^3\ \text{ft/lb} \times 1/453.6\ \text{lb/g} \times 0.3239\ \text{cal/ft-lb}}{1.0\ \text{cal/(g)(°C)}}$$

$$\Delta T_e = \frac{0.715\ \text{°C}}{10^3\ \text{ft-lb/lb}}$$

$$\Delta T_e = \frac{1.29\ \text{°F}}{10^3\ \text{ft-lb/lb}}$$

ENERGY AND ITS TRANSFORMATIONS

GROUNDWATER POTENTIAL AND DARCY'S LAW

The detail of the preceding section gives insight into the nature of the losses incurred in the flow of a real fluid. The potential energy

$$\Phi = gz + \frac{P}{\rho_w} \tag{4.43}$$

is the dominating quantity in the flow process, where Φ is designated the potential per unit fluid mass, or simply potential. The quantity Φ is composed of two parts, elevation and pressure, and is derived from the general formula (Hubbert, 1940)

$$\Phi = gz + \int_{P_a}^{P} \frac{dP}{\rho_w} \tag{4.44}$$

where P and P_a are limiting values of pressure over the interval considered. Assuming ρ_w is constant,

$$\Phi = gz + \frac{P - P_a}{\rho_w} \tag{4.45}$$

The components of total head shown in Fig. 4.3 are expressed

$$h = z + \frac{P}{\rho_w g} \tag{4.46}$$

The pressure at the point of consideration, including atmospheric pressure P_a, is then

$$P = \rho_w g (h - z) + P_a \tag{4.47}$$

Substituting this result in Eq. (4.45) gives

$$\Phi = gz + \frac{[\rho_w g(h - z) + P_a] - P_a}{\rho_w} = gh \tag{4.48}$$

The magnitude of the quantity Φ is then the product of total head h, measured in a piezometer tube, and the acceleration of gravity.

Fluid, or groundwater, potential as commonly used in flow problems is therefore based on the following assumptions: (1) The velocity, or kinetic energy, component of fluid potential is so low that it can be omitted; (2) the fluid is viscous; and (3) water is essentially incompressible and in a steady-state condition, or

$$\int P\, dV = 0 \quad \text{and} \quad V = V_0 \tag{4.49}$$

where V is volume, and V_0 is the initial volume.

It is possible now to reexamine Darcy's empirical law derived with the aid of the apparatus in Fig. 4.1. For this problem, Darcy's law is restated

$$q = -K\frac{\partial h}{\partial L} \tag{4.50}$$

where h is total head consisting of elevation and pressure, and L is the distance over which flow takes place. In that $\Phi = gh$,

$$\frac{1}{g}\frac{\partial \Phi}{\partial L} = \frac{\partial h}{\partial L} \tag{4.51}$$

Substituting this result in Eq. (4.50),

$$q = -K\frac{1}{g}\frac{\partial \Phi}{\partial L} \tag{4.52}$$

the minus sign now indicating that discharge is in a direction opposite to the increase in potential. Substituting the five factors comprising hydraulic conductivity [Eq. (4.15)],

$$q = \frac{-N\rho_w d^2}{\mu}\frac{\partial \Phi}{\partial L} \tag{4.53}$$

The fluid potential expressed above is a potential function similar to the voltage potential in the flow of electricity and the temperature potential in the flow of heat. Equivalents of Darcy's law are easily written for the flow of electricity and heat in media that are anisotropic, with x, y, and z as principal directions

$$I_x = -\frac{1}{R_x}\frac{\partial V}{\partial x} \qquad I_y = -\frac{1}{R_y}\frac{\partial V}{\partial y} \qquad I_z = \frac{1}{R_z}\frac{\partial V}{\partial z} \tag{4.54}$$

and

$$H_x = -J_x\frac{\partial T_e}{\partial x} \qquad H_y = -J_y\frac{\partial T_e}{\partial y} \qquad H_z = -J_z\frac{\partial T_e}{\partial z} \tag{4.55}$$

where I = current flow, coulombs/sec, or amperes
    V = voltage
    R = electrical resistance
    H = heat flow, cal/sec
    T_e = temperature
    J = thermal conductivity

Equivalent quantities of the three systems are clearly evident.

GRADIENT OF POTENTIAL AND DIVERGENCE OF THE GRADIENT

The term *field* is used synonymously with both a region and the value of some physical quantity within the region. If the physical quantity is a scalar, such as potential, the field is termed a *scalar field*. If the quantity of interest is a vector, such as flow velocity, the field is a *vector field*. Suppose now that the scalar quantity $\Phi(x,y,z)$ is known at every point in a groundwater region for a three-dimensional problem. Starting at a given point, the potential increases in some directions and decreases in others, the rate of change with respect to distance depending on the particular direction examined. Such rates of change with distance are termed *directional derivatives*. It can be shown (Boas, 1966) that the direction of the largest change in Φ is perpendicular to the equipotential Φ equals a constant. This direction is of particular interest in that it coincides with the direction of groundwater movement from regions where Φ is high to regions where Φ is low. Hence, $\Phi(x,y,z)$ equal to a constant is a curve, and the gradient of Φ (or grad Φ) is a vector perpendicular to this curve when the media is isotropic. In vector notation,

$$\text{grad } \Phi = \nabla \Phi = \mathbf{i}\frac{\partial \Phi}{\partial x} + \mathbf{j}\frac{\partial \Phi}{\partial y} + \mathbf{k}\frac{\partial \Phi}{\partial z} \tag{4.56}$$

where the vectors $\mathbf{i}$, $\mathbf{j}$, and $\mathbf{k}$ are unit vectors in the x, y, z directions, respectively, and $\nabla \Phi$ (pronounced del Φ) is the accepted abbreviation for grad Φ. The symbol ∇ indicates the vector operation

$$\nabla = \mathbf{i}\frac{\partial}{\partial x} + \mathbf{j}\frac{\partial}{\partial y} + \mathbf{k}\frac{\partial}{\partial z} \tag{4.57}$$

The discussions above have been concerned with $\nabla \Phi$, where Φ is a scalar quantity and $\nabla \Phi$ is a vector function. We shall now be concerned with the results of a vector operation ∇ on a vector function $\mathbf{q}$. The divergence of $\mathbf{q}$ (abbreviated div $\mathbf{q}$) is defined as the dot product

$$\nabla \cdot \mathbf{q} = \text{div } \mathbf{q} = \frac{\partial q_x}{\partial x} + \frac{\partial q_y}{\partial y} + \frac{\partial q_z}{\partial z} \tag{4.58}$$

where q_x, q_y, and q_z are three components of $\mathbf{q}$. As the quantity $\nabla \Phi$ of Eq. (4.56) is a vector function, $\mathbf{q}$ can be set equal to $\nabla \Phi$ in Eq. (4.58) and

$$\nabla \cdot \nabla \Phi = \text{div grad } \Phi \tag{4.59}$$

This is written $\nabla^2 \Phi$ and is termed the laplacian of Φ. From Eqs. (4.56) and (4.57)

$$\nabla^2 \Phi = \text{div grad } \Phi = \frac{\partial}{\partial x}\frac{\partial \Phi}{\partial x} + \frac{\partial}{\partial y}\frac{\partial \Phi}{\partial y} + \frac{\partial}{\partial z}\frac{\partial \Phi}{\partial z} \tag{4.60}$$

or

$$\nabla^2 \Phi = \frac{\partial^2 \Phi}{\partial x^2} + \frac{\partial^2 \Phi}{\partial y^2} + \frac{\partial^2 \Phi}{\partial z^2} \tag{4.61}$$

The definition of divergence of grad Φ, then, lies in the definition of a partial differential equation. The divergence of a flow has the simple physical meaning, per unit volume, as the excess of the flow which leaves a small volume to the flow which penetrates it. As will be demonstrated later,

$$\nabla^2 \Phi = 0 \qquad (4.62)$$

represents the Laplace equation, and

$$\nabla^2 \Phi = \frac{1}{a} \frac{\partial \Phi}{\partial t} \qquad (4.63)$$

where a is a constant, represents the diffusion equation. These are important mathematical expressions used extensively in distributed-parameter problems of groundwater flow.

4.2 MEASUREMENTS OF POTENTIAL ENERGY AND INTERPRETATIONS

The subsurface position of groundwater flow precludes direct observation of movement and requires the use of other observations relating to movement. Among these observations, groundwater potential is the most reliable. If the medium in which the potential is measured is viewed on a macroscopic scale, some parts will be of high permeability and others of low permeability, and the resulting flow pattern will depend to a large degree on the spatial arrangement of these materials.

CONDITIONS FOR ISOTROPIC AND HOMOGENEOUS FLOW

Expressed symbolically, the line designated $\Phi(x,y)$ equals a constant and is termed an *equipotential line,* where Φ is the potential and x and y describe its position in space. The line designated $\Psi(x,y)$ equals a constant is called a *flow line,* and represents the path followed by a particle of water. A flow line is drawn in such a way that its direction at a point (that is, the direction of the tangent to the point) is in the same direction as the flow at that point. At any one point, the flow can have but one direction; thus, only one flow line can pass through each point of the field. It follows that flow lines never intersect.

Since, by definition, the potential is constant along a given equipotential line, there can be no component of flow tangential to that line, and flow lines and equipotential lines must intersect at some angle greater than zero. The conditions under which they intersect at right angles in two-dimensional flow are demonstrated below with the aid of Fig. 4.6. In these calculations, it is assumed that the influence of gravity does not vary significantly over a region, so that total head [Eq. (4.48)] provides an adequate measure of the potential energy of the water at a point.

ENERGY AND ITS TRANSFORMATIONS

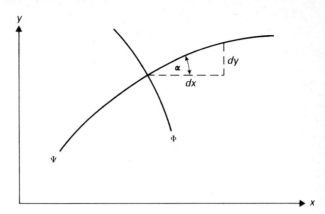

Fig. 4.6 Intersecting flow line and equipotential line

For the flow line of Fig. 4.6, the tangent of the angle α is expressed

$$\tan \alpha = \frac{dy}{dx} = \frac{q_y}{q_x} = \frac{k_y \, (\partial h/\partial y)}{k_x \, (\partial h/\partial x)} \tag{4.64}$$

for a general anisotropic medium. If k_x equals k_y,

$$\tan \alpha = \frac{dy}{dx} = \frac{\partial h}{\partial y} \frac{\partial x}{\partial h} \tag{4.65}$$

For the equipotential line, the statement that the head along it does not change in value is given as

$$\frac{\partial h}{\partial x} dx + \frac{\partial h}{\partial y} dy = 0 \tag{4.66}$$

or

$$\frac{dy}{dx} = -\frac{\partial h}{\partial x} \frac{\partial y}{\partial h} \tag{4.67}$$

If two lines are mutually perpendicular, the slope of one equals the negative reciprocal of the slope of the other. Testing this condition for the slopes expressed in Eqs. (4.65) and (4.67),

$$\frac{\partial h}{\partial y} \frac{\partial x}{\partial h} = -\frac{1}{(-\partial h/\partial x)(\partial y/\partial h)} \tag{4.68}$$

or

$$\frac{\partial h}{\partial y} \frac{\partial x}{\partial h} = \frac{\partial h}{\partial y} \frac{\partial x}{\partial h} \tag{4.69}$$

and the lines are proved perpendicular.

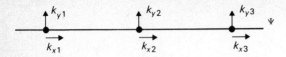

Fig. 4.7 Diagram illustrating point permeabilities along a flow line.

A porous medium is isotropic with respect to permeability if it is equally permeable to flow in all directions. For the hypothetical flow line of Fig. 4.7, this condition is fulfilled if $k_{x1} = k_{y1}$, $k_{x2} = k_{y2}$, $k_{x3} = k_{y3}$, and not fulfilled (i.e., is anisotropic) when $k_{x1} \neq k_{y1}$, $k_{x2} \neq k_{y2}$, etc. A porous medium is homogeneous with respect to permeability if the permeabilities are the same from point to point or, from Fig. 4.7, if $k_{x1} = k_{x2} = k_{x3}$ and $k_{y1} = k_{y2} = k_{y3}$. Nonconcurrence with this condition implies nonhomogeneity. A medium that is both isotropic and homogeneous is characterized by point permeabilities that are equal in all directions, as well as from point to point; that is, $k_{x1} = k_{y1} = k_{x2} = k_{y2} =$, etc. A flow pattern depicting this condition has flow lines and equipotential lines intersecting at right angles (a condition of isotropy) to form a pattern of squares or curvilinear squares (a condition of homogeneity).

POTENTIOMETRIC SURFACE

Measurements of hydraulic head obtained from water wells may be contoured on a map, each individual data point representing a measure of the potential energy of the water in the rock unit to which it refers. The hypsometric surface depicted is referred to as a *potentiometric map*. Another term in common usage is *piezometric surface,* defined by Meinzer (1923) as an imaginary surface that everywhere coincides with the static level of water in the aquifer. Generally, however, the aquifer is characterized by more than one piezometric surface.

A cross-sectional view of flow in the xz plane and a cutaway view of the water table's projection on a horizontal surface are shown in Fig. 4.8. Water from different depths along any vertical will rise to different levels, each set of levels for a given depth having its own characteristic projection. A piezometric surface for any depth is thus a two-dimensional projection of a three-dimensional field (Hubbert, 1940), there being a unique projection for each depth. If a potentiometric map is drawn from data obtained from wells of different depths, the surface obtained is a composite of potential measurements.

In spite of these limitations, potentiometric maps are used to good advantage when the flow is two dimensional in the xy plane. Figure 4.9 is

such a map, characteristic of a nonhomogeneous region. The following statements hold:
1. Head losses between adjacent pairs of lines of equal head are equal.
2. The hydraulic gradient varies inversely with distance between lines of equal head.

If inflow for any section is just balanced by outflow, the relative steepness of the hydraulic gradient reflects the overall resistance offered to fluid flow, a consequence of Darcy's law. In Fig. 4.9, the permeability in the region between h_4 and h_3 is about twice as great as between any other set of equipotential lines. Relatively uniform permeability gives rise to equidistant spacing, as between h_1 and h_3. These points are best demonstrated by equating the quantity of flow in a unit of time entering one of the rectangles or squares with the quantity of flow in a unit of time leaving the rectangle or square, or

$$\Delta q_1 = k_1 \frac{\Delta h}{L_1} a_1 = \Delta q_2 = k_2 \frac{\Delta h}{L_2} a_2 \tag{4.70}$$

As a_1 equals a_2,

$$\frac{k_1}{k_2} = \frac{L_1}{L_2} \tag{4.71}$$

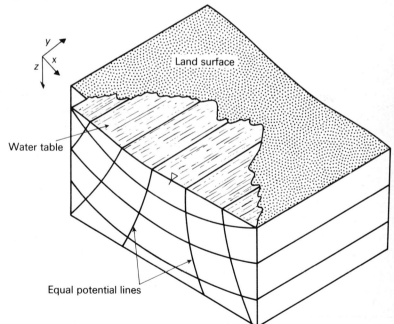

Fig. 4.8 Cross-sectional view of a flow pattern, with a cutaway view of the water table's projection on a horizontal surface.

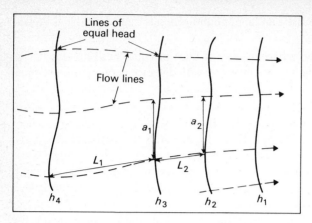

Fig. 4.9 Potentiometric map and intersecting flow lines for a nonhomogeneous aquifer.

Hence, the flow pattern for an isotropic, nonhomogeneous medium consists of lines intersecting at right angles to form a pattern of rectangles or curvilinear rectangles, the ratio of the lengths of sides of the rectangles equal to the ratio of the respective permeabilities. It follows that the conditions of homogeneity are satisfied when L_1 equals L_2, and the flow pattern consists of lines intersecting to form a pattern of squares.

If an area is one of widespread irrigation or some other form of uniform recharge, the amount of water in the system may increase in the direction of flow. If the aquifer is homogeneous, this results in a decrease in distance between equipotential lines (Fig. 4.10a). A decrease in flow in a homogeneous aquifer, as will occur in an area of uniform natural discharge, results in an increase in the distance between equipotential lines in the direction of flow (Fig. 4.10b). If the permeability of a nonhomogeneous aquifer is known, the increase or decrease of flow can be easily computed by measuring the increase or decrease of the hydraulic gradient.

When permeability varies in the direction of flow and water is either added to or removed from the system in the same direction, the potential configuration will be controlled by both of these factors. For example, the two conditions, (1) that of continuous loss of water in the direction of flow and (2) that of decreasing permeability in the direction of flow, have opposing effects on the hydraulic gradient. The first results in a decrease in gradient, and the second results in an increase in gradient. If the loss of flow is proportional to the decrease in permeability, the resulting surface will approach the configuration of a homogeneous aquifer. Darcy's law is again the basis for these conclusions.

ENERGY AND ITS TRANSFORMATIONS

GRADIENT OF HYDRAULIC HEAD: VERTICAL SECTIONS

As the rate of change of potential with distance is a directional derivative whose value depends on the direction examined, a potentially useful and readily accessible direction in which measurements may be obtained is vertically downward. Emphasis now is placed on measurements obtained during the drilling of boreholes or water wells that relate to the vertical component of flow. The rate of change in hydraulic head with respect to change in elevation in a single borehole or well is defined

$$\frac{dh}{dz} = \lim_{z_1 \to z_2} \frac{\Delta h}{\Delta z}$$

where

$$\Delta h = h_1 - h_2 \quad \text{and} \quad \Delta z = z_1 - z_2$$

for $z_1 > z_2$. The gradient of head has useful geometrical and physical meanings which when analyzed, should provide considerable insight into the vertical configuration of groundwater flow.

Consider, for example, the following experiment. A piezometer is installed at a given depth in a homogeneous, isotropic medium. After sufficient time has elapsed, a stable water level is observed and measured.

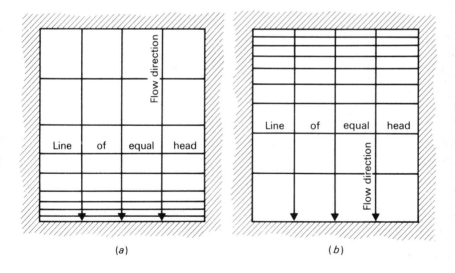

Fig. 4.10 Schematic diagrams showing potentiometric surfaces and intersecting flow lines in a homogeneous, isotropic aquifer characterized by (*a*) an increase in flow in the direction of flow and (*b*) a decrease in flow in the direction of flow. (*Figure 4.10a after Skibitzke and Da Costa, 1962.*)

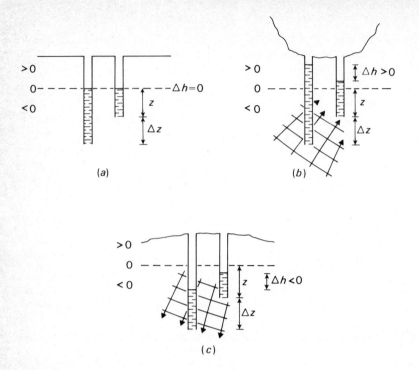

Fig. 4.11 Schematic diagrams of (*a*) normal, or hydrostatic, pressure, (*b*) an area of upward flow, and (*c*) an area of downward flow.

The piezometer is then placed deeper in the system and temporarily terminated until a new equilibrium level is attained. What sort of changes would be expected? Consider several cases.

The first to be considered is one of normal pressure, or with perfect coincidence of the length of water column which can be supported by the pressure in any given stratum and the depth of that stratum below the top of the zone of saturation (Fig. 4.11*a*). At this point, the situation is described completely by hydrostatics, the pressure-producing mechanism being the weight of the superincumbent body of water. As there are no shear stresses in a fluid at rest, the descriptive term *normal pressure for a given depth* emphasizes that only normal forces are present. This hydrostatic case is a good reference with which dynamic heads may be compared. The convention followed is that of designating the water-table position as zero, above the water table as greater than zero, and below the water table as less than zero.

For the hydrostatic condition, dh/dz equals zero, as shown in the plot of Fig. 4.12*a*. An identical plot would result if the equipotential lines were vertical, as in lateral flow (Fig. 4.4). Hence, dh/dz equals zero for both

ENERGY AND ITS TRANSFORMATIONS

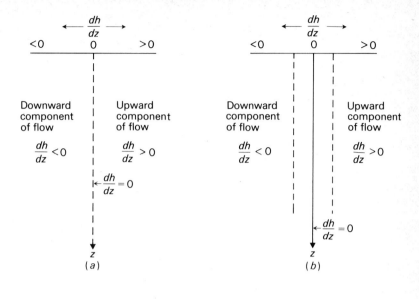

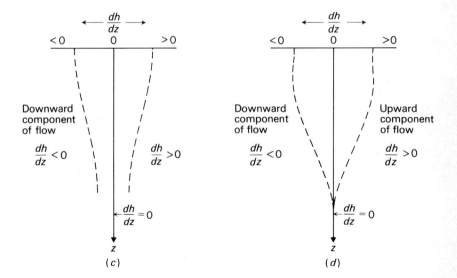

Fig. 4.12 Flow system information determined from a borehole in uniform material: (a) hydrostatic or lateral flow conditions, (b) conditions of upward or downward flow components when the gradient is uniform with depth, (c) a decrease in active circulation with increasing depth, and (d) relative stagnation or lateral flow conditions encountered at depth.

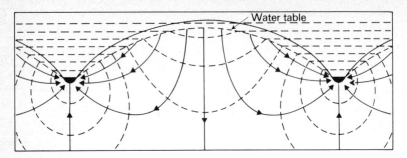

Fig. 4.13 Flow pattern in a uniformly permeable material between the sources distributed over the air-water interface and the valley sinks. (*After Hubbert, 1940. Used with permission of the University of Chicago Press.*)

stagnation (hydrostatics) and lateral flow. This is generally the assumption made in construction and interpretation of piezometric maps.

It follows that plots of dh/dz for upward and downward components of flow in Figs. 4.11b and 4.11c are greater than and less than zero, respectively, when viewed from the zero designation of the water table. If the gradient is uniform with depth, dh/dz versus z would plot as approximate straight lines parallel to the $dh/dz = 0$ line (Fig. 4.12b). However, because active circulation often decreases with increasing depth, it is not unexpected to discover a general convergence of dh/dz toward the zero line (Fig. 4.12c). The limiting condition occurs when the value of the gradient varies from greater than or less than zero to zero, which suggests that penetration has proceeded from regions with either upward or downward components of flow to regions of either relative stagnation or lateral flow (Fig. 4.12d). If there is relative stagnation, the effective lower boundary of the flow system is delineated.

CONFINED AND UNCONFINED FLOW

Although valid only for the isotropic, homogeneous condition, the above discussion should adequately serve as an introduction to the groundwater flow cell, defined as unconfined flow bounded by adjacent topographic highs (Mifflin, 1968) and used synonymously with Hubbert's (1940) flow patterns in uniformly permeable material (Fig. 4.13). As water offers no resistance to deformation, its movement from high elevations to lower elevations would, under closed conditions, result in the drainage of the water contained in topographic highs and so producing a flat surface of minimum potential energy (the hydrostatic condition). This tendency is opposed by the continuous replenishment of water derived from the atmosphere. The result of this movement-renewal process is the flow pattern in Fig. 4.13, the upper surface in which is a subdued replica of topography.

ENERGY AND ITS TRANSFORMATIONS

It has been suggested that for valleys having lower-order streams only, no concentration of discharge at the stream occurs (Toth, 1962). Instead, the discharge is distributed over an area between a so-called midline and the valley bottom (Fig. 4.14). The midline is seen to be a near-vertical equipotential line, which is present also in the flow pattern of Fig. 4.13, although masked somewhat by the exaggerated vertical scale. Flow lines intersecting such a line or group of lines must be horizontal, or nearly so, which gives rise to a region of lateral flow. By inspection of Figs. 4.13 and 4.14, flow lines up-gradient from the region of lateral flow are directed downward with respect to the water table, and flow lines down-gradient from the region of lateral flow are directed upward. These areas then correspond to the postulated downward and upward flow conditions of Fig. 4.11, and they are referred to as *recharge* and *discharge areas,* respectively. The idealized flow pattern can be described in three parts: the recharge area, or source; the area of lateral flow; and the discharge area, or sink. These designations are based not only on the areas where the processes of recharge and discharge take place, but on the direction of flow lines with respect to the water table in the zone of saturated flow (Toth, 1962). These concepts do not by any means represent new ideas, but were first introduced by Meinzer (1917) to describe the flow system in Big Smoky Valley, Nevada.

The unconfined flow patterns of Figs 4.13 and 4.14 are typically associated with flow in an ideal homogeneous and isotropic environment. Deviations from this idealized geologic setting give rise to other types of

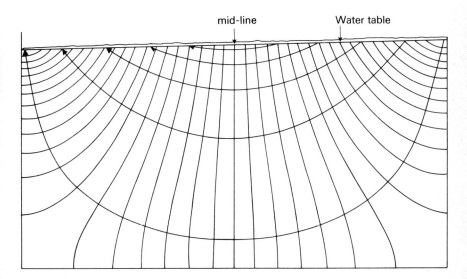

Fig. 4.14 Symmetrical flow pattern with respect to the midline between the valley bottom and the groundwater divide. (*After Toth*, 1962.)

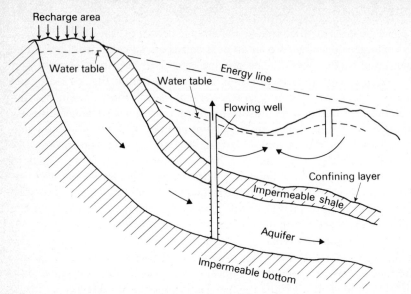

Fig. 4.15 Cross section of a confined groundwater basin.

flow. At the other extreme, for example, is Chamberlin's (1885) concept of a confined basin, exemplified by dipping beds with replenishment at outcrop and by marked changes in lithology in interlayered units (Fig. 4.15). If a well is placed in this system, water will rise in the well to the elevation of the water table in the intake area, minus any friction losses incurred from the point of intake to the point of withdrawal. The dynamic mechanism required to maintain the high-pressure system is the continuous replenishment by precipitation. The pressure-producing mechanism is assumed to be the hydrostatic weight of the body of water extending from the water table in the outcrop area to the discharge area. Clearly, a piezometric surface of the flow field must now be completely independent of surface topography. This situation exemplifies complete confinement.

It is advantageous to consider the confined system postulated by Chamberlin and the unconfined system described by Hubbert as end members of a complex isomorphic series of degrees of confinement. What is consistently found in nature is a system of flow that possesses distinct characteristics of each of the end members. For example, some degree of hydraulic continuity is a chief characteristic of almost all water-bearing rocks, which suggests that the potential distribution with depth is affected by the potential distribution of the overlying water table. On the other hand, the flow field may possess characteristics of a confined system whenever the flow is refracted when emerging from a low-permeability unit such that it proceeds tangentially, or nearly so, to the bottom surface of the unit.

ENERGY AND ITS TRANSFORMATIONS

Flow-line refraction is merely the bending of flow lines in such a way that fluid mass is conserved when water flows across a boundary between strata of different permeability. One way to look at this phenomenon is provided by Darcy's law: all factors equal, the higher the permeability, the smaller the area required to pass a given volume of water in a given time. Analytically, a refraction occurs such that the permeability ratio of the two units equals the ratio of the tangents of the angles the flow lines make with the normal to the boundary. From Fig. 4.16a,

$$\frac{k_1}{k_2} = \frac{\tan a_1}{\tan a_2} \qquad (4.72)$$

In Fig. 4.16b, the hydraulic gradient in the lower layer is steepened to accommodate the flow crossing the boundary. In Fig. 4.16c, the hydraulic gradient in the lower layer is relatively flat. This phenomenon has been described by Casagrande (1937) and Hubbert (1940).

In gross detail, Fig. 4.17 represents an unconfined system in that there exists a hydraulic continuum in the vertical direction. Flow lines in the high-permeability unit are such that they approach a position parallel to the bottom surface of the low-permeability unit. Note that there is little or no difference in piezometric head along an imaginary vertical line passing through the unit of high permeability. If this imaginary line is extended upward, it crosses numerous equipotential lines in the low-permeability unit and relatively few equipotential lines in the unit of medium permeability. It follows that if a well is drilled in this system, a large increase in the piezometric head will be noted when the hole first enters the unit of high permeability. The reason is that static levels there can establish themselves more rapidly than in other parts of the section. This is commonly observed during drilling and is often attributed to confinement of water under pressure. In reality, it is the movement of water through the unit of low permeability that

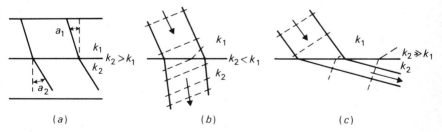

Fig. 4.16 Diagrams of flow-line refraction and conditions at the boundaries between materials of differing permeability. (*Fig. 4.16b and c after Casagrande*, 1937. Reprinted from the Journal of the New England Water Works Association, vol. 51, no. 2, 1937.)

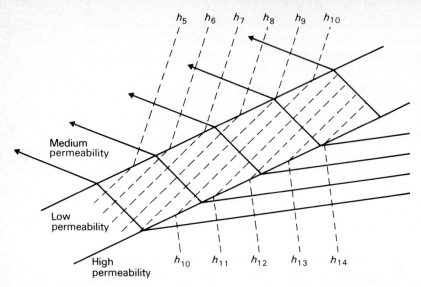

Fig. 4.17 Flow-line refraction in an unconfined system of layered permeability contrasts. (*Modified after Hubbert*, 1940. *Used with permission of the University of Chicago Press.*)

gives rise to the large change in head, which suggests that the conditions implied by the term *confinement* will arise when a unit of low permeability overlies a unit of high permeability.

In summary, the following statements are offered:

1. Many groundwater systems are essentially unconfined in that the potential distribution within them is affected by the potential distribution of the water table in the overlying rocks. These systems, or parts of them, often exhibit conditions implied by the term *confinement* because of large permeability differences between adjacent units.
2. A confined system is one in which the potential distribution is not affected by the potential distribution of the overlying water table. The piezometric surface of the flow field is completely independent of surface topography.

4.3 OTHER FORMS OF ENERGY

Since the gain of heat compensates for the loss of mechanical energy in groundwater flow, heat must be accepted as another form of energy to satisfy the conditions of the energy conservation law. This means that any phenomenon capable of producing heat is likewise a form of energy. A

ENERGY AND ITS TRANSFORMATIONS

chemical reaction, for example, may liberate heat, as may the flow of electricity through a wire. As groundwater moves from areas where energy is high to areas where it is low, its movement in response to thermal, electrical, or chemical gradients is suited to analysis as energy.

Analysis of this type requires further application of thermodynamic principles, and those of free energy in particular, a topic which will be dealt with in considerable detail later. Suffice it to say at this time that free energy is analogous to potential energy in that it represents the work that must be done on a system at constant pressure and temperature to bring it from some state A to some state B. Another way to look at free energy is as follows: The decrease in free energy in going from state A to state B equals the work done isobarically and isothermally by the system on its surroundings. Hence, free energy is a thermodynamic potential, quite analogous to the mechanical concept of potential. As with potential energy, data selection is purely arbitrary. For example, one may be interested in the work done in going from a state of pure water to a state of a definite concentration of dissolved constituents in the water, or from a state of certain dissolved concentration to a state of different concentration.

In general, the following statements are true:

1. Free energy increases with increases in hydrostatic pressure.
2. Free energy decreases with increases in concentration of dissolved material.
3. Free energy decreases with increasing temperature.

CHEMICAL OSMOSIS

One apparatus used to demonstrate the phenomenon of chemical osmosis is shown in Fig. 4.18, where a column of relatively pure water is

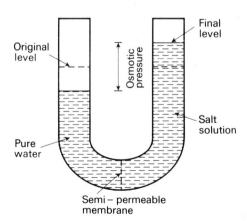

Fig. 4.18 U-tube demonstration of osmosis.

hydrostatically balanced by a column of saline water through a semipermeable membrane. Because movement takes place from the pure-water side to the saline-water side, the pure water must somehow possess more energy than the saline solution. As this is not mechanically evident, the reasons must lie in the statement given above: Free energy decreases with an increase in salinity concentration. As a consequence of water movement, the saline water becomes somewhat diluted and its salinity decreases. As the system is presumed closed, the mass transfer of water causes an increase in hydrostatic pressure of the saline solution and a decrease in hydrostatic pressure of the pure solution. In terms of energy, the free energy of the saline solution increases in accordance with a decrease in salinity and an increase in hydrostatic pressure. The free energy of the pure water decreases in accordance with a decrease in hydrostatic pressure. Movement will cease when the free energies of the pure and saline water are equal.

The initial conditions of hydrostatic balance in Fig. 4.18 can be expressed

$$F_{sp} > F_{ss}$$

where F is free energy, and the subscripts sp and ss refer to pure and saline water, respectively. The free energy consists of that due to hydrostatic pressure P and that due to chemical energy, designated Π. As energy terms are additive,

$$P_{sp} + \Pi_{sp} > P_{ss} + \Pi_{ss}$$

Recognizing that P_{sp} equals P_{ss},

$$\Pi_{sp} > \Pi_{ss}$$

by the amount that F_{sp} exceeds F_{ss}.

If, for the moment, the sign of the energy terms is ignored, the terminal, or equilibrium, condition is expressed

$$P_{sp} + \Delta P_{sp} + \Pi_{sp} = P_{ss} + \Delta P_{ss} + \Pi_{ss} + \Delta \Pi_{ss} \tag{4.73}$$

As energy is neither created nor destroyed, but merely transferred, the change in free energy of the pure water at equilibrium is equal in magnitude but opposite in sign to the free-energy change of the saline solution; that is,

$$-\Delta F_{sp} = \Delta F_{ss} \tag{4.74}$$

at equilibrium. Collecting appropriate terms from Eq. (4.73),

$$-\Delta P_{sp} = \Delta \Pi_{ss} + \Delta P_{ss} \tag{4.75}$$

Ignoring the subscripts for the two-column system,

$$\Delta \Pi = -2\Delta P \tag{4.76}$$

ENERGY AND ITS TRANSFORMATIONS

The energy term Π is called the *osmotic pressure* and is defined as the hydrostatic pressure that must be placed on a concentrated solution to establish equilibrium between it and the pure solvent, that is, to raise the free energy of the concentrated solution to that of pure water. For example, from the initial conditions of Fig. 4.18 and the developments above, no net movement could have taken place if the pressure of the saline solution was raised by $2\Delta P$. From this definition,

$$\Pi = P_{ss} - P_{sp} \tag{4.77}$$

where P_{ss} is the hydrostatic pressure of the saline solution, and P_{sp} the pressure of the pure solution to satisfy the condition ΔF equals zero.

The field counterpart of the apparatus in Fig. 4.18 does not involve a pure solvent and a concentrated solution, but salinities of different concentration. The general example is as follows: If the salinities of the water in two aquifers separated by a thin layer of clay or shale are unequal, a net movement of water may occur from the aquifer of low salinity to the aquifer of high salinity. During this process, the salt ions of the more diluted aquifer cannot move freely across the shale membrane. As a consequence, the dissolved constituents of the originally diluted aquifer become more concentrated and its salinity increases. The dissolved constituents of the originally salty aquifer become diluted. This movement can be opposed if pressure can be applied to the fluid of high salinity. This balancing pressure may be provided by the hydrostatic head that is developed.

The phenomenon of chemical osmosis is called upon to explain certain pressure and salinity anomalies in sedimentary basins. In Fig. 4.19, two aquifers are separated by a semipermeable membrane. The aquifers have characteristic heads Δh_I and Δh_II, and the activities of their waters are denoted by $\alpha^\mathrm{I}_{H_2O}$ and $\alpha^\mathrm{II}_{H_2O}$. Activities of water are inversely related to

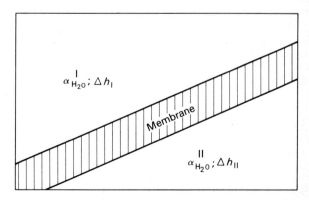

Fig. 4.19 Shale membrane separating two aquifers. (*After Hanshaw and Zen, 1965.*)

salinities, so that if $\alpha^{II}_{H_2O} > \alpha^{I}_{H_2O}$, aquifer I is the saltier of the two. If $\Delta h_{II} > \Delta h_{I}$, the water will of course flow from aquifer II to aquifer I. As the salt ions cannot move freely across the membrane, the water in aquifer II gets saltier and the water in aquifer I gets fresher. As free energy increases with decreasing salinity, a counterforce is set up and the flow may cease when the increased free energy of the water in aquifer I balances the head difference between the aquifers.

The process described above is called *salt filtering,* or *ultrafiltration* (Back and Hanshaw, 1965), a mechanism often called upon to explain the occurrence of brines in deep sedimentary basins. Bredehoeft and others (1963) recognized that brines are usually found (1) in deeper parts of formations that contain fresh water at shallow levels near the outcrop areas, (2) in close proximity to evaporites or other soluble strata, (3) in formations near saline surface-water bodies with hydraulic conditions that are favorable for brine encroachment, and (4) in formations that have been subject to migration of brines from one or more of these sources. The mechanisms for the formation of brines in categories 2 through 4 are obvious. Category 1 is more difficult to explain. Because many brines are as much as six times more concentrated than normal sea water, the suggestion that original marine water was trapped in the sediment is generally ruled out. The osmotic mechanism described above was proposed by DeSitter (1947) to account for the occurrence of brines in basins which are not subject to migration of brines from other sources and which do not have extensive saline deposits.

Returning now to Fig. 4.19, if there is no appreciable difference in Δh_I and Δh_{II}, and if $\alpha^{II}_{H_2O} > \alpha^{I}_{H_2O}$, water may flow from aquifer II to I. If such flow takes place in a poorly transmissive environment, the net effect may approximate that of the closed system of Fig. 4.18, where no flow is contributed from the outside. Under these conditions, the pressure of the fluid in aquifer II may decrease in response to osmotic withdrawal, and the pressure in aquifer I may increase. Areas of high or low fluid pressures often constitute anomalies in certain environments that may be logically explained by chemical osmosis (Berry, 1959; Berry and Hanshaw, 1960; Back and Hanshaw, 1965).

CAPILLARITY

When the lower part of a dry, porous material comes in contact with a saturated material, the water rises to a certain height above the top of the saturated sample. The driving force responsible for the rise is termed *surface tension,* a force acting parallel to the surface of the water in all directions because of an unbalanced molecular attraction of the water at the boundary. The tensional nature of these forces can be compared with those set up in a stretched membrane.

Capillarity results from a combination of the surface tension of a liquid and the ability of certain liquids to wet the surfaces with which they come in contact. This wetting (or wetability) causes a curvature of the liquid surface, giving a contact angle between liquid and solid different than 90°. The idealized system commonly utilized to examine the phenomenon is a water-containing vessel and a capillary tube. When the tube is inserted in the water, the water rises to a height h_c (Fig. 4.20). The meniscus, or curved surface at the top of the tube, is in contact with the walls of the tube at some angle α, the value of the contact angle depending on the wall material and the liquid. For a water-glass system, α is taken as zero.

Points A and C of Fig. 4.20 are at atmospheric pressure (14.7 lb/in.²). A fundamental law of hydrostatics states that the pressure intensity in a fluid at rest can vary in the vertical direction only. Hence, point B is also at atmospheric pressure, and the pressure at point D must be less than atmospheric by an amount equal to the weight of the water column above B, or by the amount $h_c \gamma_w$. At a height of 34 ft, the value of $h_c \gamma_w$ is 14.7 lb/in.² A capillary rise of 34 ft suggests that the pressure just below the meniscus is zero. Practically speaking, zero water pressure at any point within a water column is impossible.

The height of capillary rise in small-diameter tubes can be determined by considering the water column BC isolated from the tube. The weight of the water column is the product of capillary rise h_c, the circular area of the tube ΠR^2, and the unit weight of water. This weight is assumed to be carried by the surface tension acting on the circumference of the tube with which the water is in contact, or $2\Pi R$ times the surface tension T_s. Equating these relations,

$$h_c \Pi R^2 \gamma_w = 2\Pi R T_s \qquad (4.78)$$

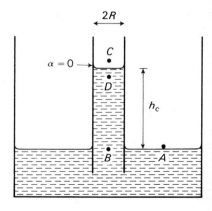

Fig. 4.20 Capillary rise in a glass tube.

or

$$h_c = \frac{2T_s}{\gamma_w R} \qquad (4.79)$$

For water at ordinary temperatures against glass, the surface tension is about 75 dynes/cm, or 4.2×10^{-4} lb/in. Equation (4.79) reduces to

$$h_c = \frac{2.32 \times 10^{-2}}{R} \qquad (4.80)$$

where h_c and R are expressed in inches.

The fact that the height of capillary rise is inversely proportional to the diameter of passage permits us to account for a greater capillary rise in fine-grained materials than in coarse-grained materials. Beyond this widely accepted generalization, very little of the capillary-tube theories can be applied with consistency to materials found in nature. The actual influence of the diameter of passage, for example, is difficult to evaluate. When water is drawn upward from the zone of saturation, the maximum diameter of passage controls the process. When water percolates downward, the minimum diameter of passage is the controlling factor. Hence, the complexities due to variable width of passage make the development of tenable theories of capillarity extremely difficult.

Edlefsen and Anderson (1943) gave the following account of capillary rise in terms of free energy. Recognizing that soil moisture moves spontaneously from the saturated zone to the unsaturated zone, they concluded the free energy of wet soils is higher than the free energy of dry soils. In capillary tubes, the free energy of water just under the meniscus at height h above the free water surface is

$$\Delta F = -h\gamma_w \qquad (4.81)$$

the negative sign representing a state of tension, or negative hydrostatic pressure with respect to datum. The water in the vessel is pure, with a free energy of zero.

ELECTRO AND THERMOOSMOSIS

The movement of water across a semipermeable membrane may also take place in response to an electrical potential across the membrane, or to a temperature gradient. In general, water movement in response to these energy gradients is largely speculative and of minor concern. In the experimental example of electroosmosis, two electrodes are installed in a saturated mass and a direct current applied. There results a movement of water from the positive electrode to the negative electrode. This phenomenon is associated with exchangeable ions which, under the influence of an electrical current, are positively charged and move toward the negative electrode. They are replaced by positive hydrogen ions produced by electrolysis of water. In nature, a membrane may develop a potential difference when it separates two electrolyte solutions containing cation species at different concentrations (Back and Hanshaw, 1965).

Thermoosmosis is the movement of water in response to a temperature gradient. Several mechanisms have been proposed for this phenomenon, including both transport via the vapor phase and viscous flow due to the gradient of vapor pressure.

4.4 EXTENSIONS FOR TWO MISCIBLE FLUIDS: SEA–WATER INTRUSION IN COASTAL AQUIFERS

Under natural conditions, the flow of fresh water toward the sea limits the landward encroachment of sea water. With development of groundwater supplies and subsequent lowering of the water table or piezometric surface below mean sea level, the dynamic balance between fresh and sea water is disturbed, permitting the sea water to intrude usable parts of the aquifer (Fig. 4.21). This phenomenon has been reported in several parts of the world, including The Netherlands (Liefrinck, 1930), Japan (Senio, 1951), Israel (Schmorak, 1967), and the United States (Barksdale, 1940; Leggette, 1947; Bennett and Meyer, 1952).

Sea-water intrusion is another example of open-system behavior in response to outside disturbances, and it can be controlled by maintaining the water table or piezometric surface at or near the coast. Of the five methods of control suggested by Banks and Rictor (1953), four utilize this method. The four are:

1. A reduction in pumpage or rearrangement of the pumping pattern
2. Artificial recharge
3. Establishment of a pumping trough along the coast, thereby limiting the area of intrusion to the trough
4. Formation of a pressure ridge along the coast

The fifth method has a subsurface barrier designed to reduce the permeability of the media to the flow of salt water.

Field observations in coastal areas suggest the presence of a mixing zone, or zone of dispersion, where miscible sea and fresh water interfuse about their boundary. This phenomenon is best documented in the Biscayne aquifer near Miami, Florida (Kohout, 1960; Kohout and Klein, 1967). In this area, the saltwater front is dynamically stable as much as 8 miles seaward of the position computed by methods to be discussed in this section. There is ample evidence to suggest that the zone of diffusion between the salt and fresh water moves seaward during periods of heavy recharge, and inland

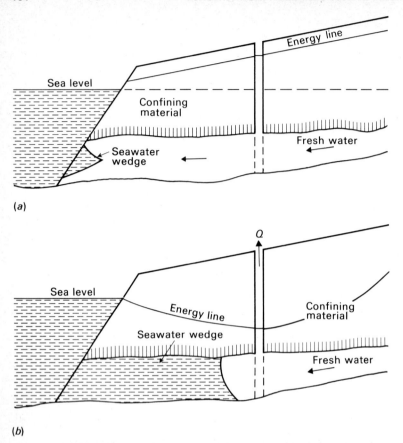

Fig. 4.21 Hydraulic conditions near a coastline (*a*) not subject to sea-water intrusion, and (*b*) subject to seawater-intrusion with an advancing sea-water wedge.

during periods of low freshwater head (Kohout and Klein, 1967). During the periods of inland flow, the cyclic flow of salt water in the zone of dispersion tends to limit the extent to which sea water invades the aquifer. A typical cross section of the zone of dispersion is shown in Fig. 4.22. A generalized theory of the phenomenon is given by Bear and Bachmat (1967).

HYDROSTATICS OF THE GHYBEN–HERZBERG RELATION

At about the turn of the century, two investigators working independently along the European coast recognized that salt water occurred underground, not at sea level, but at a depth below sea level equivalent to approximately 40 times the height of fresh water above sea level (Ghyben, 1889; Herzberg, 1901). The analytical explanation of this phenomenon is referred to as the

ENERGY AND ITS TRANSFORMATIONS

Ghyben-Herzberg formula and is derived through simple hydrostatics. A necessary condition for application of hydrostatic principles to this situation is the existence of two segregated fluids with a common interface. With this condition, the weight of a unit vertical column of fresh water extending from the water table to the interface is balanced by the weight of unit vertical column of sea water extending from sea level to the same depth as the point on the interface; that is, the weight of the column of fresh water of length $h_f + z$ equals the weight of the column of salt water of length z (Fig. 4.23).

By designating ρ_f and ρ_s as the densities of fresh and salt water, respectively, the condition of hydrostatic balance is expressed

$$\rho_s g z = \rho_f g (h_f + z) \tag{4.82}$$

or

$$z = \frac{\rho_f}{\rho_s - \rho_f} h_f \tag{4.83}$$

where z is the depth below sea level to a point on the interface. If the density of fresh water is taken as 1.0 and sea water as 1.025,

$$z = 40h \tag{4.84}$$

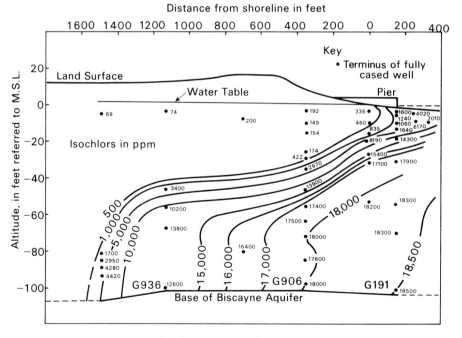

Fig. 4.22 Cross section showing the zone of diffusion. (*After Kohout and Klein, 1967.*)

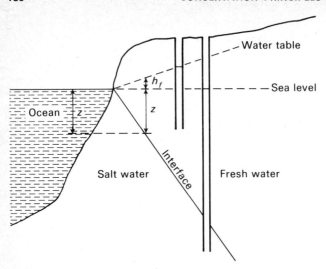

Fig. 4.23 Hydrostatic conditions of the Ghyben-Herzberg relation.

as confirmed by Ghyben and Herzberg by observation. This means, where this condition is approximately correct, a freshwater level of 20 ft above sea level corresponds to 800 ft of fresh water below sea level. Stated in another way, a lowering of the water table by 5 ft will cause a 200-ft rise of salt water.

The relation cited above is, at best, only approximately correct in that groundwater is not at rest but continually discharged into the sea. No such provision is taken into account in the hydrostatic theory. The coincidence at the coastline of a zero freshwater and saltwater head, along with the assumption of a rigid interface across which no flow can occur, effectively closes the system at its discharge point. For the undisturbed state, with low potential gradients, this approximation may be adequate; when influenced by pumping, the head of the intruded sea water in the aquifer may often be well below sea level, and Eq. (4.83) inadequate.

ELEVATION AND SLOPE OF THE INTERFACE UNDER DYNAMIC EQUILIBRIUM

The general case of dynamic equilibrium between two fluids assumed to be immiscible was investigated by Hubbert (1940; 1953). The interface was assumed to segregate the fluids rigidly, but to be common to both and subject to adjustment with change in the state of the flow of either fluid. Each configuration, however, represented a position of dynamic equilibrium.

When two homogeneous fluids occupy adjacent regions in space, each is characterized by its own potential, which may be given (Section 4.1) by

$$\Phi_1 = gz + \frac{P}{\rho_1}$$

$$\Phi_2 = gz + \frac{P}{\rho_2}$$
(4.85)

where ρ_1 and ρ_2 represent the respective densities, with $\rho_2 > \rho_1$. To examine the elevation of the interface at any point, the first of Eqs. (4.85) is solved for the pressure P and substituted in the second equation. This gives

$$\Phi_2 = gz + \frac{(\Phi_1 - gz)}{\rho_2}\rho_1$$
(4.86)

Solving for z,

$$z = \frac{1}{g}\left(\frac{\rho_2}{\rho_2 - \rho_1}\Phi_2 - \frac{\rho_1}{\rho_2 - \rho_1}\Phi_1\right)$$
(4.87)

Equation (4.87) gives the elevation of any point in either of the two regions in Fig. 4.24 for which Φ_1 and Φ_2 are known, including a point on the interface (Hubbert, 1940). Given any two of the quantities z, Φ_1, and Φ_2, the third can be uniquely determined with Eq. (4.87).

By substituting the subscripts f and s representing fresh and salt water, respectively and recognizing that $\Phi = gh$, Eq. (4.87) becomes

$$z = \frac{\rho_s}{\rho_s - \rho_f}h_s - \frac{\rho_f}{\rho_s - \rho_f}h_f$$
(4.88)

The second term on the right-hand side of Eq. (4.88) is identical with the Ghyben-Herzberg formula. When the head of the intruded sea water, h_s, is at sea level, the two equations give identical results. The more negative

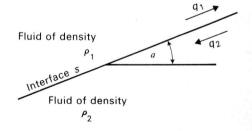

Fig. 4.24 Slope of the interface between two immiscible fluids. (*After Hubbert, 1940. Used with permission of the University of Chicago Press.*)

h_s becomes, the greater the discrepancy between computations based on hydrostatic versus dynamic conditions.

Perlmutter and others (1959) have applied Eq. (4.88) to determine the depth of a theoretical freshwater-saltwater contact within a zone of diffusion. In this calculation, z is the altitude of a point on the interface, h_s is the altitude of the water level in a well filled with salty water of density ρ_s, supposedly terminated at the interface, and h_f is the altitude of the water level in a well filled with fresh water of density ρ_f, also supposedly terminated at the interface. In a practical situation, however, the heads used are at some vertical distance apart, h_f in a freshwater environment and h_s in a saltwater environment (Fig. 4.25). The underlying assumption is that the extension, or reduction, as the case may be, of either well to the interface will not alter the head values used in the computation. Hence, the computation is valid only when there is negligible vertical movement of water between the two measurement points of Fig. 4.25.

Carrying out the calculations of Perlmutter et al. (1959), with h_s taken as 6 ft below sea level, h_f as 2.7 ft above sea level, and ρ_s as 1.018,

$$z = \frac{1.018}{1.018 - 1.000}(-6) - \frac{1.000}{1.018 - 1.000}(2.7) = -489 \text{ ft}$$

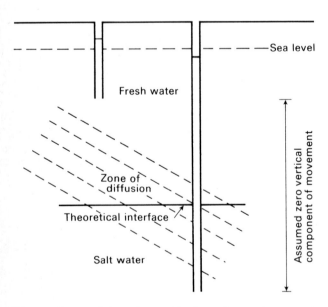

Fig. 4.25 Conditions for application of equations to the theoretical interface in a zone of diffusion on Long Island, New York.

ENERGY AND ITS TRANSFORMATIONS

Applying these data to the Ghyben-Herzberg formula,

$$z = \frac{1.018}{1.018 - 1.000} (2.7) = -152.5 \text{ ft}$$

The elevation computed by Eq. (4.88) was shown to fall well within the diffused zone. The Ghyben-Herzberg computation was well above the zone.

The problem of vertical flow between measurement points was investigated by Lusczynski (1961). If Δh is taken as the head loss between the points, the depth of the interface below sea level is expressed

$$z = \frac{\rho_f \Delta h}{\rho_s - \rho_f} + \frac{\rho_s h_s}{\rho_s - \rho_f} - \frac{\rho_f h_f}{\rho_s - \rho_f} \tag{4.89}$$

which differs from Eq. (4.88) by the magnitude of the first term. For example, if Δh in the above example is about -0.5 ft, the new depth of the interface is about 517 ft below sea level. Equation (4.89) is an important special case of some general results derived by Lusczynski (1961), and it differs from Hubbert's treatment in that the system is realistically taken as open with respect to vertical flow components.

If we return now to Eq. (4.87), the slope of the interface, whose angle is a, is determined at any point (Hubbert, 1940):

$$\sin a = \frac{\partial z}{\partial s} = \frac{1}{g} \left(\frac{\rho_2}{\rho_2 - \rho_1} \frac{\partial \Phi_2}{\partial s} - \frac{\rho_1}{\rho_2 - \rho_1} \frac{\partial \Phi_1}{\partial s} \right) \tag{4.90}$$

In general, Φ_1 and Φ_2 are not known. However, in that Darcy's law can be expressed (Section 4.1)

$$q = -K \frac{1}{g} \frac{\partial \Phi}{\partial L} \tag{4.91}$$

we get

$$\frac{\partial \Phi}{\partial L} = \frac{\partial \Phi}{\partial s} = -\frac{gq}{K} \tag{4.92}$$

Substitution in Eq. (4.90) gives

$$\sin a = \frac{\partial z}{\partial s} = -\frac{\rho_2}{\rho_2 - \rho_1} \frac{q_2}{K_2} + \frac{\rho_1}{\rho_2 - \rho_1} \frac{q_1}{K_1} \tag{4.93}$$

If the denser salt water is assumed to be static, q_2 becomes zero and

$$\sin a = \frac{\partial z}{\partial s} = \frac{\rho_1}{\rho_2 - \rho_1} \frac{q_1}{K_1} \tag{4.94}$$

As $\rho_2 > \rho_1$, so $\sin a > 0$, and the slope of the interface will increase upward in the direction of flow. If both fluids are static, $\sin a$ equals zero, and the equilibrium position is horizontal. If both fluids are in motion, the slope of the interface is found by solving Eq. (4.93).

The dynamic relations between two fluids of different density are applicable not only to saltwater-freshwater conditions, but to other fluids assumed to be immiscible, such as oil and water (Hubbert, 1953). The well-known gravitational theory treating oil that overlies water in anticlinal structures is based on the premise that both fluids are static, which results in a horizontal interface. On the other hand, if the water is moving and the oil is static, the interface is tilted in accordance with Eq. (4.93), with q_1 equal to zero.

Another approach to the saltwater problem is the analysis of Glover (1959), where the flow through the seepage surface above sea level is assumed negligible (Fig. 4.26). The interface between the fresh water and sea water is determined from the equation

$$y^2 - \frac{2Q}{\gamma k} x - \frac{Q^2}{\gamma^2 k^2} = 0 \tag{4.95}$$

where y is the distance measured vertically downward from sea level, in feet; x is the distance measured horizontally landward from the shoreline, in feet; Q is the freshwater flow per unit length of shoreline, in feet per second; and γ is the excess of the specific gravity of sea water over that of fresh water.

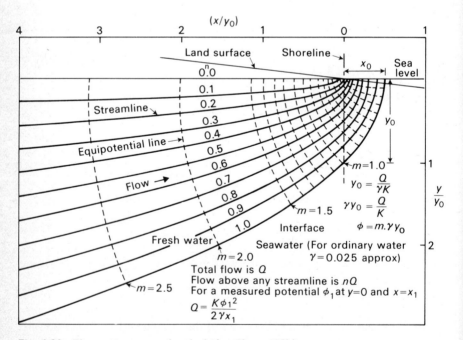

Fig. 4.26 Flow pattern near a beach. (*After Glover, 1959.*)

ENERGY AND ITS TRANSFORMATIONS

The width of the freshwater discharge gap is given by

$$x_o = \frac{Q}{2\gamma k} \tag{4.96}$$

and the potential along the line $y = 0$ is

$$\Phi_o = \left(\frac{2\gamma Q x}{k}\right)^{\frac{1}{2}} \tag{4.97}$$

Example 4.5 Assume a horizontal interface between salt water and fresh water under conditions of dynamic equilibrium.

1. If the specific discharge is the same for both fluids, which medium (salt water or fresh water) has a higher K?
2. If K for each medium is the same, which fluid has a greater specific discharge?

In answer to question 1, $q_2 = q_1 = q$ and $\sin a = 0$. From Eq. (4.93),

$$0 = -\frac{\rho_2}{\rho_2 - \rho_1}\frac{q}{K_2} + \frac{\rho_1}{\rho_2 - \rho_1}\frac{q}{K_1} \quad \text{or} \quad \frac{\rho_1}{\rho_2 - \rho_1}\frac{1}{K_1} = \frac{\rho_2}{\rho_2 - \rho_1}\frac{1}{K_2}$$

This gives

$$\frac{K_2}{K_1} = \frac{\rho_2}{\rho_1} \quad \text{or} \quad K_2 = \frac{\rho_2}{\rho_1} K_1$$

As $\rho_2 > \rho_1$, so $K_2 > K_1$.

In answer to question 2, $K_2 = K_1 = K$ and $\sin a = 0$.

$$0 = -\frac{\rho_2}{\rho_2 - \rho_1}\frac{q_2}{K} + \frac{\rho_1}{\rho_2 - \rho_1}\frac{q_1}{K}$$

Following the procedures above,

$$q_1 = \frac{\rho_2}{\rho_1} q_2$$

As $\rho_2 > \rho_1$, so $q_1 > q_2$.

4.5 ENERGY CONCEPTS APPLIED TO CHEMICAL THERMODYNAMICS

For flow processes that involve a transfer of mechanical to thermal energy, the tendency is to move from states of order to states of disorder, the term *increasing disorder* used synonymously with increased entropy. The concept of potential was introduced to describe and deal with such changes. A spontaneous change in such systems was demonstrated to be characterized by a decrease in potential energy and an increase in entropy, or unavailable energy. That is, $\Phi_2 - \Phi_1 < 0$ and $s_2^* - s_1^* > 0$ in the direction of flow describe the same condition, where Φ is potential energy and s^* is entropy.

The term *free energy* was introduced as thermodynamic potential similar to potential energy in that it represents the work that must be done on a system at constant pressure and temperature to bring the system from some state A to some other state B. If F_A represents the absolute initial free energy and F_B the absolute value of free energy after the system has undergone some change, say, chemically,

$$\Delta F = F_B - F_A \tag{4.98}$$

or

$$\Delta F < 0$$

That is, for spontaneous chemical changes at constant temperature and pressure, free energy in a system will always decrease. The symbol ΔF is used to indicate free energy, and is similar in meaning to Φ as a measure of available energy. Free energy has meaning mainly as a relative quantity, not an absolute one.

In most groundwater systems, a tendency toward equilibrium between dissolved ions and solid minerals is promoted by a large surface area of mineral in contact with a slowly moving water. For this section, free energy is a measure of the tendency, or energy available, for reaction between this solid mineral phase and the water. By analogy with the concept of potential where $\Delta \Phi < 0$, $\Delta s^* > 0$ characterizes the condition (and direction) in which mechanical flow takes place, $\Delta F < 0$, $\Delta s^* > 0$ characterizes a similar condition and direction under which spontaneous chemical changes take place. If $\Delta F < 0$ and $\Delta s^* > 0$, the mineral phase is going into solution faster than the rate of precipitation from solution, and the water is undersaturated with some mineral species. If $\Delta F = 0$ and $\Delta s^* = 0$, the system is in equilibrium, and the liquid phase is saturated. If $\Delta F > 0$, the water is supersaturated with a particular mineral species, and the rate of deposition theoretically exceeds the rate of solution; that is, the process tends to proceed in the opposite direction. The last condition, however, requires the admittance of outside energy. The magnitude of ΔF serves as a measure of how far a given mixture is from equilibrium, the term being used in the sense that solution and precipitation rates are equal, resulting in neither decrease nor increase in a particular dissolved constituent. Thus, the energy distribution in rock-water systems is such that the tendency is to strive toward equilibrium, this position corresponding with the direction of maximum entropy. In fact, this equilibrium-seeking tendency may serve as a chemical interpretation of the second law. The second law is thus related to equilibrium when it is realized that work can be obtained from a system only when the system is not already at equilibrium.

An important distinction should be made between the concept of equilibrium that is arrived at from conservation considerations and is to be applied to chemical thermodynamics, and the concept of dynamic equilibrium of the steady state. Equilibrium in a chemical thermodynamic sense per-

ENERGY AND ITS TRANSFORMATIONS

tains to a characteristic of a closed-svstem, where its attainment is synonymous with the attainment of maximum entropy. That is, given an initial amount of free energy, equilibrium implies the degradation of this energy to the point where no further work can be done, and entropy expresses the degree to which this is true. Dynamic equilibrium, on the other hand, is a characteristic reserved exclusively for the open system in which matter, or energy, is continually being added and removed at certain points. Although energy may be degraded in the system as a result of irreversibilities, outside energy is added so that the net rate of increase of entropy is zero (Prigogine, 1967). Theoretical developments in regional groundwater flow are based on dynamic equilibrium concepts (Section 6.1). The equations applied in chemical thermodynamics, although applied in practice to open systems, are based on closed-system concepts. This approach is fully justified, however, in that the kinetics driving the system toward chemical equilibrium are faster than the rate of flow through the system.

Free energy is measured in calories, or kilocalories, and is expressed

$$\Delta F = \Delta H - T\Delta s^* \tag{4.99}$$

where ΔH is the change in heat content associated with the transformation of reactants into products, or change in enthalpy; T is temperature; and Δs^* is the change in entropy. If heat is absorbed in the reaction, then the products contain more energy than the reactants, ΔH is positive, and the reaction is endothermic. If heat is liberated, then ΔH is negative, the products contain less energy than the reactants, and the reaction is exothermic. If one reaction liberates heat, the reverse reaction absorbs heat and, in accordance with the law of conservation of energy, the amounts of heat are equal. Values for ΔH and Δs^* are generally found in physiochemical tables. The dependence of free energy on temperature and pressure is demonstrated in most texts on physical chemistry.

EQUILIBRIUM STUDIES

Equilibrium studies demonstrate the influence of certain mineral species on the chemical character of the water, and they have been applied to limestone-dolomite terrain (Hsu, 1963; Back, Cherry, and Hanshaw, 1966) and to iron-bearing continental deposits (Barnes and Back, 1964; Back and Barnes, 1965). The condition for equilibrium is, of course, that free energy equal zero.

From our understanding of free energy, a reversible reaction will proceed in one direction until it reaches a point where it "stops" in the sense that a balance between the two reactants has been achieved. If the rate of reaction is directly proportional to the concentration of each reacting substance, the driving forces at equilibrium are sensibly equal, and may be related to a constant of proportionality, or equilibrium constant. For example, the reaction

$$aA + bB \leftrightharpoons yY + zZ$$

represents equilibrium in accordance with the conservation of mass in that a moles of A with b moles of B are just balanced by y moles of Y and z moles of Z, or

$$aA + bB = yY + zZ \tag{4.100}$$

where A, B, Y, and Z are chemical formulas, and a, b, y, and z are coefficients. When very dilute solutions are considered, the equilibrium constant K is

$$K = \frac{m_Y{}^y m_Z{}^z}{m_A{}^a m_B{}^b} \tag{4.101}$$

where m_I is the concentration of the Ith component of the reaction. For more concentrated solutions,

$$K = \frac{a_Y{}^y a_Z{}^z}{a_A{}^a a_B{}^b} \qquad a_I = \gamma_I m_I \tag{4.102}$$

when dealing with activities a, or effective concentrations of substances, where γ_I is the activity coefficient. In general, Eq. (4.102) is used in hydrological studies, and a method of conversion from parts per million or equivalents per million to activities will be given shortly.

A value for K as given by Eq. (4.102) can be determined from a chemical analysis of water for a specified reaction. The calculated value represents a particular state of equilibration of ions at a given instant, which is referred to as the *ionic activity product* K_{iap}. If the free energy of the system is less than zero, work can still be obtained from the system, and K_{iap} is less than K_{eq}. For the equilibrium state, entropy is a maximum, and K_{iap} equals K_{eq}. Hence, we may characterize the three significant states of undersaturation, saturation (equilibrium), and oversaturation as

$$\Delta F < 0 \quad \text{and} \quad \Delta s^* > 0 \qquad K_{iap} < K_{eq} \qquad \text{undersaturation}$$
$$\Delta F = 0 \quad \text{and} \quad \Delta s^* = 0 \qquad K_{iap} = K_{eq} \qquad \text{saturation}$$
$$\Delta F > 0 \quad \text{and} \quad \Delta s^* < 0 \qquad K_{iap} > K_{eq} \qquad \text{oversaturation}$$

As mentioned earlier, the tendency of natural systems is to strive for equilibrium from both below and above saturation.

It follows that equilibrium studies hinge on the ability to calculate values of K_{iap} and to measure theoretical or laboratory values of K_{eq} for given reactions. The free-energy change accompanying the formation of 1 mole from its elements is the difference between the free energy of formation of the products and the free energy of formation of the reactants

$$\Delta F° = \Delta F_{products} - \Delta F_{reactants} \tag{4.103}$$

By convention, the superscript of $\Delta F°$ means that free-energy change for the reaction is calculated at 25°C and 1 atmosphere pressure, and it is referred to as the *standard* free-energy change. In the sense that the equilibrium constant has meaning only with respect to free energy, we expect a relation between free energy and the constant. The relation holding for all states is given by

$$\Delta F = \Delta F° + RT \ln K \tag{4.104}$$

and for the equilibrium state ($\Delta F = 0$) by

$$\Delta F° = -RT \ln K \tag{4.105}$$

where R is the gas constant, and T is absolute temperature.

Equation (4.105) represents an important thermodynamic relation. The immediately apparent application is to the calculation of the direction of a reaction as well as the final equilibrium state which the reaction will eventually attain. Laboratory determinations of $\Delta F°$ can be substituted in Eq. (4.105), and K_{eq} determined. It only remains to compare values of K_{iap} calculated from field data with laboratory determinations of K_{eq}. On the other hand, calculated values of K_{iap} can be substituted in Eq. (4.104), $\Delta F°$ can be determined as above, and ΔF can be ascertained. The sign of ΔF then yields identical information.

In that K_{iap} is determined at field temperature, and $\Delta F°$ (and consequently K_{eq}) at 25°C, the normal procedure is to change free-energy values from the reference temperature to the temperature of the water. The relation required is the Van't Hoff equation

$$\frac{d \ln K}{dt} = \frac{\Delta H}{RT^2} \tag{4.106}$$

where ΔH is the heat of reaction. When ΔH is positive (an endothermic reaction), the differential term is positive, and K increases with increasing temperature.

Latimer (1952) gives several values of $\Delta F°$ and ΔH. Examples of calculation are given by Back (1961).

The statement that a system is or is not in equilibrium is relative only with respect to reactions being considered. In general, studies conducted tend to investigate whether or not the water is in equilibrium with one or a few particular mineral species. Hence, the statement that equilibrium does or does not exist implies that ΔF does or does not equal zero for the water-mineral system investigated, not the total mineral assemblage. In monomineralic terrain, best approximated by limestone or dolomite, the studies logically reflect departure from equilibrium between the groundwater and calcite and dolomite. For multimineralic terrain, the equilibrium calculations for a particular mineral species are influenced by the ionic concentration of the other species present.

Figure 4.27 shows some results of a typical equilibrium study in Florida (Back, Cherry, and Hanshaw, 1966). The figure depicts the ratio $K_{iap}/K_{eq} \times 100$, showing the location and areal extent of saturation and oversaturation. Where water is undersaturated with respect to a mineral, the mineral is being dissolved. Because of kinetic effects, it is not possible to say that where supersaturated, the mineral is being precipitated. The results of this study enabled the authors to identify the principal area of recharge for the aquifer and its lack of coincidence with piezometric highs. Undersaturation in the

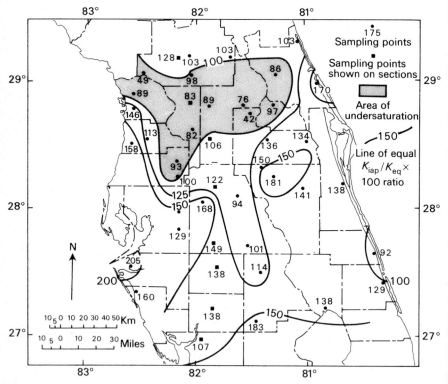

Fig. 4.27 Map of areas of undersaturation and supersaturation with respect to calcite. (*After Back, Cherry, and Hanshaw,* 1966.)

ENERGY AND ITS TRANSFORMATIONS

recharge area at depths approaching 1,000 feet suggests the deep formation of solution channels and caves.

Example 4.5 Given a water sample with the ionic composition shown here in tabular form, determine the saturation state with respect to calcite and dolomite.

Ion	Ca^{++}	Mg^{++}	Na^+	K^+	HCO_3^-	SO_4^{--}	Cl^-	At pH	At temp
Ppm	93.9	22.9	19.1	5.0	334.0	85.0	9.0	7.200	22.0°C

1. The analysis here is given in parts per million (ppm). To use these data in equilibrium calculations, it is necessary to convert the concentrations into moles per liter (which is the approximate equivalent of molality m) with the following relation:

$$\text{Molality} = \frac{\text{ppm} \times 10^{-3}}{\text{formula wt, in grams}}$$

This gives:

Ion	Ca^{++}	Mg^{++}	N^+	K^+	HCO_3^-
m	2.3×10^{-3}	0.94×10^{-3}	0.83×10^{-3}	0.13×10^{-3}	5.5×10^{-3}

Ion	SO_4^{--}	Cl^-
m	0.89×10^{-3}	0.25×10^{-3}

2. In order to represent the variation of the activity coefficient with concentration, especially in the presence of added electrolytes, a quantity called *ionic strength* must be calculated. It is a measure of the intensity of the electrical field due to the ions in solution. Ionic strength is defined as half the sum of the terms obtained by multiplying the molality of each ion in the solution by the square of its valance,

$$I = \sum_{i=1}^{a} m_i z_i^2$$

Thus, if a solution contains a number of ionic species indicated by the subscripts 1, 2, 3, . . . , so that their respective molalities are $m_1, m_2, m_3, \ldots$ and the corresponding valances are $z_1, z_2, z_3, \ldots$, then the ionic strength is given by

$$I = 0.5(m_1 z_1^2 + m_2 z_2^2 + m_3 z_3^2 + \cdots)$$

the sum being taken for all the ions present. For this example,

$$I = 0.5[(2.3 \times 10^{-3})(2^2) + (0.94 \times 10^{-3})(2^2) + (0.83 \times 10^{-3})(1^2)$$
$$+ \cdots + (0.25 \times 10^{-3})(1^2)]$$

$$I = 0.0117$$

3. Physical chemists have been able to establish approximate values of activity coefficient γ_i as a function of ionic strength. The Debye-Hückel equation,

$$-\log \gamma_i = \frac{A z_i^2 (I)^{1/2}}{1 + a_i^\circ B(I)^{1/2}}$$

gives values of the activity coefficient γ_i which are adequate for solutions of ionic strength less than 0.1, which corresponds to fresh or slightly brackish groundwater. Values of A and B are temperature-dependent constants, and values of a are the effective diameters of the ions in solution determined experimentally. Garrels and Christ (1965) list these constants. For this example, the activity coefficients are:

Ion	Ca	HCO$_3$	Mg
γ	0.659	0.905	0.676

4. Two of the most important carbonate dissolution reactions are:

Calcite: $CaCO_3 \rightleftharpoons Ca^{++} + CO_3^{--}$
Dolomite: $CaMg(CO_3)_2 \rightleftharpoons Ca^{++} + Mg^{++} + 2CO_3^{--}$

Writing in terms of the dissociation constants,

$$K^t_{CaCO_3} = a_{Ca^{++}} \cdot a_{CO_3^{--}}$$

and

$$K^t_{CaMg(CO_3)_2} = a_{Ca^{++}} \cdot a_{Mg^{++}} \cdot a^2_{CO_3^{--}}$$

where the superscript t refers to the temperature of groundwater. In natural groundwater, the CO_3^{--} is also governed by the equilibrium relation,

$$HCO_3^- \rightleftharpoons CO_3^{--} + H^+$$

where

$$K^t_{HCO_3^-} = \frac{a_{CO_3^{--}} \cdot a_{H^+}}{a_{HCO_3^-}}$$

so that,

$$a_{CO_3^{--}} = \frac{K_{HCO_3^-} \cdot a_{HCO_3^-}}{a_{H^+}}$$

This value for a is substituted in the equations expressed in terms of dissociation constants, which gives the working equations

Calcite: $$K^t_{CaCO_3} = \frac{a_{Ca^{++}} \cdot a_{HCO_3^-} \cdot K_{HCO_3^-}}{a_{H^+}}$$

Dolomite: $$K^t_{CaMg(CO_3)_2} = a_{Ca^{++}} \cdot a_{Mg^{++}} \left(\frac{a_{HCO_3^-} \cdot K^t_{HCO_3^-}}{a_{H^+}} \right)^2$$

For most calculations, a_{H^+} is taken as equivalent to 10^{-pH}. The equilibrium constants for calcite and dolomite at 22°C are, from laboratory determinations:

Constant	K_{CaCO_3}	$K_{CaMg(CO_3)_2}$	K_{HCO_3}
Value	0.415×10^{-8}	0.117×10^{-16}	0.439×10^{-10}

5. Percent saturation:

Calcite: $\dfrac{100}{K_{CaCO_3}} \dfrac{a_{Ca^{++}} \cdot K_{HCO_3^-} \cdot a_{HCO_3^-}}{a_{H^+}}$

$= \dfrac{100}{0.415 \times 10^{-8}} 0.659(0.23 \times 10^{-2})(0.439 \times 10^{-10})$

$\times \dfrac{0.905(0.55 \times 10^{-2})}{10^{-7.200}}$

$= 128$

Dolomite: $\dfrac{100}{K_{CaMg(CO_3)_2}} a_{Ca^{++}} \cdot a_{Mg^{++}} \left(\dfrac{K_{HCO_3} \cdot a_{HCO_3^-}}{a_{H^+}} \right)^2$

$= \dfrac{100}{0.117 \times 10^{-16}} [0.659(0.23 \times 10^{-2})][0.676(0.94 \times 10^{-3})]$

$\times \left\{ \dfrac{(0.439 \times 10^{-10})[0.950(0.55 \times 10^{-2})]}{10^{-7.200}} \right\}^2$

$= 100.6$

Thus, the sample is supersaturated with respect to calcite, and saturated with respect to dolomite.

OXIDATION–REDUCTION POTENTIAL

Oxidation potential, also called *oxidation-reduction potential,* permits another method of studying tendency to react in natural environments, where the reaction involves an exchange of electrons. Oxidation means an increase in positive valance, or a loss in electrons, and reduction means a decrease in positive valance, or a gain in electrons. When the process of electron transfer is conducted in a laboratory, say, along a wire connecting two elements, a potential difference E exists between the reacting substances. The symbol $E°$, termed the *standard potential,* is measured for reactions that take place at the usual arbitrary standard, that is, 25°C with all substances at unit activity.

Table 4.1 lists various half-reactions with their oxidation potentials (a half-reaction is an equation showing only a reduction process). The double arrow signifies that, under appropriate conditions, the half-reaction can go in either direction. The voltage cited applies only to forward reactions. For reverse reactions, the voltage sign changes.

The forward reaction in Table 4.1 is one of oxidation in which the reducing agent, shown on the left, is oxidized. The table is arranged so that reducing agents are listed in order of decreasing strength; that is, Li is the best reducing agent since it has the highest tendency to give off electrons, and F^- is the worst since it has the least tendency to part with electrons. A list of reducing agents arranged in decreasing order is called the *electromotive series.* The magnitude of the potential is a measure of the relative tendency of the half-reaction to occur.

Table 4.1 Some half-reactions and their oxidation potentials
(*After M. J. Sienko and R. A. Plane, 1966 and with permission of McGraw-Hill Book Co.*)

	Half-reaction	Potential volts
Li(s)	$Li^+ + e^-$	+3.05
Na(s)	$Na^+ + e^-$	+2.71
Mg(s)	$Mg^{++} + 2e^-$	+2.37
Al(s)	$Al^{+2} + 3e^-$	+1.66
Zn(s)	$Zn^{++} + 2e^-$	+0.76
Fe(s)	$Fe^{++} + 2e^-$	+0.44
$2Br^-$	$Br_2 + 2e^-$	−1.09
$2H_2O$	$O_2(g) + 4H^+ + 4e^-$	−1.23
$2Cl^-$	$Cl_2(g) + 2e^-$	−1.36
$4H_2O + Mn^{++}$	$MnO_4 + 8H^+ + 5e^-$	−1.51
$2F^-$	$F_2(g) + 2e^-$	−2.87

As the magnitude of the potential indicates relative tendency to react, it is related to free energy. The relation is

$$\Delta F° = nE°f \tag{4.107}$$

where n is the number of electrons that shift in the reaction, and f is the Faraday constant. Combining Eqs. (4.107) and (4.105),

$$F° = \frac{-RT}{nf} \ln K \tag{4.108}$$

Equation (4.107) can also be combined with Eq. (4.104), which gives
$$\Delta F = nE°f + RT \ln K \tag{4.109}$$
or

$$\frac{\Delta F}{nf} = E = E° + \frac{RT}{nf} \ln K \tag{4.110}$$

which is a form of the expression called the *Nernst equation*.

The ability of a natural environment to promote an oxidizing or reducing reaction is measured through its redox potential, termed *Eh*, and sufficiently defined by the Nernst equation

$$Eh = E° + \frac{RT}{nf} \ln K \tag{4.111}$$

In simple terms, *Eh* is a measure of the energy required to remove electrons from ions in a given chemical environment. A detailed description of one sampling technique is given by Back and Barnes (1961). Baas-Becking et al. (1960) summarize the result of over 5,000 measurements.

ENERGY AND ITS TRANSFORMATIONS

The pH of natural groundwaters, or hydrogen ion concentration, affects the solubility of minerals, and therefore exerts a certain influence on the oxidation potential of the environment. Krauskopf (1967) pointed out that redox potential measures the ability of the environment to supply electrons for an oxidizing agent, or take up electrons from a reducing agent, just as pH of an environment measures the environment's ability to supply protons (hydrogen ions) to a base, or take up protons from an acid. For this reason, both *Eh* and pH have a profound effect on concentrations of ions in groundwater systems, and both are required to relate oxidizing and reducing environments to the occurrence of elements that exist in two or more valance states. The *Eh*-pH diagram serves this purpose.

The chemical behavior of iron in groundwater systems has been extensively investigated by these methods (Hem and Cropper, 1959; Back and Barnes, 1961; Barnes and Back, 1964; Back and Barnes, 1965). Figure 4.28

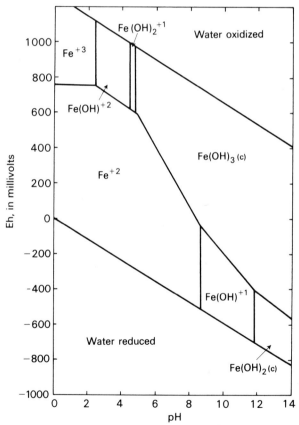

Fig. 4.28 Stability diagram for aqueous ferric-ferrous system. (*After Back and Barnes*, 1965.)

demonstrates the stability aspects of various forms of iron. The limiting conditions of pH and *Eh* on the stability of water is marked by the two straight lines with negative slopes. Within this field, the forms of iron that predominate are located with respect to *Eh* and pH. The upper part of the diagram represents increased oxidizing conditions; and the lower part, increased reducing conditions. For example, Fe^{+3} is the stable ion for high oxidizing conditions and low pH. At a lower *Eh*, it gains an electron and becomes reduced in accordance with

$$Fe^{+2} = Fe^{+3} + e^- \tag{4.112}$$

where e is the electron gained. From left to right in the diagram, the pH increases, and hydrogen ions become relatively abundant. The subscript c indicates that the most stable form of iron is crystalline, and certain environments are characterized by a low amount of iron in solution.

4.6 CONCLUDING STATEMENT

Many theoretical aspects of groundwater hydrology are based largely on an understanding of the controls of energy and its transformations in porous flow. In field problems, it is the spatial distribution of energy and entropy production that are of interest, because (1) each observation or measurement is made with reference to a point in the flow field, and (2) all processes of interest are irreversible. Energy thus serves as a unifying concept which relates to both the physical and chemical processes that take place in a groundwater system. As an entity, it is permanent and constant; it merely changes form.

1. The basis for evaluating the distribution of mechanical energy is the potential theory developed by Hubbert in a number of papers. A loss in potential is a loss in available energy and may be thought of as an increase in entropy, or unavailable energy. This may be generalized into a principle of conservation of energy, which states that in a closed system, the total energy remains constant. The loss of potential in an irreversible flow process is converted to heat, which is absorbed by the system. Hence, one form of energy is transformed into another, with the total of the two energies remaining constant. Darcy's law is an expression for the conservation of energy in a mechanical flow process.
2. In a similar fashion, a given quantity of dissolved constituents may appear in the groundwater flow, but a given amount of free energy disappears. Hence, an increase in entropy accompanies the irreversible process of dissolving minerals in a groundwater flow system. The concept of dissipation of usable energy is expressed by the second law of thermodynamics.

3. When a single fluid is being dealt with, only a single potential need be considered. With several fluids, there are as many different potentials as there are fluids. The idea of potential energy can then be extended to two or three segregated fluids, or, more importantly, to the boundary surfaces between them.
4. There are numerous other energy transformations within the subsurface portion of the hydrologic cycle. A chemical reaction may result in products that contain more energy, or less energy, than the reactants, depending on whether heat is absorbed or liberated. On the other hand, heat may actually perform work in transporting a fluid by the vapor phase. The role of heat of friction in driving chemical reactions in porous flow is poorly understood. The mechanical problems of isolation and measurement and the sensitivity requirements of the measuring devices limit our ability to determine the intensity of these processes in groundwater flow. Their effects are mostly speculative, but thought to be of minor importance.

PROBLEMS AND DISCUSSION QUESTIONS

4.1 Why must Darcy's law be corrected for actual pore space when used to determine flow velocity?

4.2 The specific discharge as expressed by Darcy's law (a vector quantity) is equal to the product of a scalar and a vector quantity. The same holds true for the equivalents of Darcy's law expressed by Eqs. (4.54) and (4.55). Describe the physical meaning of the scalar and vector quantities in question.

4.3 Combine Eqs. (4.7) and (4.12) to obtain an expression for the parameter K. (Assume that $q = v$.) How does this expression differ from Eq. (4.14)? Would you expect that both the shape factor and the porosity are dependent upon the same geologic factors, such as grain size and structural arrangement?

4.4 The various types of apparatus which are used to measure permeability in the laboratory are called *permeameters*. In the falling-head permeameter, the apparatus is similar to that shown in Fig. 4.1, with the quantity of percolating water measured indirectly by observing the rate of fall of the water level in a standpipe. Let a be area of the standpipe, and $-dh/dt$ the velocity of fall (the negative sign indicates that the head is decreasing with increased time). Let h_0 be the measured head at time t equals zero, and h_1 be the head after the lapse of an interval of time t_1. For a sample of length L and cross-sectional area A,

$$Q = -a\frac{dh}{dt} = K\frac{h}{L}A$$

Prove that

$$K = 2.3\frac{aL}{At_1}\log\frac{h_0}{h_1}$$

4.5 The transmissivity of the aquifer shown in Fig. 4.2, of thickness m and length L, is 1,000 ft^2/day.

a. Specifically, what does this indicate?
   b. What does the product Ti indicate, where T is transmissivity and i is the actual hydraulic gradient?
   c. Prove that TiL equals KiA, where A is the total cross-sectional area of flow and i is the hydraulic gradient.
4.6 Explain: Equation (4.30) is a special case of Eq. (4.38); Eq. (4.33) is a special case of Eq. (4.41).
4.7 Generally for an anisotropic medium [Eq. (4.64)], the hydraulic gradient and the direction of groundwater movement are not collinear. With anisotropy, it seems reasonable to suggest a relation between a directionally dependent permeability and the directional properties of sediment fabric. Directionally dependent quantities may sometimes be represented by symmetric tensors—quantities which possess not only magnitude and direction, but magnitude which varies with direction. What are some directionally dependent sedimentary textural characteristics which may be correlated with the tensor concept of permeability? What are some nondirectional characteristics which may be correlated with the scalar concept of permeability?
4.8 A family of flow lines is constructed on a piezometric map of an isotropic, nonhomogeneous aquifer. Intersecting lines extending across an entire length of flow in one region of the flow net form a pattern of near-perfect squares. In other regions, intersecting lines form a pattern of rectangles with different ratios for the sides. Describe a method by which the relative permeabilities may be determined over the entire region.
4.9 For a homogeneous, isotropic aquifer, prove that

$$Q = \frac{nf}{nd} KmH$$

where Q is the flow rate L^3/T through a full thickness of aquifer, nf is the number of flow channels as determined from a flow net constructed from a piezometric surface, nd is the number of potential drops over the entire flow field as determined from a piezometric map, K is the conductivity L/T, while m is the saturated thickness L, and H is the total drop in head L across the flow field.
4.10 Under what conditions is the following correct for two diverging flow lines?

$$\frac{a_1}{L_1} = \frac{a_2}{L_2}$$

where a and L are defined in Fig. 4.9, with $a_2 > a_1$ and $L_1 > L_2$.
4.11 If the rocks composing the earth's upper crust were of one homogeneous, isotropic lithology extending from the earth's surface to great depths, the only flow systems capable of forming would be unconfined. Explain.
4.12 Experimental evidence reported in Chapter 6 suggests that for highly compacted shale membranes having an abundance of clay minerals, osmotic pressures across the shale can be approximately 12 to 15 lb/in.^{2} for each 1,000-ppm difference in water salinity. How many feet of head does this represent?
4.13 The concepts presented on chemical osmosis suggest that the measurement of a hydraulic gradient across a shale membrane separating aquifers of different salinities does not necessarily mean that water is moving in the direction of the gradient, or at least it does not mean that water is moving at a rate that may be readily predicted with Darcy's law. Explain.
4.14 In what one fundamental way does the Ghyben-Herzberg equilibrium concept

ENERGY AND ITS TRANSFORMATIONS

differ from Hubbert's dynamic equilibrium concept? What implications could this have when applied to actual problems?

4.15 Groundwater of uniform density moves in the direction of decreasing head. When density of the fluid varies from point to point, hydraulic head may not indicate directly the direction of movement. One way to circumvent this problem is to convert the measured head of density ρ_s to an equivalent freshwater head with density ρ_f (which is unity).
 a. Explain how this conversion may be made.
 b. Calculate an equivalent freshwater head for the following data: from a well penetrating brackish water of density 1.05, with sea level as datum, $z = 3,000$ ft and $P/\rho_s g = 1,000$ ft. (Answer: $h_f = 4,050$ ft.)

4.16 Assume an interface between two fluids of different density in a homogeneous, isotropic medium. Comment on whether or not the following statements are true. Include the reasoning you apply in arriving at your conclusion.
 a. A sloping interface always indicates movement of one or both of the fluids.
 b. A horizontal interface is possible only if $q_1 > q_2$ (assume that $\rho_2 > \rho_1$).
 c. If one of the fluids is a liquid hydrocarbon and the other is denser salt water in motion, $\sin \alpha < 0$ when the hydrocarbon is trapped in an anticlinal structure and the interface slopes downward in the direction of the moving water. The slope will be steeper with increasing q_2.
 d. If one of the fluids is a gaseous hydrocarbon and the other is a moving denser salt water, $\sin \alpha < 0$ when the hydrocarbon is trapped in an anticlinal structure and the interface slopes downward in the direction of the moving water. The slope, however, will be less steep than in part c above.

4.17 A water sample has the following ionic composition:

Ion	Ca	Mg	Na	K	HCO$_3$	SO$_4$	Cl	pH	Temp.
ppm	19.7	8.3	30.0	1.0	179.0	8.4	12.5	8.145	10°C

Given that the activity coefficients for Ca, Mg, and HCO$_3$ are 0.76, 0.77, and 0.94, respectively, and that the equilibrium constants for calcite, dolomite, and HCO$_3$ are 0.47×10^{-8}, 0.19×10^{-16}, and 0.33×10^{-10}, respectively, calculate the saturation state of the sample with respect to calcite and dolomite. (Answer: calcite, 101.4, and dolomite, 87.9.)

4.18 Empirical studies suggest that the highest concentrations of iron are in waters of low oxidation potential ($+100$ to -20 millivolt). Would you expect a relation between the distribution of Eh (and therefore iron) and the groundwater flow pattern? Explain.

REFERENCES

Baas-Becking, L. G. M., I. R. Kaplan, and D. Moore: Limits of the natural environment in terms of pH and oxidation-reduction potentials, *J. Geol.*, vol. 68, pp. 243–284, 1960.

Back, W.: Calcium carbonate saturation in groundwater from routine analyses, *U.S. Geol. Surv., Water Supply Papers*, 1535-D, 1961.

—— and I. Barnes: Equipment for field measurement of electrochemical potentials, *U.S. Geol. Surv., Profess. Papers*, 424-C, pp. 366–368, 1961.

—— and ——: Electrochemical potential and iron bearing waters related to groundwater flow patterns, *U.S. Geol. Surv., Profess. Papers*, 498-C, 1965.

Back, W., R. Cherry, and B. Hanshaw: Chemical equilibrium between the water and minerals of a carbonate aquifer, *Natl. Speleol. Soc. Bull.,* vol. 28, July, 1966.

—— and B. Hanshaw: Chemical geohydrology, in V. T. Chow (ed.), "Advances in Hydroscience," Academic Press, Inc., New York, vol. 2, pp. 49–109, 1965.

Banks, H. O., and R. C. Richter: Sea water intrusion into groundwater basins bordering the California coast and inland bays, *Trans. Amer. Geophys. Union,* vol. 34, pp. 575–582, 1953.

Barksdale, H. C.: The contamination of groundwater by salt water near Parlin, New Jersey, *Trans. Amer. Geophys. Union,* vol. 21, pp. 471–474, 1940.

Barnes, I., and W. Back: Geochemistry of iron rich groundwater of southern Maryland, *J. Geol.,* vol. 72, pp. 435–447, 1964.

Bear, J., and Y. Bachmat: A generalized theory on hydrodynamic dispersion in porous media, *Intern. Assoc. Sci. Hydrol., Symp. Haifa,* Publ. 72, pp. 7–16, 1967.

Bennett, R. R., and R. R. Meyer: Geology and groundwater resources of the Baltimore area, *Maryland Board Nat. Resources, Dept. Geology, Mines, Water Resources, Bull.* 4, 1952.

Berry, F., Hydrodynamics and geochemistry of the Jurassic and Cretaceous systems in the San Juan basin, northwestern New Mexico and southwestern Colorado, unpubl'd. Ph.d. thesis, School of Mineral Sciences, Stanford Univ., 1959.

—— and B. Hanshaw: Geologic evidence suggesting membrane properties of shales, *Proc. Intern. Geol. Congr., 21st, Copenhagen,* Abstract vol., 1960.

Boas, M.: "Mathematical Methods in the Physical Sciences," John Wiley & Sons, Inc., New York, 1966.

Bredehoeft, J., and others: Possible mechanism for concentration of brines in subsurface formations, *Bull. Amer. Assoc. Petrol. Geologists,* vol. 47, pp. 257–269, 1963.

Carman, P. C.: "Flow of Gases through Porous Media," The Academic Press, Inc., New York, 1956.

Casagrande, A.: Seepage through dams, *J. New England Water Works Assoc.,* vol. 51, pp. 131–172, 1937.

Chamberlin, T. C.: The requisite and qualifying conditions of artesian wells, *U.S. Geol. Surv., 5th Ann. Rept.,* pp. 131–173, 1885.

Darcy, H. P.-G.: Les fontaines publiques de la ville de Dijon, Victor Dalmont, Paris, 1856.

DeSitter, L. U.: Diagenesis of oil-field brines, *Bull. Amer. Assoc. Petrol. Geologists,* vol. 31, no. 11, pp. 2030–2040, 1947.

Edlefsen, N. E., and A. B. C. Anderson: Thermodynamics of soil moisture, *Hilgardia,* vol. 15, pp. 31–297, 1943.

Ferris, J. G., and others: Theory of aquifer tests, *U.S. Geol. Surv., Water Supply Papers,* 1536-E, pp. 69–174, 1962.

Fischel, V. C.: Further tests of permeability with low hydraulic gradients, *Trans. Amer. Geophys. Union,* vol. 16, pp. 499–503, 1935.

Garrels, R. M., and C. L. Christ: Solutions, Minerals and Equilibria, Harper and Row, Publishers, Inc., New York, 1965.

Ghyben, W. B.: Notes in verband met de Voorgenomen Put boring Nabji Amsterdam, *Tijdschr. Koninhijk, Inst. Ingrs.,* The Hague, 1899.

Glover, R. E.: The pattern of fresh water flow in a coastal aquifer, *J. Geophys. Res.,* vol. 64, pp. 457–459, 1959.

Hanshaw, B., and E. Zen: Osmotic equilibrium and overthrust faulting, *Bull. Geol. Soc. Amer.,* vol. 76, pp. 1379–1386, 1965.

Hem, J. D., and W. H. Cropper: Survey of ferrous-ferric chemical equilibria and redox potentials, *U.S. Geol. Surv., Water Supply Papers,* 1459-A, pp. 1–31, 1959.

Herzberg, B.: Die Wasserversovgung einiger Nordseebaser, *J. Gasbeleucht. und Wasserversov.*, vol. 44, pp. 815–819, 1901.
Hsu, K. J.: Solubility of dolomite and composition of Florida groundwaters, *J. Hydrol.*, vol. 1, pp. 288–310, 1963.
Hubbert, M. K.: The theory of groundwater motion, *J. Geol.*, vol. 48, pp. 785–944, 1940.
———: Entrapment of petroleum under hydrodynamic conditions, *Bull. Amer. Assoc. Petrol. Geologists*, vol. 37, pp. 1944–2026, 1953.
———: Darcy's law and the field equations of the flow of underground fluids, *Trans. AIME*, vol. 207, pp. 222–239, 1956.
King, F. H.: Principles and conditions of the movements of groundwater, *U.S. Geol. Surv., 19th Ann. Rept.*, pt. 2, Washington, D.C., pp. 59–294, 1899.
Kohout, F. A.: Flow pattern of fresh water and salt water in the Biscayne aquifer of the Miami area, Florida, *Intern. Assoc. Sci. Hydrol., Comm. Subter. Waters*, publ. 52, pp. 440–448, 1960.
——— and H. Klein: Effect of pulse recharge on the zone of diffusion in the Biscayne aquifer, *Intern. Assoc. Sci. Hydrol., Symp. Haifa*, publ. 72, pp. 252–270, 1967.
Kozeny, J.: Uber Kapillare Leitung des Wassers in Boden, *Sitzber. Wiener Akad. Wiss.*, vol. 136, pt. 2, 1927.
Krauskopf, K. B.: "Introduction to Geochemistry," McGraw-Hill Book Company, New York, 1967.
Latimer, W. M.: "The Oxidation States of the Elements and Their Potentials in Aqueous Solutions," Prentice-Hall, Inc., Englewood Cliffs, N.J., 1952.
Leggette, R. M.: Salt water encroachment in the Lloyd Sand on Long Island, New York, *Water Works Eng.*, vol. 100, pp. 1076–1079 and 1107–1109, 1947.
Leonards, G. A.: Engineering properties of soils, in G. A. Leonards (ed.), "Foundation Engineering," McGraw-Hill Book Company, New York, pp. 66–240, 1962.
Liefrinck, F. A.: Water supply problems in Holland, *Public Works*, vol. 61, no. 9, pp. 19–20, 65–66, and 69, 1930.
Lusczynski, N. J.: Head and flow of groundwater of variable density, *J. Geophys. Res.*, vol. 66, no. 12, pp. 4247–4256, 1961.
Meinzer, O. E.: Geology and water resources of Big Smokey, Clayton, and Alkali Spring valleys, Nevada, *U.S. Geol. Surv., Water Supply Papers*, 423, 1917.
———, Outline of groundwater in hydrology with definitions, *U.S. Geol. Surv., Water Supply Papers*, 494, 1923.
——— and V. C. Fischel: Tests of permeability with low hydraulic gradients, *Trans. Amer. Geophys. Union*, vol. 15, pp. 405–409, 1934.
Mifflin, M. D.: Delineation of groundwater flow systems in Nevada, *Desert Res. Inst., Tech. Rept.*, Series H-W, no. 4, Reno, Nev., 1968.
Muskat, M., and H. G. Botset: Flow of gases in porous media, *Physics*, vol. 1, no. 1, 1931.
Perlmutter, N. M., J. J. Geraghty, and J. E. Upson: The relation between fresh and salty groundwater in southern Nassau and southeastern Queens Counties, Long Island, New York, *Econ. Geol.*, vol. 54, pp. 416–435, 1959.
Prigogine, I.: "Introduction to Thermodynamics of Irreversible Processes," Interscience Publishers, a division of John Wiley & Sons, Inc., New York, 1967.
Reynolds, O.: An experimental investigation of the circumstances which determine whether the motion of water shall be direct or sinuous and the law of resistance in parallel channels, *Phil. Trans. Roy. Soc. London*, vol. 174, 1883.
Schmorak, S.: Salt water encroachment in the Coastal Plain of Israel, *Intern. Assoc. Sci. Hydrol., Symp. Haifa*, publ. 72, pp. 305–318, 1967.

Senio, K.: On the groundwater near the seashore, *Assemblée generale de Bruxelles*, *Assoc. intern. hydrol. sci.*, vol. 2, pp. 175–177, 1951.

Sienko, M. J., and R. A. Plane: "Chemistry," 3d ed., McGraw-Hill Book Company, New York, 1966.

Skibitzke, H. E., and J. A. DaCosta: The groundwater flow system in the Snake River plain, Idaho, An idealized analysis, *U.S. Geol. Surv., Water Supply Papers*, 1536-D, 1962.

Slichter, C. S.: Theoretical investigation of the motion of groundwaters, *U.S. Geol. Surv., 19th Ann. Rept.*, pt. 2, Washington, D.C., pp. 295–384, 1899.

———: Field measurements of the rate of movement of underground water, *U.S. Geol. Surv., Water Supply Papers*, 140, Washington, D.C., pp. 9–85, 1905.

Stearns, N. D.: Laboratory tests on physical properties of water bearing materials, *U.S. Geol. Surv., Water Supply Papers*, 596, Washington, D.C., pp. 121–176, 1928.

Streeter, V. L.: "Fluid Mechanics," 4th ed., McGraw-Hill Book Company, New York, 1966.

Todd, D. K.: "Groundwater Hydrology," John Wiley & Sons, Inc., New York, 1959.

Toth, J.: A theory of groundwater motion in small drainage basins in central Alberta, Canada, *J. Geophys. Res.*, vol. 67, no. 11, pp. 4375–4387, 1962.

5
Compressibility, Elasticity, and Main Equations of Flow

If hydraulic conductivity is important to an understanding of the ease with which water moves through an aquifer, compressibility is equally important to an understanding of the manner in which aquifer systems respond to natural and man-made stresses. The term *compressibility* refers to the relations at a point between changes in the pressure exerted by the overlying rocks or enclosed fluid and corresponding changes in porosity. An elastic compression is one which disappears completely upon release of the stress which caused it. Like all basic phenomena, these concepts are deceptively simple, and recognition of their hydrologic implications did not occur all at once, but over a period of several years and through the efforts of numerous investigators.

Previous to Meinzer's (1928) classic treatment of compressibility and elasticity of confined aquifers, it was generally assumed that confined aquifers were rigid incompressible bodies, and changes in fluid pressure were not accompanied by changes in pore volume. It is clear that Meinzer fully disagreed with this principle as early as 1925, on the basis of studies of part of the Dakota sandstone (Meinzer and Hard, 1925). The excess in the volume of water pumped over that from natural replenishment for a 38-year period could not be accounted for by fluid compressibility alone and, in the absence of local dewatering and transmission of water from the outcrop area, the interstitial pore space of the sandstone was presumably reduced to the extent of the unaccounted-for fluid volume. The problem, then, was to demonstrate that aquifers compressed when the fluid pressure was reduced, and expanded when the fluid pressure was increased. Several lines of evidence were explored, including the response of water levels in wells to ocean tides and passing trains, and subsidence of the land surface.

Meinzer's ideas provided extensive insight into the behavior of elastic aquifers and led, about a decade later, to Theis' equation for the unsteady drawdown in the vicinity of a discharging well supplied by groundwater storage (Theis, 1935). Jacob (1940) extended the significance of these ideas by deriving a differential equation for compressible flow in elastic aquifers. Among other things, this equation stated that stored water is derived from (1) elastic compression of the aquifer, and (2) expansion of the water itself, both ideas fostered by Meinzer 15 years earlier. One of the more interesting aspects of the efforts of Theis and Jacob is that Jacob's equation, which conventionally leads to Theis' solution for a prescribed set of initial and boundary conditions, was derived after the solution itself was known.

Jacob (1940) also pointed out the significance of compression of clay bodies in interbedded aquifers. He stated that, in response to a reduction in fluid pressure, a third source of water is derived from compression of adjacent and included clay beds. As clays are generally more compressible than coarser clastics, the third source was regarded as the chief one. However, the low permeability of the clay units prevents immediate release of their contained water, which gives rise to a time lag of indefinite duration between the lowering of fluid pressure and the initial appearance of the water. The time lag is well described by consolidation theory in soil mechanics theory.

In recent years, considerable research has been conducted in the area of both low- and high-permeability compression. Lohman (1961) combined Hooke's law with Jacob's elastic component of storage and arrived at an expression for the elastic compression of aquifers. In the same year, Poland (1961) derived an expression for the component of storage derived from permanent compression of clay beds. With a completely different purpose in mind, Hantush (1960) incorporated the release of water from compressible confining layers in equations describing the unsteady flow of groundwater to a discharging well.

5.1 EFFECTIVE STRESS AND NATURAL BODY FORCES

The concept of effective stress has been used in formulating a theory for landslides (Terzaghi, 1950), for establishing a mechanism for overthrust faulting (Hubbert and Rubey, 1959), for investigating the phenomenon of land subsidence caused by fluid withdrawals (Domenico and Mifflin, 1965), and for solving problems of well construction in quicksand (Johnson Drillers Journal, 1965). It has been found equally useful for establishing the theoretical basis for fluctuations of water levels in confined aquifers subject to barometric, tidal, and other forms of loading. Most important, the ideas of effective stress were utilized in analysis of the flow of water in elastic aquifers (Jacob, 1940). A description of this principle is thus fundamental to an understanding of compressibility of rock units, and their "storativity" in particular.

In order for a sandstone, such as the Dakota, to undergo volumetric compression, there must be an increase in the grain-to-grain pressures within the matrix; oppositely, in order for it to expand, there must be a decrease in grain-to-grain pressure. Without such pressure changes, no volumetric changes can occur. A simple analogy to this process has been given in several places by Terzaghi (Terzaghi and Peck, 1948), one modification of which incorporates a spring, a watertight piston, and a cylinder.

In Fig. 5.1a, a spring under a load σ has a characteristic length z. If the spring and piston are placed in a watertight cylinder filled with water below the base of the piston, the spring supports the load σ and the water is under the pressure of its own weight, as demonstrated by the imaginary manometer tube in Fig. 5.1b. In Fig. 5.1c, an additional load $\Delta\sigma$ is placed on the system. In that water cannot escape from the cylinder, the spring cannot compress, and the additional load must be borne by the water. This is again demonstrated by the imaginary manometer, which shows the fluid pressure in excess of hydrostatic pressure. The term *excess* pressure is used in the sense that the pressure characteristic of any depth exceeds hydrostatic pressure.

As conservation laws of fluid mass state that no change in fluid pressure can occur except by loss (or gain) of water, a hermetically sealed cylinder will maintain an excess fluid pressure indefinitely. If some of the water is allowed to escape, the pressure of the water is lowered, and the spring compresses in response to the additional load it must support (Fig. 5.1d). Hence, there has been a transfer of stress from the fluid to the spring. When the excess pressure is completely dissipated, hydrostatic conditions once more prevail, and the stress transfer to the spring is complete (Fig. 5.1e).

The analogy to be made is that in any porous water-filled sediment, there are pressures in the solid phase by virtue of the points of contact, the resilient grain structure represented by the spring; and there are pressures due to the contained water, represented by the watertight cylinder. The former are referred to as *intergranular pressures,* or *effective stresses,* and the latter as *pore-water pressures,* or *neutral stresses.* The total vertical stress acting on a horizontal plane at any depth is resolved into these neutral and effective components

$$\sigma = \bar{\sigma} + P \tag{5.1}$$

where σ is the total vertical stress, $\bar{\sigma}$ is effective stress, and P is neutral stress. For the analogy cited above, $\Delta\sigma$ equals ΔP at the initial instant of loading, and $\Delta\sigma$ equals $\Delta\bar{\sigma}$ at the terminal condition. Intermediate between these extremes, total vertical stress is always in balance with the sum of the effective and neutral stresses. Neutral stresses, whatever their magnitude, act on all

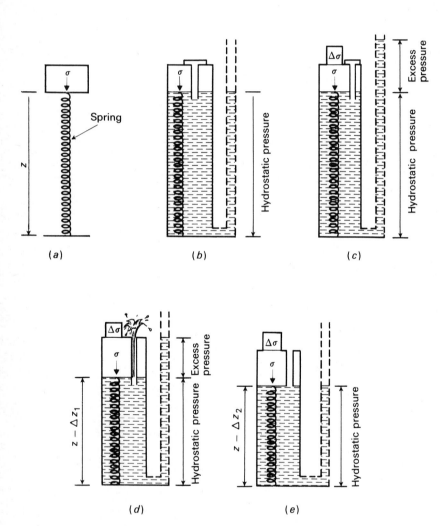

Fig. 5.1 Piston and spring analogy showing the transfer of the support for the added load from water pressure to the spring.

COMPRESSIBILITY, ELASTICITY, AND MAIN EQUATIONS OF FLOW

sides of the granular particles, but do not cause the particles to press against each other. All measurable effects, such as compression, distortion, and a change in shearing resistance, are due exclusively to changes in effective stress.

Although the concept of effective stress was first stated in relation to consolidation of clays, it is equally important to studies of high-permeability units. The concept was devised by Terzaghi (1925) and appeared in elaborate detail in the same year in which the work of Meinzer and Hard appeared. As applied to a water-saturated clay specimen in a vertical cylinder, with a means for escape of the fluid at both ends, a vertical load will cause an immediate rise of the water pressures within the clay. These pressures will then decline—more rapidly at both ends than in the middle—as water escapes from the clay sample. There follows an attendant increase in effective stress and decrease in vertical height. For materials of high permeability, the time of excess-pressure dissipation at a point may be quite rapid.

To grasp the full significance of this concept, it is important to remember that any disturbance tending toward disruption of the equilibrium condition expressed by Eq. (5.1) is offset by a corresponding change in the resolved stresses on the right-hand side. For example, the load exerted by a passing train represents an increase in the total vertical stress in much the same manner as the additional load in the spring-piston-cylinder analogy cited above. To achieve a new equilibrium, there has to be a compensating increase in one or both of the components on the right-hand side of Eq. (5.1). On the other hand, a decrease in pressure caused by pumping a confined aquifer has no effect on the total stress. Instead, that part of the load formerly carried by the grain structure must increase in proportion to the decrease in fluid pressure. Expressed mathematically, σ is constant, and, from Eq. (5.1),

$$d\bar{\sigma} = -dP \tag{5.2}$$

the negative sign indicating that a decrease in fluid pressure is accompanied by an increase in intergranular pressure. Here we postulate the manner by which water is removed from storage in the vicinity of wells in confined aquifer systems. Owing to the reduced pressure, the water expands to the extent permitted by its elasticity. At the same time, compression of the aquifer matrix results in a porosity decline, which releases water from the saturated volume. When pumping is stopped and the pressure is restored, the aquifer grains return to their original position, and the water is compressed. If aquifer pressure is not allowed to recover, but instead declines steadily, there is a progressive transfer of stress to the aquifer matrix, causing its compression.

It is possible now to derive general relations between effective and

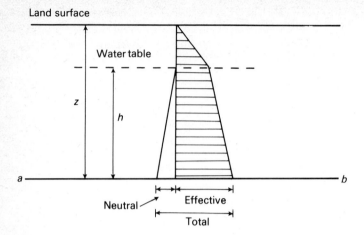

Fig. 5.2 Diagram of total, effective, and neutral stresses for a hydrostatic case.

neutral stresses in saturated materials. Figure 5.2 illustrates a simple hydrostatic case that can be applied with little error to unconfined lateral flow. The total vertical stress acting on plane *ab* at depth z is

$$\sigma = \rho_s g(z - h) + \rho_m g h \tag{5.3}$$

where ρ_s is the density of solids, and ρ_m is the density of the saturated mass. The water above plane *ab* is continuous for a distance h, and the neutral stress at any point on *ab* is given by

$$P = \rho_w g h \tag{5.4}$$

where ρ_w is the density of water. Effective stress is easily calculated from Eq. (5.1),

$$\bar{\sigma} = \rho_s g(z - h) + \rho_m g h - \rho_w g h \tag{5.5}$$

or

$$\bar{\sigma} = \rho_s g(z - h) + g h(\rho_m - \rho_w) \tag{5.6}$$

where the quantity $g(\rho_m - \rho_w)$ is the buoyant, or submerged, unit weight. If the water table coincides with land surface, z equals h, and

$$\bar{\sigma} = g z(\rho_m - \rho_w) \tag{5.7}$$

Figure 5.3a illustrates conditions in a confined aquifer where the pressure of the fluid with depth no longer can be approximated with a hydrostatic model. The term *artesian* may be employed here in that flowing wells will occur. For the conditions shown in Fig. 5.3a, the potential of the fluid increases with increasing depth through the confining layer, but not through the aquifer. The water has a steady-state component of upward movement through the confining layer. The total vertical stress acting on plane *ab* is again resolved into a neutral and effective component, the neutral stress now consisting of hydrostatic and excess pressure,

$$P = \rho_w g(H + h) \tag{5.8}$$

Total vertical stress is

$$\sigma = \rho_m g H \tag{5.9}$$

and effective stress is again the difference between total and neutral,

$$\bar{\sigma} = gH(\rho_m - \rho_w) - \rho_w g h \tag{5.10}$$

Hence, when compared with the hydrostatic case where the water table coincides with land surface, intergranular pressures are less by the amount of the excess pressure $\rho_w g h$.

If pumpage reduces h, as in Fig. 5.3b, there must follow a compensating increase in $\bar{\sigma}$, both in the aquifer material and in the low-permeability unit. The attendant deformations of the aquifer accompany the instantaneous removal of water from the aquifer, and are generally assumed to disappear upon recovery of the pressure to its original value. This is the essence of elastic compression. Deformation of the low-permeability unit also occurs with vertical movement of water out of the unit, but this entails

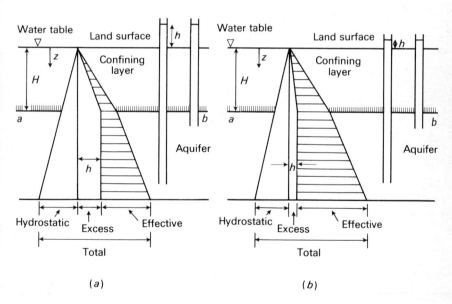

Fig. 5.3 Resolution of the total vertical stress in an aquifer and overlying confining layer for (*a*) undisturbed conditions and (*b*) some time after pumping reduces the head *h*.

a considerable time lag. These deformations are generally not recoverable upon recovery of the pressure. This is the essence of consolidation theory. Note also that the increase in effective stress within the aquifer is assumed to be not only instantaneous with the lowering of artesian head, but equal for all depths, or

$$\Delta \bar{\sigma} = \rho_w g \Delta h \qquad (5.11)$$

For the low-permeability unit, the change in effective stress varies from a maximum at the bottom to zero at the top, and entails a considerable time lag. For a layer of thickness H, the change in effective stress as a function of depth z and time t is expressed

$$\Delta \bar{\sigma}(z,t) = \rho_w g \Delta h(z,t) \qquad (5.12)$$

where $\Delta h(z,t)$ varies from Δh, at z equals H for t equals zero, to zero, where z equals zero for all times.

5.2 STORATIVITY OF ELASTIC AQUIFERS

The differential equation governing the flow of water in an elastic aquifer was derived several years ago by Jacob (1940; 1950) and, although questioned in recent years (DeWeist, 1966), is still considered a valid approximation. The point of contention is readily understood if one examines the procedure of Jacob: In accounting for the conservation of fluid mass in a control volume, the volume was first considered nondeforming to calculate the net inward mass flux, and then was allowed to deform to compute the rate of change of fluid mass in the volume. Derivations by Cooper (1966) based on a deforming coordinate system showed that Jacob's expression for what is now generally referred to as *specific storage* is essentially correct when the vertical coordinate z is taken as the deforming coordinate. The following derivations are based on Jacob's (1940; 1950) original work, with emphasis given to the role and interpretation of effective stress in the calculations.

The total weight of water W_w, within an aquifer of porosity n, thickness m, and surface area A is

$$W_w = \rho_w g n A m \qquad (5.13)$$

Recognizing that $W_w = M_w g$,

$$M_w = \rho_w n A m \qquad (5.14)$$

where M_w is the total fluid mass. The mass of contained water at time t in an elemental volume of this aquifer, whose dimensions are $\Delta x \, \Delta y \, \Delta z$, is expressed

$$\Delta M_w = \rho_w n \, \Delta x \, \Delta y \, \Delta z \qquad (5.15)$$

COMPRESSIBILITY, ELASTICITY, AND MAIN EQUATIONS OF FLOW

If the hydrostatic pressure P within this volume is decreased by ΔP, then fluid density is decreased by $\Delta \rho_w$, and the effective stress is increased by $\Delta \bar{\sigma}$. This compresses the elemental volume, which decreases the porosity by Δn. Clearly, in this process, water is expelled from the elemental volume. The time rate of change of fluid mass within the element in response to a change in fluid pressure is thus expressed

$$\frac{\partial(\Delta M_w)}{\partial t} = \frac{\partial(n\rho_w \Delta x \Delta y \Delta z)}{\partial t} \qquad (5.16)$$

On the assumption that compression is negligible in the Δx and Δy direction, whereas changes in the vertical direction are measurable with time, Eq. (5.16) becomes

$$\frac{\partial(\Delta M_w)}{\partial t} = \left(\rho_w n \frac{\partial(\Delta z)}{\partial t} + \rho_w \Delta z \frac{\partial n}{\partial t} + \Delta z\, n \frac{\partial \rho_w}{\partial t}\right) \Delta x \Delta y \qquad (5.17)$$

The three terms of this equation deal with the time rate of change of the element's vertical dimension and porosity, and time rate of change of fluid density. As the first two terms are related to vertical compression, they may be expressed in terms of intergranular pressure and, ultimately, fluid pressure [Eq. (5.2)]. The third term may also be expressed in terms of fluid pressure by invoking certain density-pressure relations.

The first term of Eq. (5.17) is related to the vertical compressibility α, which, by definition, is

$$\alpha = \frac{1}{E_s} \qquad (5.18)$$

where E_s is the bulk modulus of compression of the elemental skeleton, further defined as a stress-strain ratio

$$E_s = \frac{\Delta\text{ stress}}{\Delta\text{ strain}} = \frac{\Delta P}{-\Delta V/\Delta V_0} \qquad (5.19)$$

where ΔV is the change in elemental volume, and ΔV_0 is original volume. Since the change in Δx and Δy is assumed to be negligible, $\Delta V/\Delta V_0$ may be adequately represented by either $\Delta z/\Delta z_0$ or $d(\Delta z)/\Delta z$. The change in pressure ΔP is the change in intergranular pressure, or $\Delta \bar{\sigma}$. Making appropriate substitutions,

$$E_s = \frac{1}{\alpha} = \frac{d\bar{\sigma}}{-d(\Delta z)/\Delta z} \qquad (5.20)$$

or

$$d(\Delta z) = -\alpha \Delta z\, d\bar{\sigma} \qquad (5.21)$$

Expressed in terms of the time derivative,

$$\frac{\partial(\Delta z)}{\partial t} = -\alpha \Delta z \frac{\partial \bar{\sigma}}{\partial t} \qquad (5.22)$$

The right-hand side of Eq. (5.22) is substituted in the first term of Eq. (5.17), which gives

$$\rho_w n \frac{\partial(\Delta z)}{\partial t} = -\rho_w n \alpha \Delta z \frac{\partial \bar{\sigma}}{\partial t} \qquad (5.23)$$

The second term of Eq. (5.17) expresses the change in elemental porosity, which is also related to vertical compression. As the elemental volume consists of solids and voids, an assignment of unity to the total volume results in the following expression for the volume of solids:

$$V_s = (1 - n) \Delta x \, \Delta y \, \Delta z \qquad (5.24)$$

As V_s will remain constant with time (the change in elemental volume is assumed to be totally accounted for by the change in pore volume),

$$dV_s = d[(1 - n) \Delta x \, \Delta y \, \Delta z] = 0 \qquad (5.25)$$

Since only the Δz dimension is of concern,

$$d[(1 - n) \Delta z] = 0 \qquad (5.26)$$

Carrying out the differentiation,

$$\Delta z \, d(1 - n) + (1 - n) \, d(\Delta z) = -\Delta z \, dn + (1 - n) \, d(\Delta z) = 0 \qquad (5.27)$$

Solving for the derivative of porosity,

$$dn = \frac{1 - n}{\Delta z} d(\Delta z) \qquad (5.28)$$

This may be expressed in terms of a partial derivative with respect to time,

$$\frac{\partial n}{\partial t} = \frac{1 - n}{\Delta z} \frac{\partial(\Delta z)}{\partial t} \qquad (5.29)$$

Substituting Eq. (5.22) in Eq. (5.29),

$$\frac{\partial n}{\partial t} = (1 - n)\left(-\alpha \frac{\partial \bar{\sigma}}{\partial t}\right) \qquad (5.30)$$

Substituting this expression in the second term of Eq. (5.17),

$$\rho_w \Delta z \frac{\partial n}{\partial t} = -\rho_w \Delta z \, \alpha (1 - n) \frac{\partial \bar{\sigma}}{\partial t} \qquad (5.31)$$

Hence, the first and second terms of the equation of transient change in fluid

COMPRESSIBILITY, ELASTICITY, AND MAIN EQUATIONS OF FLOW   219

mass have been expressed in terms of the time rate of change of intergranular pressures.

The third term of Eq. (5.17) refers to changes in fluid density with time, and may be related to the compressibility of the fluid, β. By definition,

$$\beta = \frac{1}{E_w} \tag{5.32}$$

where E_w is the bulk modulus of compression of the fluid, or change in fluid volume for a corresponding change in fluid pressure, defined as

$$E_w = \frac{dP}{-d(\Delta V)/\Delta V} = \frac{1}{\beta} \tag{5.33}$$

Rearranging terms,

$$d(\Delta V) = -\beta \, \Delta V \, dP \tag{5.34}$$

The conservation-of-mass principle requires that for different states of volume and pressure, the product of density and volume must be constant for each state; that is,

$$\rho_1 \Delta V_1 = \rho_2 \Delta V_2 \tag{5.35}$$

or

$$\rho_w \, d(\Delta V) + \Delta V \, d\rho_w = 0 \tag{5.36}$$

Substituting Eq. (5.34) in (5.36) and canceling common terms,

$$\rho_w \beta \, dP = d\rho_w \tag{5.37}$$

Expressing again the time derivative,

$$\rho_w \beta \frac{dP}{dt} = \frac{d\rho_w}{dt} \tag{5.38}$$

The third term of Eq. (5.17) then becomes

$$\Delta z \, n \frac{\partial \rho_w}{\partial t} = \Delta z \, n \rho_w \beta \frac{\partial P}{\partial t} \tag{5.39}$$

Substituting Eqs. (5.23), (5.31), and (5.39) in (5.17),

$$\frac{\partial (\Delta M_w)}{\partial t} = \left[-\rho_w n \alpha \, \Delta z \frac{\partial \bar{\sigma}}{\partial t} - \rho_w \, \Delta z \, \alpha (1-n) \frac{\partial \bar{\sigma}}{\partial t} + \rho_w \, \Delta z \, n \beta \frac{\partial P}{\partial t} \right] \Delta x \, \Delta y \tag{5.40}$$

Since $d\bar{\sigma} = -dP$ [Eq. (5.2)], substitution and collection of terms gives

$$\frac{\partial (\Delta M_w)}{\partial t} = \Delta x \, \Delta y \, \Delta z \, [n \rho_w \alpha + \rho_w (1-n) \alpha + n \rho_w \beta] \frac{\partial P}{\partial t} \tag{5.41}$$

or

$$\frac{\partial(\Delta M_w)}{\partial t} = \rho_w(\alpha + n\beta)\,\Delta x\,\Delta y\,\Delta z\,\frac{\partial P}{\partial t} \quad (5.42)$$

It is desirable now to arrive at an expression containing total head, or the sum of pressure and elevation head. Taking the derivative of the expression for total head [Eq. (4.46)],

$$\frac{\partial h}{\partial t} = \frac{\partial z}{\partial t} + \frac{1}{\rho_w g}\frac{\partial P}{\partial t} \quad (5.43)$$

As z is invariant with respect to time, $\partial z/\partial t$ equals zero, and

$$\frac{\partial P}{\partial t} = \rho_w g\,\frac{\partial h}{\partial t} \quad (5.44)$$

Substituting this result in Eq. (5.42) gives

$$\frac{\partial(\Delta M_w)}{\partial t} = \rho_w(\alpha\rho_w g + n\beta\rho_w g)\,\Delta x\,\Delta y\,\Delta z\,\frac{\partial h}{\partial t} \quad (5.45)$$

By inspection, the quantity in the parentheses now has the dimensions $1/L$, and is termed the *specific storage* S_s, defined by Hantush (1964) as the volume of water that a unit volume of aquifer releases from storage because of expansion of the water and compression of the grains under a unit decline in average head within the unit volume. For a fully saturated aquifer of thickness m,

$$S = S_s m = m\rho_w g(\alpha + n\beta) \quad (5.46)$$

where S is the storativity, or coefficient of storage, defined as the volume of water an aquifer releases from, or takes into, storage per unit surface area of aquifer per unit change in the component of head normal to that surface. Stated in another way, the storativity is equal to the volume of water (measured outside the aquifer) removed from each vertical column of aquifer of height m and unit basal area when the head declines by one unit. It is therefore a ratio of a volume of water to a volume of aquifer, is dimensionless, and is less than 1.

Figure 5.4 illustrates the concept of storativity for confined and unconfined conditions. For the unconfined condition, it is the volume of water drained from the x portion of the aquifer, and it is approximately equal to specific yield. For confined conditions, no dewatering occurs, and the volume of water released is attributed to ompression of the granular structure and expansion of the water itself.

The following example may illustrate some relations between the decline of head and aquifer compressibility in confined aquifer systems. Consider a sandstone aquifer 20 ft thick extending over an area of 20×10^6

ft². Assume a 100-ft saturated thickness of shale or clay overlying the sandstone, with a saturated mass density of 4.04. The original head stands 50 ft above land surface. With the use of the notation in Fig. 5.3, the total vertical pressure acting on a unit area at the bottom of the confining layer is [Eq. (5.9)]

$$\sigma = \rho_m g H = (4.04)(32.2)(100) = 13{,}000 \text{ lb/ft}^2$$

The fluid pressure at this same point is [Eq. (5.8)]

$$P = \rho_w g (H + h) = (1.93)(32.2)(100 + 50) = 9{,}360 \text{ lb/ft}^2$$

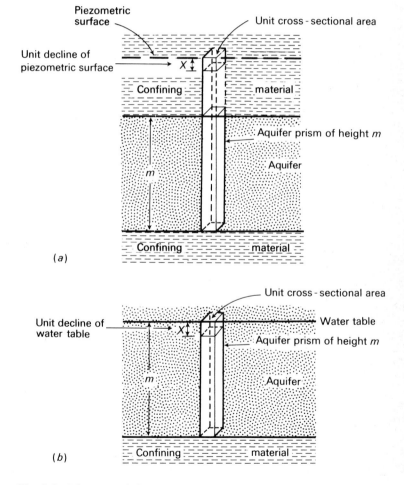

Fig. 5.4 Diagrams of the storativity for (a) confined and (b) unconfined conditions. (*After Ferris and others, 1962.*)

The intergranular pressure is the difference between the total pressure and fluid pressure, or 3,640 lb/ft². Therefore, 9,360 lb/ft² of the total pressure exerted by the confining layer is supported by the water in the aquifer, and only 3,640 lb/ft² is supported by the aquifer itself.

If, at the end of a long period of pumping, the artesian head is lowered uniformly about 40 ft, the fluid pressure in the aquifer will decrease in accordance with the following expression:

$$\rho_w g \, \Delta h = (1.93)(32.2)(40) = 2{,}496 \text{ lb/ft}^2$$

The new intergranular pressure must accordingly increase by the same amount. Hence, at the end of this pumping period, 6,864 lb/ft² of the total pressure exerted by the confining layer is supported by the water in the aquifer, and 6,136 lb/ft² by the aquifer grain structure.

According to Terzaghi (1925), the porosity of sandstone decreases 0.013 percent/1 percent load increase. The percentage load increase is the change in intergranular pressure, 2,500 lb/ft², to the original intergranular pressure, or about 70 percent. The percent of porosity decrease is easily calculated,

$$70 \times 0.013 = 0.91$$

or about 0.9 percent. If the original porosity was 10 percent, final porosity is about 9 percent. As 10 percent porosity is equivalent to 2 ft of void space for each 20-ft vertical column of aquifer, the decrease in porosity per column of aquifer is calculated to be about 0.2 ft.

From Eq. (5.20),

$$E_s = \frac{d\bar{\sigma}}{d(\Delta z)/\Delta z} = \frac{2{,}500}{0.2/20} = 250{,}000 \text{ lb/ft}^2$$

The vertical compressibility is then calculated

$$\alpha = \frac{1}{E_s} = \frac{1}{250{,}000} = 4 \times 10^{-6} \text{ ft}^2/\text{lb}$$

The component of the storativity attributed to aquifer compressibility is calculated [Eq. (5.46)]

$$S = m\rho_w g \alpha = (20)(1.93)(32.2)(4 \times 10^{-6}) = 5 \times 10^{-3}$$

With the use of the original porosity of 10 percent and the constant coefficient of fluid compressibility, 2.3×10^{-3} ft²/lb, the component of the storativity attributed to expansion of the fluid is [Eq. (5.46)]

$$S = m\rho_w g n \beta = (20)(1.93)(32.2)(0.1)(2.3 \times 10^{-8}) = 2.96 \times 10^{-6}$$

The sum of the last two calculations equals the coefficient of storage.

By using this information, it is possible now to calculate the volume of water released from storage for the total head decline over the cited area. Note that the storativity focuses attention on the volume of water released from each column of aquifer when the head is lowered 1 ft. With a 40-ft head decline, each column of aquifer releases about 0.2 ft³ of water. As there are 20×10^6 columns, the total release from storage is 4×10^6 ft³. Stated in general form,

$$V = SA \Delta h \tag{5.47}$$

where V is the volume of water, and A is the area over which the head change is effective.

5.3 ELASTIC COMPRESSION AND WATER-LEVEL FLUCTUATIONS

As a train approaches a railroad station, the downward acting pressure it exerts must be added to the total vertical pressure. Part of this load is carried by the aquifer skeleton, and part by the fluid, which causes an increase in $\bar{\sigma}$ and P proportional to the increase in σ. In response to these changes in internal stress, the water level in a well bottomed in a confined aquifer will rise owing to the increased fluid pressure, and then rapidly decline to a new position as the excess in fluid pressure is dissipated. The added weight of the train is then carried solely by the aquifer skeleton. When the train leaves the area, the effective stress decreases accordingly, which causes an expansion of the aquifer. Owing to the sudden increase in pore volume, the water level in the well declines sharply, but shortly returns to its original position (Fig. 5.5).

As in the case of passing trains, changes in atmospheric pressure must be accommodated by changes in intergranular and fluid pressure. Incremental increases or decreases in atmospheric pressure acting on a column of water in a well are respectively added to or subtracted from the pressure of the fluid. At all points distant from the well, changes in atmospheric pressure act on the confining layer which, however, has no strength to withstand or contain this applied load (Ferris et al., 1962). The effect is the same as the addition of a variable atmospheric loading pressure to that of the confining layer. The incremental stress, undiminished in amount, is borne by both the enclosed fluid and grain structure of the aquifer. It follows that as long as atmospheric pressure is undergoing change, there must exist a pressure difference between the water in the well and the water at all other points in the aquifer. When atmospheric pressure is increasing, the gradient is away from the well, which causes the water level in the well to decline. When atmospheric pressure is decreasing, the pressure gradient is toward the well, which causes the water level to rise.

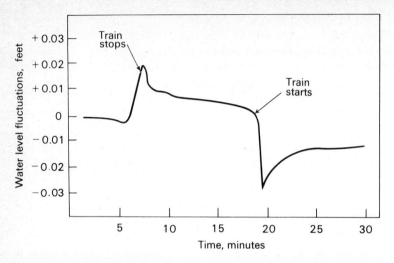

Fig. 5.5 Water-level fluctuations observed in an observation well near a railroad station. (*After Jacob*, 1939.)

The latter situation, that is, a rising water level, often presents some difficulty in understanding. Initially, it is important to recognize that the transmission of forces through confined aquifer systems is quite rapid. For example, in Fig. 5.6, there is a steady decline of the water level in a well when the atmospheric pressure is rising. When the pressure remains constant for a short period of time, the water level in the well immediately levels off and undergoes no further change. When atmospheric pressure starts to decline, the increment in stress release is fully removed from the water in the well, but only partially removed from the water at any other point in the aquifer. Hence, the pressure gradient toward the well from all directions.

Barometric fluctuations are not commonly observed in wells tapping unconfined aquifers. The reason is that changes in atmospheric pressure are transmitted equally to the column of water in a well and to the water table, owing to the direct communication between the atmosphere and the water table through the unsaturated zone. In Fig. 5.7a, the added stress is added to that of the confining layer, and is carried by both the water and the grain structure. At the well, it is carried by the water only. In Fig. 5.7b, the added stress is exerted equally on the free water surface in the well and on the water table.

The effect of a variable increment in aquifer loading for the confined condition can be examined in general terms. For aquifer loading resulting in vertical compression, Eq. (5.1) may be expressed

$$\sigma + P_e = P + \bar{\sigma} \tag{5.48}$$

COMPRESSIBILITY, ELASTICITY, AND MAIN EQUATIONS OF FLOW

where P_e is any external blanket load. As total vertical stress is constant, changes in P_e are accommodated by corresponding changes in P and $\bar{\sigma}$, or

$$dP_e = dP + d\bar{\sigma} \tag{5.49}$$

Dividing both sides of Eq. (5.49) by $d\bar{\sigma}$,

$$\frac{dP_e}{d\bar{\sigma}} = \frac{dP}{d\bar{\sigma}} + 1 \tag{5.50}$$

When an aquifer is compressed by an extensive load, the decrease in water volume ΔV_w equals the decrease in aquifer volume ΔV. As only the void volume of the aquifer is subject to change,

$$\frac{\Delta V_w}{V_w} = \frac{\Delta V}{nV} \tag{5.51}$$

which merely states that the ratio of water-volume change to total water volume equals the ratio of aquifer-volume change to total void volume of the aquifer. From equations describing the bulk modulus of compression of water and solids [Eqs. (5.20) and (5.33)],

$$\frac{\Delta V}{V} = \frac{\Delta \bar{\sigma}}{E_s} \quad \text{and} \quad \frac{\Delta V_w}{V_w} = \frac{\Delta P}{E_w} \tag{5.52}$$

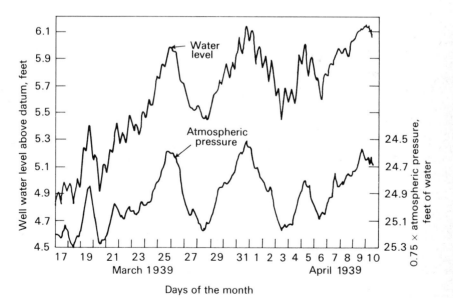

Fig. 5.6 Water-level response to changes in atmospheric pressure. (*After Robinson*, 1939.)

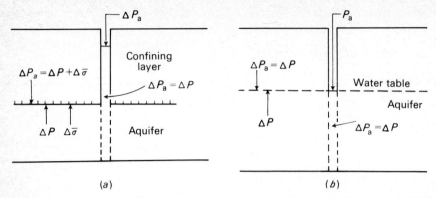

Fig. 5.7 Distribution of forces due to changes in atmospheric pressure in (*a*) a confined aquifer and (*b*) an unconfined aquifer.

Hence, when only the void volume is considered subject to compression [Eq. (5.51)],

$$\frac{\Delta P}{E_w} = \frac{\Delta \bar{\sigma}}{nE_s} \tag{5.53}$$

Rearranging terms,

$$\frac{\Delta P}{\Delta \bar{\sigma}} = \frac{E_w}{nE_s} = \frac{1/\beta}{n(1/\alpha)} = \frac{\alpha}{n\beta} \tag{5.54}$$

Substituting this result in Eq. (5.50),

$$\frac{dP_e}{d\bar{\sigma}} = \frac{\alpha}{n\beta} + 1 \tag{5.55}$$

which is quite general for an extensive load tending to compress a confined aquifer system.

Water levels in wells bottomed in aquifers subject to extensive blanket loading exhibit fluctuations that may be related to changes in vertical stress. The term *efficiency* is used to indicate how faithfully the water level reflects such changes, and may be expressed

$$E = \frac{\gamma_w \, dh}{dP_e} \tag{5.56}$$

where dh is water-level change in a well, and γ_w is the unit weight of water. When the blanket load is caused by tidal fluctuations, efficiency is referred to as *tidal efficiency*, TE, and Eq. (5.55) becomes

$$\frac{\gamma_w \, dH}{d\bar{\sigma}} = \frac{\alpha}{n\beta} + 1 \tag{5.57}$$

where H is the height of the tide. On recognizing that

$$\gamma_w \, dh = dP \tag{5.58}$$

Eq. (5.56) becomes

$$\text{TE} = \frac{dP}{\gamma_w \, dH} = \frac{dP}{d\bar{\sigma}(\alpha/n\beta + 1)} \tag{5.59}$$

Substituting Eq. (5.54),

$$\text{TE} = \frac{\alpha/n\beta}{1 + \alpha/n\beta} \tag{5.60}$$

The blanket load discussed above was assumed to act on the aquifer, but not on the column of water in a well bottomed in the aquifer. With atmospheric pressure, incremental pressures act on both the matrix and a water column in an open well. Barometric efficiency (BE) may be introduced to examine how efficiently the well behaves as a barometer. For this condition, Eqs. (5.49), (5.55), and (5.56) become

$$-d\bar{\sigma} = dP - dP_a \tag{5.61}$$

and

$$\frac{dP_a}{d\bar{\sigma}} = \frac{\alpha}{n\beta} + 1 \tag{5.62}$$

and

$$\text{BE} = \frac{\gamma_w \, dh}{dP_a} \tag{5.63}$$

respectively, where P_a is atmospheric pressure. As the combined pressure of the atmosphere and the column of water in a well is balanced by water pressure in the aquifer,

$$\gamma_w h + P_a = P \tag{5.64}$$

where h is the head above the top of the aquifer (Fig. 5.7). The change in water level is then expressed in terms of the change in effective stress [Eq. (5.61)],

$$\gamma_w \, dh = dP - dP_a = -d\bar{\sigma} \tag{5.65}$$

the minus sign indicating that the head increases with a decrease in effective stress. That is, if effective stress in the aquifer decreases in response to a decrease in atmospheric pressure, the water level in a well will rise. Substituting Eqs. (5.65) and (5.62) in Eq. (5.63),

$$\text{BE} = -\frac{d\bar{\sigma}}{d\bar{\sigma}(\alpha/n\beta + 1)} = -\frac{1}{\alpha/n\beta + 1} \tag{5.66}$$

If α equals zero, the aquifer is incompressible, and the absolute value of BE is 1 and TE is zero. A barometric efficiency of 1 suggests that the well is a perfect barometer. On the other hand, as α gets large, BE approaches zero and TE approaches 1. It follows that BE is closely related to the compressibility of water, and TE is closely related to the compressibilty of the aquifer matrix. In terms of the storativity, the percentage of storage attributable to expansion of the water is equal to BE, and that attributable to compression of the aquifer matrix is equal to TE, so that the absolute value of their sum is 1 (Hantush, 1964).

Some additional relations between the storativity of aquifers and their compressibility can be demonstrated by rearranging the expression for specific storage [Eq. (5.45)],

$$\frac{S_s}{n\beta\rho_w g} = \frac{\alpha}{n\beta} + 1 \qquad (5.67)$$

which is equal to $dP_e/d\bar{\sigma}$ [Eq. (5.55)]. In terms of the absolute value of BE [Eq. (5.66)],

$$\frac{1}{\text{BE}} = \frac{\alpha}{n\beta} + 1 = \frac{S_s}{n\beta\rho_w g} = \frac{dP_e}{d\bar{\sigma}} \qquad (5.68)$$

Numerous cases of water-level fluctuations due to atmospheric and tidal loading have been reported (Robinson, 1939; Jacob, 1941; Parker and Stringfield, 1950).

5.4 EXTENSIONS FOR LOW–PERMEABILITY UNITS

Subsidence of the land surface in many areas of the world has been ascribed to several causes: tectonic movement; solution; compaction of sedimentary materials due to static loads, vibrations, or increased density brought about by water-table lowering; and changes in reservoir pressures with loss of fluids. Noteworthy cases of appreciable subsidence attributed to the last cause have been reported in Long Beach Harbor, California (Gilluly and Grant, 1949); the San Joaquin Valley, California (Poland and Davis, 1956; Poland, 1961); the upper Gulf coastal region, Texas (Winslow and Wood, 1959); the Savannah area, Georgia (Davis et al., 1963); and Las Vegas Valley, Nevada (Domenico et al., 1964). Other prominent localities include Mexico City (Cuevas, 1936) and London (Wilson and Grace, 1942). The geologic requisites and qualifying conditions for the occurrence of subsidence are so well adapted to alluvial basins that it is likely far more occurrences of this phenomenon are taking place than have been reported. The chief reason for this is the lack of close control of benchmarks necessary to detect small changes in land-surface altitude.

COMPRESSIBILITY, ELASTICITY, AND MAIN EQUATIONS OF FLOW

Serious problems can be caused by land-surface subsidence. The normal upward force of skin friction acting on piles or well casings may be reversed, which will subject these elements to "downdrag." This may result in failure or protrusion above land surface. Gradients of canals or other conveyors of surface water may be reduced, or even reversed, and so affect the normal flow of water. Cracking of concrete or brick structures is common in subsiding areas. Subsidence in coastal areas may seriously affect structures at or near sea level; expensive remedial action may be required, such as pressurization of selected aquifer zones.

Two related phenomena are of interest in the study of land subsidence caused by fluid withdrawals: (1) total amount of vertical shortening of a column of confining layer, which corresponds to the total volume of water removed from the column, and (2) the time rate of subsidence, which corresponds to the time rate of flow out of the layer. The latter process is generally described as a progressive increase in effective stress, resulting in consolidation of a compressible stratum as it drains, as shown by the isochrone development in Fig. 5.8. Isochrones represent the degree of consolidation at various depths as a function of time. The terminal isochrone, at time t equals infinity, represents full consolidation, or ultimate subsidence, for the given head decline in the aquifer.

Given the time rate of flow, the total flow volume can be obtained by integration over limits from t equals zero to t equals infinity. However,

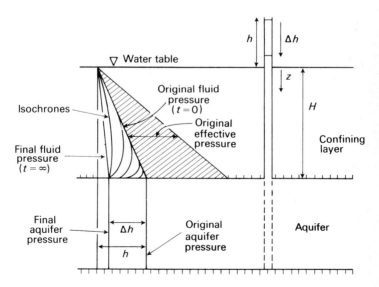

Fig. 5.8 Depth-pressure diagram of the development of isochrones in a confining layer in response to lowered aquifer pressure.

because of the wide variety of initial and boundary conditions that may be established to describe the initial pressure distribution within the clay layer as well as at its edges, several mathematical models could be described. For this reason, a general case of ultimate subsidence will be treated here. The time rate of subsidence and its relation to the final state of subsidence for a specific set of initial and boundary conditions will be reserved for treatment in the chapter on unsteady flow.

By analogy with Eqs. (5.18) and (5.19),

$$\alpha' = \frac{1}{E_c} \tag{5.69}$$

and

$$E_c = \frac{\Delta \bar{\sigma}}{\Delta H / H_0} \tag{5.70}$$

or

$$\frac{\Delta H}{H_0} = \frac{1}{E_c} \Delta \bar{\sigma} \tag{5.71}$$

where α' is vertical compressibility of a confining layer, $\Delta H/H_0$ is relative compression, and E_c is the bulk modulus. As the solids in an elemental volume are assumed to be incompressible, changes in height are fully accounted for by change in void ratio Δe, and (Fig. 5.9)

$$\frac{\Delta H}{H_0} = \frac{\Delta e}{1 + e} \tag{5.72}$$

Equation (5.72) expresses the relative compression of the element, or

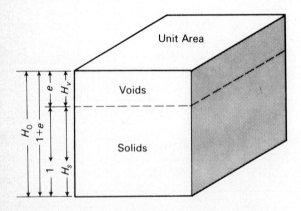

Fig. 5.9 Schematic representation of an element of confining layer.

COMPRESSIBILITY, ELASTICITY, AND MAIN EQUATIONS OF FLOW

Table 5.1 Range in values for the bulk modulus of compression, E, and specific storage γ_w/E (from data presented by Kögler and Scheidig, 1948, as modified by Domenico and Mifflin, 1965)

Material	E, lb/ft^2	γ_w/E, 1/ft
Plastic clay	1×10^4–8×10^4	6.2×10^{-3}–7.8×10^{-4}
Stiff clay	8×10^4–1.6×10^5	7.8×10^{-4}–3.9×10^{-4}
Medium-hard clay	1.6×10^5–3×10^5	3.9×10^{-4}–2.8×10^{-4}
Loose sand	2×10^5–4×10^5	3.1×10^{-4}–1.5×10^{-4}
Dense sand	1×10^6–1.6×10^6	6.2×10^{-5}–3.9×10^{-5}
Dense sandy gravel	2×10^6–4×10^6	3.1×10^{-5}–1.5×10^{-5}
Rock, fissured, jointed	3×10^6–6.5×10^7	2.1×10^{-5}–1×10^{-6}
Rock, sound	Greater than 6.25×10^7	Less than 1×10^{-6}

relative amount of water expelled per unit height. Assuming Δe is proportional to $\Delta \bar{\sigma}$,

$$\Delta e = a_v \Delta \bar{\sigma} \tag{5.73}$$

where a_v is the coefficient of compressibility, defined as the rate of change of void ratio to the rate of change of effective stress causing the deformation. In soil mechanics usage, a_v is the slope of the line obtained by plotting void ratio versus pressure for test specimens (see Section 7.7). The subscript v indicates compression in the vertical direction only.

Substituting $a_v \Delta \bar{\sigma}$ for Δe in Eq. (5.72) and equating this new expression with the right-hand side of Eq. (5.71),

$$\frac{1}{E_c} = \frac{a_v}{1+e} \tag{5.74}$$

This quantity expresses the height of a pore-water column expelled from the unit element when the effective stress is increased by one pressure unit. In that the compressibility of clay is one or two orders of magnitude larger than the compressibility of sand (Table 5.1), the term accounting for the expansion of water is ignored in this calculation. The specific storage of the element is then expressed by any of the following equalities:

$$S_s' = \rho_w g \alpha' = \frac{\rho_w g}{E_c} = \frac{a_v \rho_w g}{1+e} \tag{5.75}$$

In soil mechanics literature, $a_v \rho_w g/(1+e)$ is used almost exclusively in combination with permeability k', which gives

$$c_v = \frac{k'(1+e)}{a_v \rho_w g} \tag{5.76}$$

where c_v is termed the *coefficient of consolidation*. Clearly, this is the ratio of permeability to specific storage, the significance of which will be explored in detail later. Again, the subscript v designates vertical compression.

By definition, the specific storage of a compressible confining layer is the volume of water that a unit volume releases from storage, owing to its compression, when the average head within the volume undergoes a unit decline. The specific discharge is defined as the product of the specific storage and the actual head change within the element,

$$dq' = S'_s h(z)\, dz \tag{5.77}$$

The total volume of water extruded from a column of height H with a unit basal area is then

$$q' = S'_s \int_0^H h(z)\, dz \tag{5.78}$$

The solution of this equation is a general expression for the vertical shortening of a confining layer when steady-flow conditions are reestablished within it. If $h(z)$ is assumed to vary linearly across the clay unit,

$$h(z) = \Delta h\left(\frac{z}{H}\right) \tag{5.79}$$

where z varies from zero at the top to H at the bottom (Fig. 5.8). For a more general description of changes in aquifer pressure both above and below a confining layer (Fig. 5.10),

$$h(z) = \Delta h_1\left(\frac{z}{H}\right) + \Delta h_2\left(\frac{H-z}{H}\right) \tag{5.80}$$

Substitution in Eq. (5.78) gives

$$q' = \frac{\gamma_w}{E_c} \int_0^H \left[(\Delta h_1 - \Delta h_2)\frac{z}{H} + \Delta h_2\right] dz \tag{5.81}$$

or (Domenico and Mifflin, 1965)

$$q' = \frac{\gamma_w}{E_c}\frac{\Delta h_1 + \Delta h_2}{2} H \tag{5.82}$$

Examination of Eq. (5.82) shows that, when steady flow is reestablished, the vertical shortening of a confining layer equals the product of specific storage and the final area described by the increase in effective stress on a two-dimensional depth-pressure diagram. In Fig. 5.8, this area is a triangle of height H and base Δh_1 (Δh_2 equals zero), and in Fig. 5.10, it is a trapezoid (one-half of the sum of the bases Δh times the height). The area thus described is appropriately termed the *effective pressure area*.

The concepts given above pertain to permanent compression of low-permeability layers in confined basins, the volume of water extracted being a one-time reserve associated with permanent reduction in pore volume. Contrary to popular opinion, this process in no way reduces aquifer storage capacity in that pore-volume reduction takes place in the confining units. Indeed, the process makes available a volume of water that would otherwise not be recoverable (Poland, 1961).

Another factor to be taken into consideration is the elastic component of subsidence attributed to permanent compression of aquifers. This may be especially significant in oil field subsidence, where the shale or clay members are well compacted, and most of the subsidence is attributed to compression of the oil sands (Gilluly and Grant, 1949). By analogy with

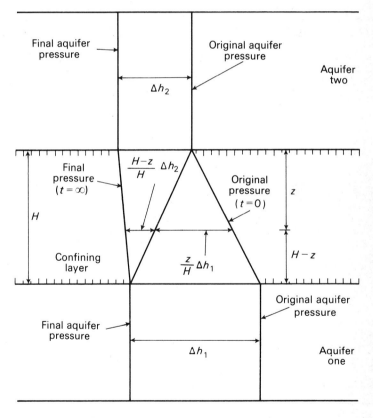

Fig. 5.10 Depth-pressure diagram for a confining layer in response to lowered pressure in adjacent aquifers.

Hooke's law, Eqs. (5.18) and (5.19) may be combined and expressed (Lohman, 1961)

$$\frac{\Delta m}{\Delta P} = \alpha m \tag{5.83}$$

where Δm is the change in thickness of an aquifer of original thickness m, ΔP is the change in artesian pressure, and α is the vertical compressibility. The coefficient of storage may be written [Eq. (5.46)]

$$\frac{S}{\rho_w g} = \alpha m + n\beta m \tag{5.84}$$

Combining Eqs. (5.83) and (5.84) and solving for Δm,

$$\Delta m = \Delta P \left(\frac{S}{\rho_w g} - \beta nm \right) \tag{5.85}$$

Given values of S, n, and m, and the amount of pressure decline, this equation is easily solved for the amount of vertical compression of an elastic aquifer.

As all regions of a basin do not subside uniformly with pressure change, geology must be an important factor relating cause and effect. Two parameters considered fundamentally geologic are thickness of the compressible units and specific storage [Eq. (5.82)]. Utilizing estimates of the bulk modulus (Fig. 5.11a) and pressure-level declines (Fig. 5.11b), Domenico and others (1966) demonstrated that the existing pattern of surface deformation in Las Vegas Valley, Nevada (Fig. 5.11c), could be reproduced (Fig. 5.11d) without invoking structural or other tectonically oriented mechanisms. The extent of the calculated subsidence bowl, expressed in terms of specific subsidence (subsidence per foot of compressible material), is in good agreement with the actual subsidence bowl. Information obtained from the bulk-modulus map indicates that fine-grained sediments deposited in low-energy depositional environments (playas) may be as much as seven times as compressible as those characteristic of high-energy environments (alluvial fans).

Example 5.1 Equation (5.85) for elastic aquifers can be shown to be identical in principle with Eq. (5.82) for compressible confining layers. Recognizing that $\Delta P = \Delta h\, \gamma_w$ and the storativity for confining layers is $\gamma_w m / E_c$, where m is thickness,

$$\Delta m = \Delta h\, \gamma_w \left(\frac{\gamma_w m / E_c}{\gamma_w} - \beta nm \right)$$

Recognizing further that the compressibility of water is ignored in consolidation theory, β equals zero and

$$\Delta m = \frac{\gamma_w}{E_c} \Delta h\, m$$

COMPRESSIBILITY, ELASTICITY, AND MAIN EQUATIONS OF FLOW

which differs from Eq. (5.82) only in that the average head change across the confining layer is not considered. This requires that head changes in aquifers separated by a confining layer must be equal in magnitude. For example, assume that essentially no flow occurs across a confining layer located between two aquifers. Upon pumping, the heads in the adjacent aquifers within a given region are lowered about 50 ft. The specific storage of the confining layer is 1×10^{-3} ft^{-1} and its thickness is 40 ft. Maximum vertical shortening (subsidence), ignoring elastic compression of the aquifers, is calculated from Eq. (5.82) or from the equation cited above,

$$q' = \frac{\gamma_w}{E_c} \frac{2\Delta h}{2} H = \frac{\gamma_w}{E_c} \Delta h \, m = 1 \times 10^{-3} \text{ft}^{-1} \times 50 \text{ ft} \times 40 \text{ ft} = 2 \text{ ft}$$

5.5 EFFECTIVE STRESS AND RESISTANCE TO SHEAR

Aside from giving rise to measurable effects of compression, amply demonstrated by water-level fluctuations, changes in effective stress will cause changes in resistance to shear. Earth materials derive strength, and therefore resistance to shear, both from solid friction between the individual particles, and from cohesion. This is generally expressed through Coulomb's empirical equation

$$s_t = c + \bar{\sigma} \tan \theta \tag{5.86}$$

where s_t is the unit shearing strength, c is the unit cohesion, $\bar{\sigma}$ is the effective stress, and θ is the angle of internal friction. Equation (5.86) can also be expressed

$$s_t = c + (\sigma - P) \tan \theta \tag{5.87}$$

This shows that any increase in P tends to decrease resistance to shear developed by solid friction independently of its effect on the angle of internal friction. Equation (5.87) is the basis both of the fluid pressure-overthrust faulting hypothesis presented by Hubbert and Rubey (1959), and of the work of Terzaghi (1950) relating fluid pressure and landslides. The value of the shearing strength can be made small simply by increasing the value of the fluid pressure P. If P is somehow made to approach σ, the solid friction is reduced to zero, and the earth mass is in a state of incipient flotation (resistance to shear developed by cohesive force is small or altogether lacking in many granular materials). In the overthrust concept, fault blocks can be moved large distances by relatively small forces in the absence of solid friction.

For the generation of landslides, increases in P can be accounted for by excess recharge during periods of high precipitation. On the other hand, fluid pressure sufficient to negate solid friction requires very special geologic conditions (Rubey and Hubbert, 1959): (1) the presence of clay rocks, (2) interbedded sandstones, (3) large total thickness, (4) rapid sedimentation. These conditions make possible the hermetic sealing of the fluids in the

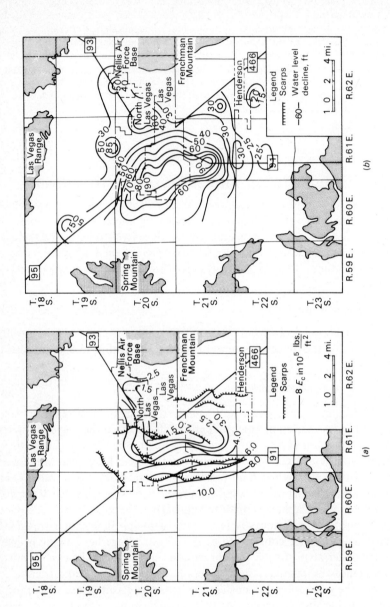

COMPRESSIBILITY, ELASTICITY, AND MAIN EQUATIONS OF FLOW 237

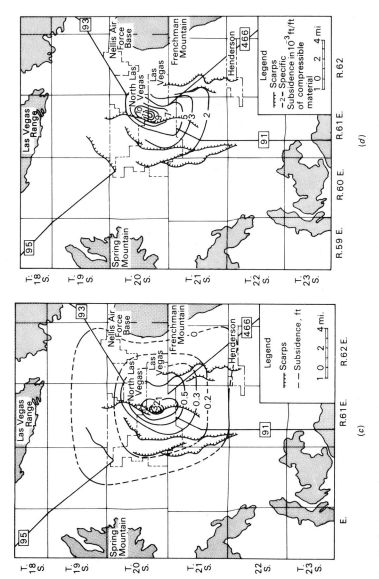

Fig. 5.11 Maps of Las Vegas Valley showing (*a*) the bulk modulus of compression, (*b*) change in total head, (*c*) actual amount of subsidence, and (*d*) a calculated bowl of subsidence (expressed in terms of specific subsidence). (*After Domenico and others*, 1966.)

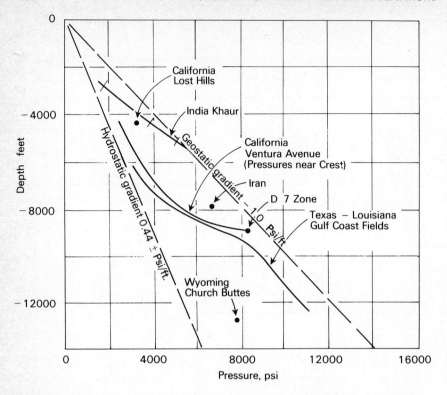

Fig. 5.12 Relation of pressure and depth in some excess-pressure oil pools. (*After Watts*, 1948. *Used with permission of the Amer. Min., Metallurgical, and Petroleum Eng. Assoc.*)

sandstone, which prevents the escape of the fluid and, therefore, preserves the fluid pressure. As in the piston-cylinder system of Fig. 5.1, with the absence of fluid loss, the added load of basin sedimentation is borne by the fluid. Fluid loss, of course, is never absent, but may be slow relative to the rate of sedimentation (Bredehoeft and Hanshaw, 1968).

Fluid pressures as great as 0.9σ have been observed in deep oil wells in many parts of the world. Figure 5.12 shows the relation between pressure and depth in some excess-pressure pools reported by Watts (1948). The geostatic gradient demonstrates the change in total vertical pressure with depth, and the hydrostatic gradient shows the theoretical change in fluid pressures for simple hydrostatics. Actual fluid pressure is seen to be considerably in excess of hydrostatic pressure, in some cases approaching the total vertical pressure.

The same concepts apply to the formation of quicksand (Fig. 5.13a), which indicates that quicksand is not a type of material, but a particular flow

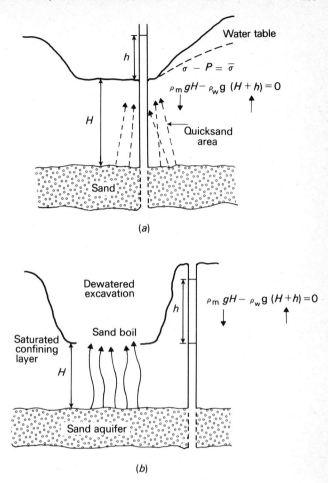

Fig. 5.13 Diagrams of (*a*) quicksand development and (*b*) "blow through" conditions in excavation.

condition by which the solid friction, or resistance to shear, is reduced to zero. The "blow through" condition reported by Ferris and others (1962) associated with excavations in confining materials is another example of fluid pressure in excess of total vertical pressure (Fig. 5.13*b*).

5.6 MAIN EQUATIONS OF FLOW

For groundwater flow, the conservation-of-mass principle requires that the rate of increase or decrease of fluid mass in an element situated in the flow field be equal to the difference between the rates of influx and efflux; that is, there can be no gain or loss in mass. For the differential control volume of dimensions dx, dy, dz (Fig. 5.14), the mass inflow rate through face $ABCD$ is

$$\text{Mass inflow rate } ABCD = \rho_w q_x \, dy \, dz \tag{5.88}$$

where q_x is the specific discharge normal to the area ΔA. The rate of outflow through face $EFGH$ is

$$\text{Mass outflow rate } EFGH = \left(\rho_w q_x + \frac{\partial}{\partial x} \rho_w q_x \, dx\right) dy \, dz \tag{5.89}$$

the derivative allowing for the possibility of nonuniform density, or specific discharge, or both. The net inward flux is the difference between inflow and outflow,

$$\text{Net inward flux} = -\frac{\partial}{\partial x} \rho_w q_x \, dx \, dy \, dz \tag{5.90}$$

Making similar calculations for sides $ABFE$, $DCGH$, $ADHE$, and $BCGF$ and adding the results,

$$\text{Net inward flux} = -\left(\frac{\partial}{\partial x} \rho_w q_x + \frac{\partial}{\partial y} \rho_w q_y + \frac{\partial}{\partial z} \rho_w q_z\right) dx \, dy \, dz \tag{5.91}$$

As the net inward flux equals the rate at which water is accumulating in the differential element, it is equivalent to the rate of change of fluid mass [Eq. (5.45)]

$$\frac{\partial(\Delta M_w)}{\partial t} = \rho_w(\alpha \rho_w g + n\beta \rho_w g) \Delta x \, \Delta y \, \Delta z \, \frac{\partial h}{\partial t} \tag{5.92}$$

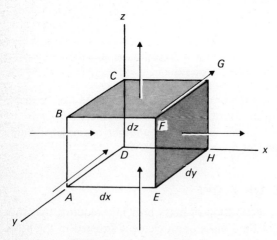

Fig. 5.14 Differential control volume.

COMPRESSIBILITY, ELASTICITY, AND MAIN EQUATIONS OF FLOW

Combining Eqs. (5.91) and (5.92) and canceling common terms,

$$-\left(\frac{\partial}{\partial x}\rho_w q_x + \frac{\partial}{\partial y}\rho_w q_y + \frac{\partial}{\partial z}\rho_w q_z\right) = \rho_w(\alpha \rho_w g + n\beta \rho_w g)\frac{\partial h}{\partial t} \quad (5.93)$$

which is the derivation credited to Jacob (1950).

By regarding both the fluid and the solid matrix as incompressible, the density ρ_w is constant, which gives

$$\frac{\partial q_x}{\partial x} + \frac{\partial q_y}{\partial y} + \frac{\partial q_z}{\partial z} = 0 \quad (5.94)$$

which is a fundamental form of the continuity equation for a rigid medium and incompressible fluid. This equation states that the net rate of change of specific discharge at any point must be zero. In other words, changes in specific discharge in one direction for the element of Fig. 5.14 must be balanced by changes of opposite sign in other directions if there is to be a zero net rate of change of water mass in the element.

DIFFERENTIAL EQUATION FOR STEADY FLOW

The generalizations of Darcy's law for flow in three coordinate directions can be further idealized for an isotropic, homogeneous medium,

$$q_x = -K\frac{\partial h}{\partial x} \qquad q_y = -K\frac{\partial h}{\partial y} \qquad q_z = -K\frac{\partial h}{\partial z} \quad (5.95)$$

Direct substitution in the continuity equation for incompressible flow in a rigid medium gives

$$\frac{\partial}{\partial x}\left(-K\frac{\partial h}{\partial x}\right) + \frac{\partial}{\partial y}\left(-K\frac{\partial h}{\partial y}\right) + \frac{\partial}{\partial z}\left(-K\frac{\partial h}{\partial z}\right) = 0 \quad (5.96)$$

which reduces to

$$\frac{\partial^2 h}{\partial x^2} + \frac{\partial^2 h}{\partial y^2} + \frac{\partial^2 h}{\partial z^2} = 0 \quad (5.97)$$

Equation (5.97) is the Laplace equation introduced in Section 4.1, and conveys the same message as the statement of continuity. That is, the sum of the components of the space rate of gradient change must be zero if the volume is to remain unchanged. For two-dimensional flow,

$$\frac{\partial^2 h}{\partial x^2} + \frac{\partial^2 h}{\partial y^2} = 0 \quad (5.98)$$

In that Darcy's law can be expressed

$$q = -K\frac{1}{g}\frac{\partial \Phi}{\partial L} \quad (5.99)$$

substitution into the continuity equation gives

$$\frac{\partial^2 \Phi}{\partial x^2} + \frac{\partial^2 \Phi}{\partial y^2} \qquad (5.100)$$

which is Laplace's equation in terms of potential gradients.

The Laplace equation occurs in other sciences based on mathematical physics. To pursue the need of gaining insight into similar physical systems, an analogous equation may be obtained by considering the flow of current in a simple one-dimensional electrical field (Karplus, 1958). In Fig. 5.15, the current I_x flows through each part of the wire, so that

$$I_{x1} = I_{x2} = I_x \qquad (5.101)$$

is an expression of continuity. The voltage drop across the element can be obtained from Ohm's law [Eq. (4.54)],

$$\Delta V = -I_x R_x \Delta x \qquad (5.102)$$

or

$$\frac{\Delta V}{\Delta x} = -I_x R_x \qquad (5.103)$$

where $\Delta V/\Delta x$ is the voltage gradient. As Δx approaches zero, the differentials become derivatives, and the rate of change of the gradient is found by differentiating both sides,

$$\frac{d^2 V}{dx^2} = 0 \qquad (5.104)$$

Equation (5.104) is analogous to the one-dimensional form of the Laplace equation for fluid flow.

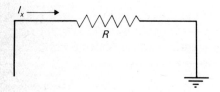

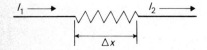

Fig. 5.15 One-dimensional resistance electric circuit.

COMPRESSIBILITY, ELASTICITY, AND MAIN EQUATIONS OF FLOW

The two-dimensional flow of fluid in a homogeneous, anisotropic medium can be expressed

$$K_x \frac{\partial^2 h}{\partial x^2} + K_y \frac{\partial^2 h}{\partial y^2} = 0 \tag{5.105}$$

where $K_x \neq K_y$. Equivalents for steady flow in a two-dimensional field of anisotropic resistance and thermal conductance are expressed

$$\frac{1}{R_x} \frac{\partial^2 V}{\partial x^2} + \frac{1}{R_y} \frac{\partial^2 V}{\partial y^2} = 0 \tag{5.106}$$

and

$$J_x \frac{\partial^2 T_e}{\partial x^2} + J_y \frac{\partial^2 T_e}{\partial y^2} = 0 \tag{5.107}$$

respectively, where V is voltage, J is thermal conductivity, and T_e is temperature.

Physical interpretation of the Laplace equation expressed for fluid, current, and heat flow remains the same. Further, the similarity of the equations, however apparent, is not the basis for the analogy exploited. They merely aid its recognition. These equations are analogous in principle, as well as form, because the laws of continuity and conservation of energy are adhered to in all cases.

DIFFERENTIAL EQUATION FOR UNSTEADY FLOW

Direct substitution of Darcy's law in Eq. (5.93) gives

$$K \left(\frac{\partial^2 h}{\partial x^2} + \frac{\partial^2 h}{\partial y^2} + \frac{\partial^2 h}{\partial z^2} \right) = (\alpha \rho_w g + n \beta \rho_w g) \frac{\partial h}{\partial t} \tag{5.108}$$

For two-dimensional flow,

$$\frac{\partial^2 h}{\partial x^2} + \frac{\partial^2 h}{\partial y^2} = \frac{S_s}{K} \frac{\partial h}{\partial t} \tag{5.109}$$

Neglecting some minor second-order terms, Cooper (1966) has verified this equation in a derivation based on a deformable coordinate system.

It is recalled that S_s is the specific storage, or volume of water that a unit volume of aquifer releases owing to an expansion of water and compression of the matrix under a unit decline in average head within the unit volume. For an aquifer of thickness m, the right-hand side of Eq. (5.109) becomes

$$\frac{\alpha \rho_w g + n \beta \rho_w g}{K} \frac{m}{m} \frac{\partial h}{\partial t} = \frac{S}{T} \frac{\partial h}{\partial t} \tag{5.110}$$

where S is the storativity and T is the transmissivity. For two-dimensional flow in a confined aquifer of uniform thickness

$$\frac{\partial^2 h}{\partial x^2} + \frac{\partial^2 h}{\partial y^2} = \frac{S}{T}\frac{\partial h}{\partial t} \tag{5.111}$$

which is the diffusion equation introduced in Section 4.1, where the constant a equals T/S.

Equation (5.111) is an approximate mathematical description of a simple aquifer in which the transmissivity and storativity undergo no areal change. It describes the configuration of the piezometric head in the simplest type of aquifer. As in the case of steady flow, an analogous equation can be derived for the flow of current in an electric field (Karplus, 1958). The electric system selected must now contain both energy dissipative elements, as in the steady-state model, and energy storage elements. The resistors function as energy dissipators by providing resistance to current flow, identical in function with permeability. The capacitor is selected as the energy storage element as it can serve adequately as a potential energy reservoir similar in function to the storage of water in an aquifer. The analogy is developed for one-dimensional flow (Fig. 5.16).

From continuity considerations,

$$I_{x3} = I_{x1} - I_{x2} \tag{5.112}$$

Recognizing that current through a capacitor is proportional to the time rate of change of voltage (just as time rate of change of storage is proportional to the rate of change of total head),

$$I_{x3} = I_{x1} - I_{x2} = C\Delta x \frac{\partial V}{\partial t} \tag{5.113}$$

where $C\Delta x$ is total capacitance between the element and ground. The difference between I_{x1} and I_{x2} is expressed as the rate of change of current per unit length,

$$I_{x1} - I_{x2} = -\frac{\partial I}{\partial x}\Delta x = C\Delta x \frac{\partial V}{\partial t} \tag{5.114}$$

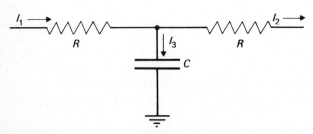

Fig. 5.16 One-dimensional resistance-capacitance electric circuit.

Ohm's law is expressed

$$I_x = -\frac{1}{R_x}\frac{\partial V}{\partial x} \qquad (5.115)$$

Taking derivatives of both sides,

$$\frac{\partial I}{\partial x} = \frac{\partial}{\partial x}\left(-\frac{1}{R_x}\frac{\partial V}{\partial x}\right) = -\frac{1}{R_x}\frac{\partial^2 V}{\partial x^2} \qquad (5.116)$$

Combining Eqs. (5.114) and (5.116),

$$\frac{\partial^2 V}{\partial x^2} = RC\frac{\partial V}{\partial t} \qquad (5.117)$$

Each resistor in an RC network represents the resistance to flow, and is analogous to the reciprocal of aquifer transmissivity. Each capacitor is analogous to the fluid storing capability of a specific part of the aquifer. From a physical point of view, the resistance to fluid flow is inversely proportional to the aquifer thickness, and the storage capacity is directly proportional to aquifer thickness. The product of circuit resistance and capacitance RC is called the *time constant*, and is defined as the time in seconds required to charge or discharge a capacitor to a certain percentage of its final value. In practical terms, it expresses the rate at which steady-state conditions are approached. The greater the resistance, or capacitance, the longer the time required to approximate steady conditions. The ratio of specific storage to conductivity is analogous to the time constant, and can be given similar interpretation. For example, the period of time required for a cone of depression forming around a battery of wells in a basin to approach steady-state conditions gets smaller as the ratio of specific storage to conductivity gets smaller. This becomes somewhat important when it is recalled that the specific storage is a variable with a range measured in orders of magnitude. For a confining layer, the time required to achieve full consolidation also depends on the ratio of specific storage to conductivity. The larger the specific storage, or the smaller the conductivity, the greater is the time to reestablish steady flow.

It follows that an equivalent equation for unsteady two-dimensional flow of heat can be expressed

$$\frac{\partial^2 T_e}{\partial x^2} + \frac{\partial^2 T_e}{\partial y^2} = \frac{1}{\kappa}\frac{\partial T_e}{\partial t} \qquad (5.118)$$

where κ is the diffusivity.

APPLICATIONS

Since the mid-1930s, there has been a remarkable increase in the application of the steady- and unsteady-flow equations to groundwater problems. These investigations have taken two principal directions (Meyboom, 1966): (1) efforts to understand the time variations in piezometric head in the vicinity of a pumped well, and (2) efforts to understand the spatial variations in piezometric head on a regional basis. The main difference in the scientific approach to the problem is, then, time and scale (Meyboom, 1966), time variations being of interest in the study of small-scale systems disturbed by pumping, and spatial variations being of interest in the large-scale study of undisturbed systems.

Contributions to the rapidly expanding fund of analytical models to describe the dynamics of unsteady flow to a well have been forthcoming since at least 1935, when Theis (1935) introduced his now classic nonequilibrium equation. Scientists from several disciplines, including soil physicists and petroleum, agricultural, and soils engineers, as well as hydrologists, have contributed to this knowledge. It is only necessary to review the appendices of the Proceedings of the Symposium on Transient Groundwater Hydraulics (1963) to be fully impressed by the size and scope of these contributions. More than 50 transient case solutions have been presented in this publication, most of which describe the piezometric head in the vicinity of a pumping well.

With steady flow, the time factor does not enter into consideration. Apart from a simple well-flow equation and numerous engineering applications of seepage through earth dams, weirs, and other engineering structures, the selected scale is regional. This approach materialized with Hubbert's (1940) treatment of the steady-state regime, and has been expanded by a considerable amount of work since 1960.

Before moving on to these applications in the next two chapters, it should be stressed that the equations as presented describe the piezometric head under relatively simplified conditions. These equations were arrived at by combining the continuity equation with Darcy's macroscopic law, and thereby providing a mathematical statement encompassing certain variables that are easily measured in the field.

5.7 CONCLUDING STATEMENT

Two main points of this chapter may be stated briefly:

1. Regardless of whether we are dealing with high-permeability sediments, or with clays, the overall behavior of natural materials under applied stresses can be described with the same set of underlying principles; only the magnitude of volume change and time of response are different.
2. Regardless of whether we are dealing with the flow of groundwater, electricity, or heat, the differential equation describing each of these phenomena is similar in form as well as principle; only the physical parameters of the equations are called different things.

COMPRESSIBILITY, ELASTICITY, AND MAIN EQUATIONS OF FLOW

The concept of effective stress is fundamental to the compression of both high- and low-permeability materials, as well as to their strength characteristics. This concept, along with the basic ideas on compressibility, provides an important connecting link between the disciplines of soil mechanics and groundwater hydrology. Indeed, numerous problems in soil mechanics are actually problems in fluid flow. The close relationship between these disciplines is often camouflaged by the use of different names and notations for identical parameters.

The discovery that the differential equation describing the flow of groundwater is identical in form and principle with the equations dealing with electricity and heat flow has more than academic significance. For an analogous set of initial and boundary conditions, a solution to one of these equations is a solution to all of them. For a hydrologic solution, it only remains to identify the hydrologic counterparts of the pertinent physical parameters and boundary conditions. It follows that textbooks treating electricity and heat flow can often be interpreted in terms of their contribution to our understanding of groundwater flow. These ideas will be fully explored in Chapter 7.

PROBLEMS AND DISCUSSION QUESTIONS

5.1 From Meinzer's paper (Compressibility and elasticity of artesian aquifers), cite and briefly discuss the lines of evidence that might be explored to demonstrate that aquifers are compressible and elastic.

5.2 Derive a companion equation for Eq. (5.10), where the hydraulic head is below land surface. Assume that the water table is at land surface, as depicted in Fig. 5.3, and that the hydraulic head is above the top of the aquifer. [Answer: $\bar{\sigma} = gH(\rho_m - \rho_w) + \rho_w gh$.]

5.3 The storativity of an elastic artesian aquifer is 3×10^{-4}. The porosity of the aquifer is 0.3, the compressibility of water is 2.36×10^{-8} ft²/lb, and the thickness of the aquifer is 200 ft.
    a. How many cubic feet of water are removed from storage under an area of 1×10^6 ft² when the head declines 1 ft? (Answer: 300 ft³.)
    b. Ascertain the storativity if the aquifer matrix is incompressible. (Answer: 8.8×10^{-5}.)
    c. What is the value for the compressibility of the aquifer? (Answer: 1.7×10^{-8} ft²/lb.)

5.4 Consider a well pumping from storage at a constant rate from an extensive elastic aquifer.
    a. The rate of head decline times the storativity summed up over the area over which the head decline is effective equals _____?
    b. At any given time, the total head decline, summed up over the area over which the head change is effective, times the storativity equals _____?

5.5 Prove that $S \Delta h A_1 = \Delta m A_2$, where S is the storativity of an elastic aquifer (ignoring fluid compressibility), Δh is the change in head which is effective over the area A_1, and Δm is the change in aquifer thickness, which is effective over the area A_2 (assume $A_1 = A_2$).

5.6 *a.* Prove that the percentage of storage attributable to expansion of the water in an elastic aquifer is equal to the barometric efficiency, and the percentage of storage attributable to compression of the aquifer matrix is equal to the tidal efficiency.

  b. The barometric efficiency of a well in an elastic aquifer is 0.4. If the porosity of the aquifer is 0.2 and the thickness is 300 ft, estimate the storativity. (Answer: $S = 2.2 \times 10^{-4}$.)

5.7 Arrange the following rock types in decreasing order of their relative *long-term* storativity. Assume that all rock units are of equal thickness and water occurs under confined conditions.

  (a) Unconsolidated sand and gravel, clean
  (b) Unconsolidated sand and gravel with interbedded soft clay
  (c) Consolidated sandstone, clean
  (d) Consolidated sandstone and interbedded soft shale
  (e) Limestone, dense

5.8 The equation describing total settlement in basins subjected to fluid withdrawals [Eq. (5.82)] does not include the application of an external load. With the aid of Fig. 5.9, derive an equivalent expression for total settlement ΔH caused by an application of an external load $\Delta \bar{\sigma}$. Identify or point out three terms of this equation and their equivalents in Eq. (5.82). (Answer: $\Delta H = \Delta \bar{\sigma} H / E_c$.)

5.9 An elastic aquifer is overlain by 100 ft of clayey material. The aquifer has a storativity of 5×10^{-4} and a porosity of 0.3 and is 200 ft thick. The clayey material has a specific storage of 4×10^{-4} ft^{-1}.

  a. Calculate the elastic compression of the aquifer for a 100-ft decline in head. (Answer: 0.042 ft.)

  b. Calculate the inelastic compression of the clayey material for a 100-ft decline in head. (Answer: 2 ft.)

  c. Ascertain the bulk modulus of compression of the aquifer. (Answer: 3×10^7 lb/ft^2.)

5.10 A clay layer extending from ground surface to a depth of 50 ft is underlain by an artesian sand. A piezometer drilled to the top of the aquifer registers a head of 20 ft above land surface. The saturated unit weight of the clay is 125 lb/ft^3. Assume a water table at land surface.

  a. What are the values of the total, neutral, and effective pressures at the bottom of the clay layer? (Answer: $\sigma = 6{,}250$ lb/ft^2; $p = 4{,}368$ lb/ft^2; $\bar{\sigma} = 1{,}882$ lb/ft^2.)

  b. If a trench is dug to a depth of 20 ft, what are the total, neutral, and effective pressures at the bottom of the clay layer beneath the trench? Assume the trench is filled with water to land surface. (Answer: $\sigma = 4{,}998$ lb/ft^2; $p = 4{,}368$ lb/ft^2; $\bar{\sigma} = 630$ lb/ft^2.)

  c. Is it possible to maintain the stability of the bottom of the trench in question *b* if the trench is dewatered?

  d. What is the approximate minimum depth of a trench excavated in the clay layer that will cause a quick condition or failure of its bottom? Assume the trench to be filled with water. (Answer: 30 ft.)

5.11 The differential equation for unsteady flow has been a point of controversy in recent years. Compare and briefly discuss the derivation and assumptions by Jacob with the derivation by Cooper based on a deforming coordinate system.

REFERENCES

Bredehoeft, J., and B. Hanshaw: On the maintenance of anomalous fluid pressures, *Bull. Geol. Soc. Amer.*, vol. 79, pp. 1097–1106, 1968.

Cooper, H. H.: The equation of groundwater flow in fixed and deforming coordinates, *J. Geophys. Res.*, vol. 71, pp. 4785–4790, 1966.

Cuevas, J. A.: Foundation conditions in Mexico City, *Proc. Intern. Conf. Soil Mech.*, vol. 3, Cambridge, Mass., 1936.

Davis, G. H., J. B. Small, and H. B. Counts: Land subsidence related to decline of artesian pressure in the Ocala limestone at Savannah, Georgia, Eng. Geol. Case Histories, vol. 4, *Geol. Soc. Amer.*, pp. 1–8, 1963.

DeWeist, R. J. M.: On the storage coefficient and the equations of groundwater flow, *J. Geophys. Res.*, vol. 71, pp. 1117–1122, 1966.

Domenico, P. A., and M. D. Mifflin: Water from low permeability sediments and land subsidence, *Water Resources Res.*, vol. 4, pp. 563–576, 1965.

———, ———, and A. L. Mindling: Geologic controls on land subsidence in Las Vegas Valley, *Proc. Ann. Eng. Geol. Soils Eng. Symp.*, 4th, Moscow, Idaho, pp. 113–121, 1966.

———, D. A. Stephenson, and G. B. Maxey: Groundwater in Las Vegas Valley, *Desert Res. Inst., Tech. Rept.* 7, Reno, Nevada, 1964.

Ferris, J. G., and others: Theory of aquifer tests, *U.S. Geol. Surv., Water Supply Papers*, 1536-E, pp. 69–174, 1962.

Gilluly, J., and U. S. Grant: Subsidence in Long Beach Harbor area, California, *Bull. Geol. Soc. Amer.*, vol. 60, pp. 461–521, 1949.

Hantush, M. S.: Modification of the theory of leaky aquifers, *J. Geophys. Res.*, vol. 65, pp. 3713–3725, 1960.

———: Hydraulics of Wells, in V. T. Chow (ed.), "Advances in Hydroscience," vol. 1, Academic Press, New York, pp. 281–432, 1964.

Hubbert, M. K., and W. W. Rubey: Role of fluid pressures in mechanics of overthrust faulting, Part I: Mechanics of fluid filled porous solids and its application to overthrust faulting, *Bull. Geol. Soc. Amer.*, vol. 70, pp. 115–166, 1959.

Jacob, C. E.: Fluctuations in artesian pressure produced by passing railroad trains as shown in a well on Long Island, New York, *Trans. Amer. Geophys. Union*, vol. 20, pp. 666–674, 1939.

———: On the flow of water in an elastic artesian aquifer, *Trans. Amer. Geophys. Union*, vol. 2, pp. 574–586, 1940.

———: Notes on the elasticity of the Lloyd Sand on Long Island, New York, *Trans. Amer. Geophys. Union*, vol. 22, pp. 783–787, 1941.

———: Flow of groundwater, in H. Rouse (ed.), "Engineering Hydraulics," John Wiley & Sons, Inc., New York, pp. 321–386, 1950.

Johnson Drillers Journal, Series on quicksand, July–August, September–October, November–December, 1965.

Karplus, W. J.: "Analog Simulation," McGraw-Hill Book Company, New York, 1958.

Kögler, F., and A. Scheidig: "Baugrund Und Bauwerk," Wilhelm Ernst und Sohn, Berlin, 1948.

Lohman, S. W.: Compression of elastic artesian aquifers, *U.S. Geol. Surv., Profess. Papers*, 424-B, pp. 47–48, 1961.

Meinzer, O. E.: Compressibility and elasticity of artesian aquifers, *Econ. Geol.*, vol. 23, pp. 263–291, 1928.

——— and H. H. Hard: The artesian water supply of the Dakota sandstone in North Dakota, with special reference to the Edgeley quadrangle, *U.S. Geol. Surv., Water Supply Papers*, 520-E, pp. 73–95, 1925.

Meyboom, P.: Current trends in hydrogeology, *Earth Sci. Revs.,* vol. 2, no. 4, pp. 345–364, 1966.

Parker, G. G., and V. T. Stringfield: Effects of earthquakes, rains, tides, winds, and atmospheric pressure changes on the water in the geologic formations of southern Florida, *Econ. Geo.,* vol. 45, pp. 441–460, 1950.

Poland, J. F.: The coefficient of storage in a region of major subsidence caused by compaction of an aquifer system, *U.S. Geol. Surv., Profess. Papers,* 424-B, pp. 52–54, 1961.

────── and G. H. Davis: Subsidence of the land surface in Tulare-Wasco (Delano) and Los Banos–Kettleman City areas, San Joaquin Valley, California, *Trans. Amer. Geophys. Union,* vol. 37, pp. 287–296, 1956.

Robinson, T. W.: Earth tides shown by fluctuations of water levels in wells in New Mexico and Iowa, *Trans. Amer. Geophys. Union,* vol. 20, pp. 656–666, 1939.

Rubey, W. W., and M. K. Hubbert: Role of fluid pressure in mechanics of overthrust faulting, Part II: Overthrust belt in geosyncline area of western Wyoming in light of fluid pressure hypothesis, *Bull. Geol. Soc. Amer.,* vol. 70, pp. 168–205, 1959.

Symposium on Transient Groundwater Hydraulics, Colorado State University, Fort Collins, July 25–27, 1963.

Terzaghi, K.: "Erdbaumechanic auf Bodenphysikalischer Grundlage," Franz Deuticke, Vienna, 1925.

──────: Mechanism of landslides, in "Berkey Volume" (Application of geology to engineering practice), *Geol. Soc. of Amer. Publ.,* pp. 83–123, 1950.

────── and R. B. Peck: "Soil Mechanics in Engineering Practice," John Wiley & Sons, Inc., New York, 1948.

Theis, C. V.: The relation between the lowering of the piezometric surface and the rate and duration of discharge of a well using groundwater storage, *Trans. Amer. Geophys. Union,* vol. 2, pp. 519–524, 1935.

Watts, E. V.: Some aspects of high pressure in the D7 Zone of the Ventura Avenue Field, *AIME, Petrol. Div.,* Vol. 174, pp. 191–200, 1948.

Wilson, G., and H. Grace: The settlement of London due to underdrainage of the London clay, *J. Inst. Civil Engrs. (London),* vol. 19, pp. 100–127, 1942.

Winslow, A. G., and L. A. Wood: Relation of land subsidence to groundwater withdrawals in the upper Gulf coastal region of Texas, *AIME, Mining Div., Mining Eng.,* pp. 1030–1034, 1959.

part three

The Groundwater Basin as a Distributed-parameter System

6
Regional Groundwater Flow: Theoretical Models of the Steady-state and Related Field Observations

The point of view taken in this chapter is that flow systems often exhibit properties or phenomena whose definite interrelations can be substantiated (1) when the drainage basin is accepted as the smallest permissible unit for study, and (2) when the complete flow pattern is accepted as the denominator common to both the system and its observable properties. The methodology adopted is to define a flow system through a suitably chosen dynamic property, capable of being measured at individual points, which when compared with other system properties and their spatial distributions, demonstrates the manner in which these properties follow from and interrelate with it. The chosen property is groundwater potential, and its spatial distribution constitutes the flow pattern. By relinquishing the lumped-parameter concept described in Section 1.3, related system properties, such as chemical character, temperature, and surface manifestations of flow, may be studied with regard to their location within the flow field.

According to this concept, hydrologic parameters and phenomena are viewed in terms of a single system of flow rather than independent events or measurements. This attitude is prejudiced in favor of order, regularity, and wholeness, and proponents of this approach are attracted to elements of flow systems which are measurable and which may prove to satisfy these criteria. A review article by Meyboom (1966a) organizes some of the pertinent literature on the subject, and presents a sound case in support of this attitude. The mutual acceptance by other groups is acknowledged, as witnessed by the statement attributed to Thomas and Leopold (1964):

> Research projects in hydrogeology should be considered as supporting activities of the overall objective of defining numerically the hydrogeologic system and the underlying flow patterns and superimposed chemical systems.

Two basic approaches are employed, field and theoretical. The field methodology is classically geological, and was fostered originally by Meinzer, as evident from his early observational studies in the Great Basin (Meinzer, 1917; 1922; 1927). These ideas have since been rediscovered and advanced by a group of Canadian hydrologists. The theoretical approach was first presented by Hubbert (1940), and entails the formulation of an ideal system model that reliably portrays the flow pattern linking source and sink. The combined approach is the essence of flow system theory, here defined as the methodology of searching for and delineating flow systems.

A regional system can be defined as a large groundwater flow system which encompasses one or more topographic basins (Mifflin, 1968). Water enters the system at the source area, or areas, by virtue of passing through the vadose zone, and exits at the sink area, or areas, to positions outside the saturated zone of earth materials—to the atmosphere, surface water systems, or to plants. At every point within the system, each molecule of water has the potential to move toward a system sink. Water which does not move, or moves so slowly as to be virtually undetectable, is considered exterior to the region of dynamic flow. The delineation of such systems is the act of representing, portraying, or describing graphically or verbally the region of groundwater flow and interrelated processes.

6.1 THEORETICAL MODELS: SOLUTIONS OF LAPLACE'S EQUATION

Expressions stated as

$$\frac{\partial^2 h}{\partial x^2} + \frac{\partial^2 h}{\partial y^2} = 0 \tag{6.1}$$

REGIONAL GROUNDWATER FLOW

and

$$\frac{\partial^2 \Phi}{\partial x^2} + \frac{\partial^2 \Phi}{\partial y^2} = 0 \qquad (6.2)$$

are the most common forms of Laplace's equation for two-dimensional flow in an isotropic and homogeneous medium. A more compact statement of the equation is obtained by introducing the laplacian operator ∇^2 (Section 4.1), which is used to signify the sum of the second derivatives with respect to all space variables of interest. This gives

$$\nabla^2 \Phi = 0 \qquad (6.3)$$

The problem of finding a solution of Laplace's equation subject to given boundary conditions is called a *boundary-value problem*. Time is not a variable in these problems, and the complete solution may be represented by two intersection families of curves. In isotropic media, the lines are mutually perpendicular, and one family of curves determines the other. Hence, either may be taken as the complete solution.

In this section, the flow net formed by intersecting families of curves is introduced as a theoretical means for obtaining a quantitative evaluation of regional groundwater flow. A few analytical, numerical, and electrical solutions are briefly reviewed.

BOUNDARY CONDITIONS

A groundwater system may be isolated for special consideration by introducing the effect of its environment through boundary conditions. If only a part of a flow region is to be considered, it may be of importance to account for a specified flux into the region. If the flux is a subsurface flow, Darcy's law may be employed. An important special type of such a boundary condition is termed a *no flow* boundary, and is used when no flux is allowed into the region. Expressed mathematically, Φ equals a constant value on each side of a vertical or horizontal boundary of the flow field so that

$$\frac{\partial \Phi}{\partial n} = 0 \qquad (6.4)$$

where n is the direction normal to the boundary. Hence, a no-flow boundary is generated by the absence of a potential gradient. No-flow boundaries may be encountered at the contact between permeable and impermeable material, and at major topographic divides.

A second boundary condition may require a specified head along a given boundary. An important special type of this boundary is termed a *constant head* boundary, as might be required along the margins of rivers or lakes in hydraulic connection with groundwater. In all cases, the boundary indicates, or fixes, the limit or extent of the flow region of interest.

An example of a typical boundary-value problem complete with a statement of the boundary conditions is shown by the two-dimensional vertical section of Fig. 6.1. The crosshatched area represents an isotropic, homogeneous region in which Laplace's equation is to be solved. Emphasis in this chapter will be focused on vertical sections of unconfined flow of this type. Regardless of which method of solution is employed, approximations of the boundary conditions are required along the entire boundary of the flow region. Since there is no lateral flow across major topographic highs and lows (Fig. 4.13), $\partial \Phi / \partial x$ equals zero describes the condition at the lateral boundaries, where x is taken as the direction normal to the boundary. This renders the boundary "impermeable." For the impermeable bottom, $\partial \Phi / \partial z$ equals zero, where z signifies the direction normal to the boundary.

A fourth condition is required along the upper surface of flow. In the application of flow nets to engineering seepage problems, the upper surface of flow is termed a *free surface;* its form in the problem is generally treated as an unknown. In regional flow applications, the upper surface is represented by $\Phi = f(x)$, where $f(x)$ is the equation of the water-table configuration, and is known from field observations. In practice, values of Φ along the water table are likely to be a function of both time and position; however, as the flow is assumed to be steady, Φ is assumed to be a function of the space variable x. This implies that the water table is "fixed" in space, and local fluctuations are either unimportant, or have been averaged to give a mean water-table position.

Other conditions pertaining to certain types of boundaries will be discussed as required.

ANALYTICAL SOLUTIONS

Whereas Hubbert (1940) arrived at the flow pattern of Fig. 4.13 by recognizing

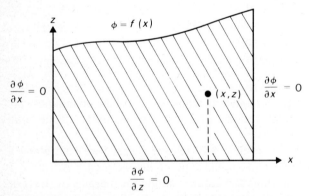

Fig. 6.1 Two-dimensional, unconfined, homogeneous, isotropic region with conditions at the boundaries.

that flow lines everywhere parallel to $-\text{grad }\Phi$ form an orthogonal system with the family of equipotential surfaces, it is possible to obtain flow patterns as mathematical solutions to formal boundary-value problems. Emphasis in this section is not on the various mathematical techniques employed, but on a few solutions which bear directly on the overall objectives of this chapter. Students interested in the widely used separation-of-variables technique and the specialized Schwartz-Christoffel transformation and Green's functions methods for solving elliptical partial differential equations are referred to any standard text dealing with potential theory. Applications of fluid flow through or under engineering structures may be found in Muskat (1937), Harr (1962), and Polubarinova-Kochina (1962).

The first significant extension of Hubbert's work was presented by Toth (1962, 1963), and incorporated the following assumptions:

1. The medium is isotropic and homogeneous to a specified depth, below which there exists a horizontal impermeable basement.
2. Flow is restricted to a two-dimensional vertical section. The topography can be approximated by simple curves, such as straight lines or sine waves, and the water table is a subdued replica of topography.
3. The upper boundary of the flow field is the water table, the lower boundary is the impermeable basement, and the lateral boundaries are major groundwater divides.

Figure 6.2 shows the mathematical models established for vertical flow sections, with the two water-table configurations expressed as

$$f(x) = z_0 + c'x \tag{6.5}$$

and

$$f(x) = z_0 + c'x + a' \sin b'x \tag{6.6}$$

where the first represents a linear water table whose elevation decreases from the topographic high to the valley bottom, and the second is a sine wave superimposed upon the regional slope. Other boundary conditions are as shown in Fig. 6.1. The symbols of Fig. 6.2 and Eqs. (6.5) and (6.6) mean the following (Toth, 1962; 1963):

z_0 = elevation of the water table above datum at the valley bottom

α = angle of slope of the water table

$c' = \tan \alpha$

$a' = \dfrac{a}{\cos \alpha}$

$b' = \dfrac{b}{\cos \alpha}$

a = amplitude of sine curve
$b = 2\Pi/\lambda$ = frequency of the sine wave
λ = period of the sine wave

A separation-of-variables technique is used to solve Laplace's equation for the flow regions of Fig. 6.2, which leads to a converging infinite series that

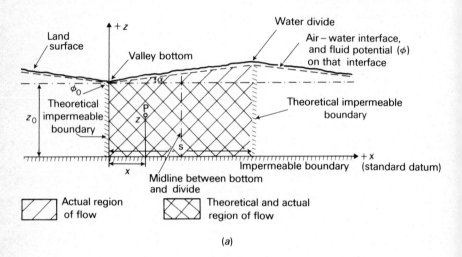

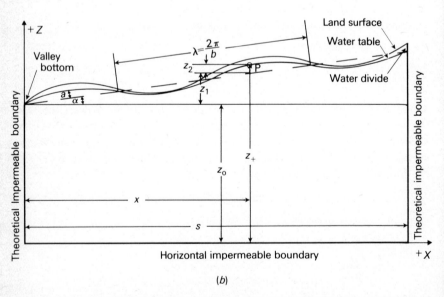

Fig. 6.2 Two-dimensional, unconfined, homogeneous, isotropic models for (a) linear and (b) sine-wave approximation of the water table. (*After Toth*, 1962, 1963.)

may be used to obtain independent values of Φ. For the linear water table, the solution is

$$\Phi = g\left(z_0 + \frac{cs}{2}\right) - \frac{4gcs}{\Pi^2} \sum_{m=0}^{\infty} \frac{\cos[(2m+1)\Pi x/s]\cosh[(2m+1)\Pi z/s]}{(2m+1)^2 \cosh[(2m+1)\Pi z_0/s]} \quad (6.7)$$

For the sine-wave approximation,

$$\Phi = g\Bigg(z_0 + \frac{c's}{2} + \frac{a'}{sb'}(1 - \cos b's) + 2\sum_{m=1}^{\infty}\left[\frac{a'b'(1 - \cos b's \cos m\Pi)}{b'^2 - m^2\Pi^2/s^2}\right. $$

$$\left.+ \frac{c's^2}{m^2\Pi^2}(\cos m\Pi - 1)\right]\cos\left(\frac{(m\Pi x/s)\cosh(m\Pi z/s)}{s\cosh(m\Pi z_0/s)}\right)\Bigg) \quad (6.8)$$

By applying Eq. (6.8) to a variety of hypothetical situations, Toth (1963) recognized a pattern formed by a certain grouping of flow lines, from which he arrived at a definition of a flow system:

> A flow system is a set of flow lines in which any two flow lines adjacent at one point of the flow region remain adjacent through the whole region; they can be intersected anywhere by an uninterrupted surface across which flow takes place in one direction only.

From this definition, an area of downward flow (recharge area) is that part of the system in which the net saturated flow of groundwater is directed away from the water table; an area of upward flow (discharge area) is that part of the system in which the net saturated flow of groundwater is directed toward the water table. Three distinct types of flow systems have been recognized (Toth, 1963) (Fig. 6.3):

1. A local system, which has its recharge area at a topographic high and its discharge area at a topographic low that are adjacent to each other. This system is virtually synonymous with the flow system portrayed in Fig. 4.13.
2. An intermediate system, which is characterized by one or more topographic highs and lows located between its recharge and discharge areas.
3. A regional system, which has its recharge area at the major topographic high and its discharge area at the bottom of the basin.

From a comparative study of variations in selected geometric parameters, such as depth to impermeable basement, slope of the valley flanks, and local relief, Toth (1963) made three pertinent conclusions regarding the conditions under which local, intermediate, and regional systems may develop. The more significant aspects of these empirical conclusions may be examined by recognizing that the water table is a subdued replica of topography, whose potential is given by

$$\Phi_{WT} = g(z_0 + c'x + a' \sin b'x) \tag{6.9}$$

where $c'x$ is the general slope of the water table, and $a' \sin b'x$ is the *local relief*. Hence, any assumptions concerning topographic configuration automatically affect the water-table configuration. For example, if local relief is negligible,

$$\Phi_{WT} = g(z_0 + c'x) \tag{6.10}$$

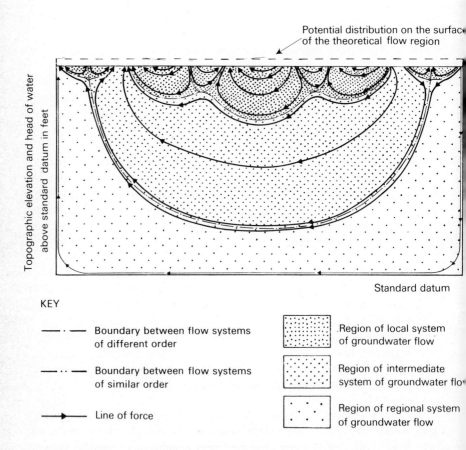

Fig. 6.3 Two-dimensional, unconfined, homogeneous, isotropic model for the distribution of local, intermediate, and regional groundwater flow systems. (*After Toth*, 1963.)

where z_0 and c' are constant. Thus,

$$\frac{\partial \Phi_{WT}}{\partial x} = g\left[\frac{\partial z_0}{\partial x} + \frac{\partial c'x}{\partial x}\right] = gc' \tag{6.11}$$

where gc' is a constant. A constant water-table gradient implies an absence of alternating highs and lows, which suggests only regional systems are present (Fig. 4.14).

From Eq. (6.9), a' is the height of local relief and $\sin b'$ is a function of x, in this case a trigonometric function with values ranging from $+1$ to -1, depending on x. It follows that

$$\frac{\partial \Phi_{WT}}{\partial x} = g\left[\frac{\partial z_0}{\partial x} + \frac{\partial c'x}{\partial x} + \frac{\partial}{\partial x}(a' \sin b'x)\right] \tag{6.12}$$

or

$$\frac{\partial \Phi_{WT}}{\partial x} = g\left[c' + a' \frac{\partial(\sin b'x)}{\partial x}\right] \tag{6.13}$$

If x_1 and x_2 are selected so that $\sin b'$ equals $+1$ at x_1 and $\sin b'$ equals -1 at x_2, then for x_1,

$$\frac{\partial \Phi_{WT}}{\partial x_1} = g(c' + a') \tag{6.14}$$

where $\partial \Phi_{WT}/\partial x_1 > 0$. At x_2,

$$\frac{\partial \Phi_{WT}}{\partial x_2} = g(c' - a') \tag{6.15}$$

and $\partial \Phi_{WT}/\partial x_2 < 0$. This suggests that the direction of movement is capable of being completely reversed as one travels in the direction of the x axis, a condition reversed for a sinusoidal water-table characteristic of numerous adjacent local systems (Fig. 6.4).

For an extended flat area, both $c'x$ and $a' \sin b'x$ approach zero, so that

$$\Phi_{WT} = gz_0 \tag{6.16}$$

and

$$\frac{\partial \Phi_{WT}}{\partial x} = g\frac{\partial z_0}{\partial x} = 0 \tag{6.17}$$

Hence, the water table is a horizontal surface, and the system is static for this extreme condition (or nearly static for very small gradients). Discharge is possible only by evapotransportation.

The conclusions derived above were first stated by Toth (1963) from his empirical study:

1. If local relief is negligible, and there is a general slope of topography, only regional systems will develop.
2. Pronounced local relief suggests that no extensive unconfined regional systems can exist across valleys of large rivers or highly elevated watersheds. The greater the relief, the deeper the local systems that develop.
3. Under extended flat areas unmarked by local relief, neither regional nor local systems can develop. Water-logged areas may develop, and the groundwater may be highly mineralized from concentration of salts.

The results obtained by programming Eq. (6.7) or (6.8) are analytically correct only for the specific conditions established in Fig. 6.2. A more general approach has been presented by Freeze and Witherspoon (1966), where n homogeneous and isotropic interlayered sections can be treated in two-dimensional form. Further, a generalized configuration for the water table is incorporated with a series of straight-line segments so that any configuration can be approximated as a function of the space variable x. This approach is not burdened by the assumption that the water table is a smooth curve that reflects exactly a smooth topography, but requires additional data, namely, the actual water-table configuration. The flow system, as modeled, is unconfined, and the conditions that are considered typical of confined flow are due to permeability differences between adjacent layers.

The boundary-value problem to be solved is actually n interrelated problems. Where n equals 3, Fig. 6.5 applies. Two boundary conditions must be satisfied at the contact between adjacent layers. The condition that Φ_n equals Φ_{n+1} ensures a continuous potential across the contact. A second condition dealing with respective permeabilities and potential gradients ensures a continuous normal component of velocity at the contact. This is merely a reiteration of the tangent refraction law [Eq. (4.72)]. Figure 6.6

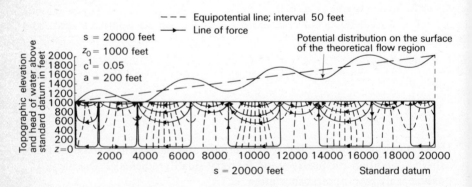

Fig. 6.4 Two-dimensional, unconfined, homogeneous, isotropic model for the distribution of adjacent local systems. (*After Toth*, 1963.)

REGIONAL GROUNDWATER FLOW

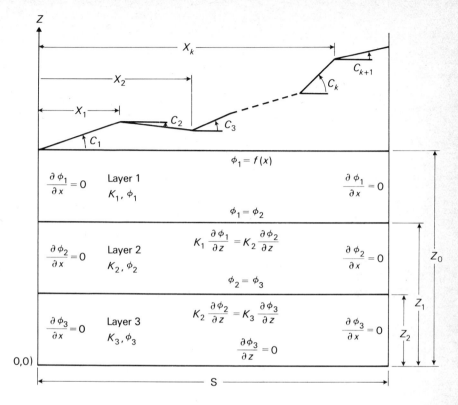

Fig. 6.5 Two-dimensional, unconfined model with n homogeneous, isotropic layers. (*After Freeze and Witherspoon*, 1966.)

demonstrates the effect of the second boundary, as well as the influence of assumed variations in basin geology.

The unwieldy nature of the solution precludes its presentation here, and the interested reader may refer to the original paper by Freeze and Witherspoon (1966) for a general treatment of the problem.

NUMERICAL APPROXIMATIONS: METHOD OF FINITE DIFFERENCES

One of the most commonly used numerical methods for solving boundary-value problems is known as the *method of finite differences*. In this method, the governing differential equation is replaced by an approximating difference equation in such a manner that the budgetary requirements of the original differential equation are approximately conserved. The continuous region for which a solution is desired is replaced by an array of discrete points. This allows reduction of the problem to selection of a system of algebraic equations requiring only basic operations, such as multiplication, addition, and subtraction. Ordinarily, iterative techniques, such as those developed for digital computers, are required to solve problems of this sort.

The first step in making a groundwater system discrete is to obtain the descriptive steady-state equation as approximated for a two-dimensional grid network. The region of interest is shown in Fig. 6.7a by a net consisting of two sets of mutually perpendicular lines. The solution sought, of course, is the potential distribution in the region.

An approximation of the gradient between nodes 0 and 1 and nodes 2 and 0 (Fig. 6.7b) is given as head difference divided by distance,

$$\left(\frac{\partial h}{\partial x}\right)_0^1 \approx \frac{h_1 - h_0}{\Delta x} \qquad \left(\frac{\partial h}{\partial x}\right)_2^0 \approx \frac{h_0 - h_2}{\Delta x} \qquad (6.18)$$

The second space derivative is the rate of change of the first, or

$$\left(\frac{\partial^2 h}{\partial x^2}\right)_0 \approx \frac{\left(\frac{\partial h}{\partial x}\right)_0^1 - \left(\frac{\partial h}{\partial x}\right)_2^0}{\Delta x} \qquad (6.19)$$

Substituting Eq. (6.18),

$$\left(\frac{\partial^2 h}{\partial x^2}\right)_0 \approx \frac{1}{\Delta x^2}(h_1 + h_2 - 2h_0) \qquad (6.20)$$

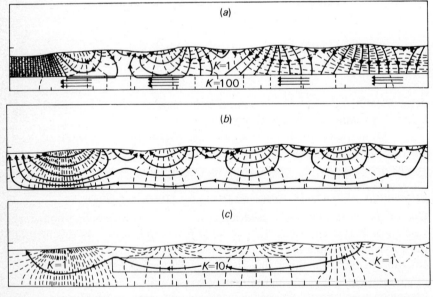

Fig. 6.6 Flow patterns of (a) interbasin flow controlled by continuous high-permeability unit, (b) interbasin flow controlled by topographic relief, and (c) interbasin flow controlled by a lenticular body of high permeability. (*After Freeze and Witherspoon, 1967.*)

REGIONAL GROUNDWATER FLOW

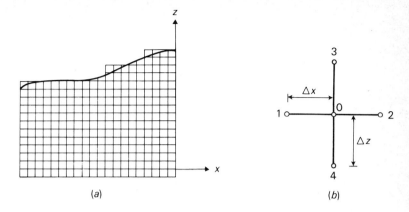

Fig. 6.7 Diagram of (a) finite grid used in numerical analysis and (b) element of the finite grid containing four nodal points.

Applying these same principles to the nodes aligned in the z direction,

$$\left(\frac{\partial^2 h}{\partial z^2}\right)_0 \approx \frac{1}{\Delta z^2}(h_3 + h_4 - 2h_0) \tag{6.21}$$

A finite difference approximation of Laplace's equation for two-dimensional flow in a homogeneous, isotropic medium is then expressed

$$\frac{\partial^2 h}{\partial x^2} + \frac{\partial^2 h}{\partial z^2} \approx \frac{1}{\Delta x^2}(h_1 + h_2 - 2h_0) + \frac{1}{\Delta z^2}(h_3 + h_4 - 2h_0) = 0 \tag{6.22}$$

Assuming the grid spacing is chosen so that Δx equals Δz,

$$h_1 + h_2 + h_3 + h_4 - 4h_0 = 0 \tag{6.23}$$

or

$$h_0 = \tfrac{1}{4}(h_1 + h_2 + h_3 + h_4) \tag{6.24}$$

Equation (6.23) or (6.24) makes it possible to obtain values of the head at all nodal points such that the finite difference equation is satisfied simultaneously at all nodes. In the case of Eq. (6.24), a satisfactory solution is obtained when the total head at all nodal points equals one-fourth of the sum of the heads at the four adjacent nodes. As a first approximation, the heads along the boundaries and at each node are estimated. All assigned values will obviously be in error, with the exception of the boundary values. Beginning at nodes adjacent to the boundary, a new value for the head at each node is computed from the assigned values at the four adjacent nodes. Equation (6.24) is used in this computation. This procedure is repeated until the computed heads converge to their limiting values.

Equation (6.23) for any node may be modified to account for the error of the assumed starting heads

$$h_1 + h_2 + h_3 + h_4 - 4h_0 = R_0 \qquad (6.25)$$

where R_0 is the residual amount representing the error at node zero for the assumed starting heads. The idea here is to remove the residual so that Eq. (6.25) converges to Eq. (6.23) (Scott, 1963). The residual can be eliminated by adding to it an equal number of opposite sign. Clearly, the elimination of a residual at any one node changes the surrounding residuals at other nodes. Further, the change of a residual by -4 changes all adjacent residuals by $+1$. The procedure that follows entails the removal of all residuals node by node, starting with the highest.

The space variable in the above development is no longer a continuous variable, but has been made discrete. The finite difference approach with its characteristic nodal pattern, then, involves the replacement of a continuous field by an assemblage of discrete, or lumped, elements in such a way that the characteristics of the original field are approximately conserved. As each nodal point represents, in fact, a specified area in space, the following assumptions are inherent in the finite difference approach: (1) Variations in the permeability parameter within the specified area represented by a nodal point can be ignored, and (2) absolute information content is assumed by assigning numerical values of permeability to nodal points in the absence of conclusive field data. Similarly, making a continuous system into a distribution of discrete lumped elements involves other approximations and attendant errors which can be kept to a minimum by careful consideration of the space between adjacent nodes. Such errors are directly proportional to the square of the chosen nodal spacing.

Although a digital computer is generally required for both analytical and numerical solutions, the procedure used for each differs markedly. In the case of analytical solutions, the solution is first derived and the computer is called upon in much the same manner as a calculator to grind out numerical values of the potential at various points within the region. These calculations are made independently of each other. With the numerical method, the solution *is* the numerical value of the potential, and the method of obtaining such values is intimately related to the computational methods used (Freeze and Witherspoon, 1966).

The more powerful method of numerical solutions is particularly adaptable to the study of the interlayered condition in Fig. 6.5, as well as a general nonhomogeneous, anisotropic flow field (Freeze, 1966). When a digital computer is used, the following three steps are suggested. First, a suitable grid must be chosen. Next, finite difference equations must be developed for interior points as well as for points on the boundaries. The third step requires a method of solution for the resulting system of equations, and a program for the computer. A few solutions of general interest have been reported (Fayers and Sheldon, 1962; Freeze and Witherspoon, 1966), and students interested in the required specialized techniques are referred to these papers, as well as to general literature on the subject (Southwell, 1946; Dusinberre, 1949; Allen, 1954; McCracken and Dorn, 1964). Digital computation for numerical solutions will be discussed in Chapter 7.

ELECTRICAL ANALOGS

The analogy between the flow of fluids, electrical current, and heat has already been discussed. The thermal analogy is not readily exploited because of the difficulty of maintaining constant temperature boundaries and equipotential lines. Of more practical concern is the analogy between the flow of fluids and the flow of electricity. Electrical analogs suitable for such analysis may be classified in three groups:

1. Conductive liquids
2. Conductive sheets and solids
3. Lumped-element analogs

The first two groups are referred to as continuous models in that every part of the prototype can be identified with a corresponding part of the analog. Conductive liquid models employ the familiar electrolytic tank, a large receptacle containing a liquid which conducts electricity. The voltage distribution along the surface of the liquid is measured in response to electric excitation of the boundaries.

Application of conductive sheet analogs to hydraulic flow or other laplacian fields requires the following (Karplus, 1958):

1. A conductive material of the same geometrical shape as the hydraulic flow field. Resistive paper is the most suitable for the flow situations described in this chapter.
2. A power source by which voltages or currents, or both, may be used to stimulate the boundary conditions.
3. A sensing device by which the voltage distribution within the conducting material may be measured.

For flow solutions in a vertical plane, the potential along the upper boundary of flow is presumed known, so that a varying potential is required there. This boundary may be simulated by applying an appropriate voltage to individual metallic disks (electrodes) inserted along the upper surface of the paper. Errors arising from simulating a continuous boundary with a series of discrete points may be avoided by employing a continuous silverpoint electrode along the upper surface (Karplus, 1958).

This type of model has been applied to determine the types, distribution, and intensity of groundwater flow systems for a study area in western Canada (Toth, 1968). The main components of the model are shown in Fig. 6.8, and include a dc power supply, a potential divider, a voltmeter and probe for making readings, and a conducting paper. As mentioned above, the conducting paper is cut to the shape of the conducting medium (vertical section), with due consideration given to the ratio of basin length to depth, and to the shape of the upper surface. The electrical potential applied to the electrodes along the upper surface is controlled by potentiometers, the desired value at each electrode being proportional to the elevation of the water table at that point. A sensor probe is used to measure electrical potentials in the interior of the model. Voltage equipotentials are easily sketched in. If we are interested only in the potential distribution and not in discharges, conversion by appropriate scaling factors to hydraulic head adds nothing to the solution, and is not required.

The model described above incorporates the assumptions of homogeneity and isotropy. Attempts to simulate the nonhomogeneous situation electrically have not been altogether successful. For the interlayered case in Fig. 6.5, for example, separate sheets of differing conductivity as well as suitable means of connecting them would be required. Scott (1963) proposed an alternative for approximate solutions to the layered condition when the permeability does not vary by more than a factor of 7 or 8. The method proposes removal of some of the paper in the less-permeable layers in

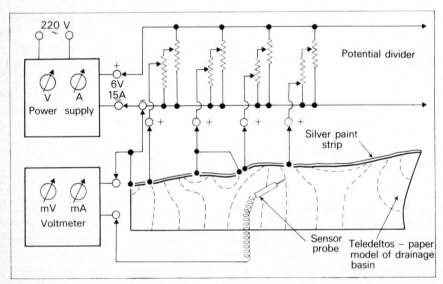

Fig. 6.8 Schematic presentation of an electric analog used to investigate unconfined, two-dimensional flow in a homogeneous, isotropic region. (*After Toth*, 1968.)

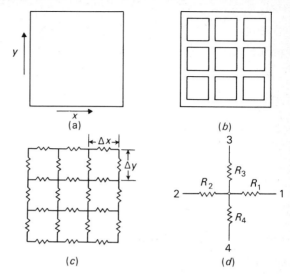

Fig. 6.9 Diagrams showing (a) a two-dimensional conductive sheet, (b) a partially discrete equivalent of the two-dimensional conductive sheet, (c) an equivalent resistive network, and (d) an element of the resistive equivalent. (*After Karplus, 1958. Used with permission of McGraw-Hill Book Co.*)

proportion to the permeability difference between the layers. The cutaway sections may be randomly punched holes or rectangles.

The physical basis for the lumped-element analog can be developed from the conductive sheet discussed above. If a network of squares are cut from the two-dimensional sheet in Fig. 6.9a, the remaining conductive paths (Fig. 6.9b) can be replaced by the resistive network in Fig. 6.9c, the resistors representing a lumped equivalent of the original field. Kirchhoff's equation, which states that the algebraic sum of the currents into any node must be zero, can be applied to the nodal diagram in Fig. 6.9d. This provides an expression analogous in form and in principle to the finite difference approximation of Laplace's equation for fluid flow.

As current is voltage divided by resistance, Kirchhoff's equation is expressed

$$\frac{V_1 - V_0}{R_1} + \frac{V_2 - V_0}{R_2} + \frac{V_3 - V_0}{R_3} + \frac{V_4 - V_0}{R_4} = 0 \qquad (6.26)$$

If the field is presumed to be uniform in its resistive properties, the resistors are equal, which gives

$$V_1 + V_2 + V_3 + V_4 - 4V_0 = 0 \qquad (6.27)$$

which is the finite difference form of Laplace's equation governing the steady flow of electricity, and is the electrical equivalent of Eq. (6.23). Hence, the continuous electrical field can be made discrete in much the same manner as the hydrological field in the method of finite differences, and a continuous region can be replaced by a distribution of lumped elements. This means that the resistive sheet in Fig. 6.8 can be duplicated with a network of resistors connected as shown in Fig. 6.9c, but in the shape of the prototype field. This, in effect, associates each electrical element with a specific area of the prototype. The term *lumped* is appropriate here because this approach, although designed to investigate the distributed system, has elements of the black-box approach in that variations in system parameters within a specified area of the prototype field (say, as represented by one resistor) are ignored. For nonhomogeneous systems, it is thus possible to lump the system into elements which are effectively homogeneous in themselves.

Termination of the resistors at the two lateral and one vertical boundaries adequately accounts for three no-flow boundary conditions. The electric potential is applied to the "water table" at the nodes along the upper surface, and again is controlled by potentiometers, or even fixed resistors. Further, the layered case no longer represents a major problem. If permeability of the individual interlayered units does not vary greatly, resistors may be chosen in proportion to prototype permeability, which gives an interlayered model that can be used for approximate solutions.

The lumped-element analog will be discussed in greater detail in the section on simulation.

SUMMARY AND ADDITIONAL INFORMATION

This section suggests it is possible to obtain a picture of regional flow for any basin for which the pertinent data are available. Data requirements include permeability distribution and geometry of the basin boundary. Although the ideal homogeneous case is contrary to field observations, it is merely a *condition* in the mathematical development, not an assumption necessary to the general validity of the theory. Further, the steady-state solutions do not deny the existence of a fluctuating upper boundary of flow. The main argument is that the effect on flow patterns will be small if (1) the fluctuating zone is small compared with total saturated thickness, and (2) the relative configuration of the water table is unchanged throughout the cycles of fluctuation. Given that these conditions are reasonably satisfied, the value of the information provided by the flow net can best be appreciated if one is interested in the distribution of recharge and discharge areas along the water table (Fig. 6.3), the depth and lateral extent of local systems in hummocky terrain (Fig. 6.4), and the degree to which the zones of high permeability act as major conductors of water and their overall influence on the flow pattern (Fig. 6.6). In summarizing the factors that influence these items, Freeze and Witherspoon (1967) cite the following as the most important:

1. The three most influential factors affecting the potential distribution are
   a. The ratio of basin depth to lateral extent.
   b. The configuration of the water table.
   c. Variations in permeability.
2. A major valley will tend to concentrate discharge in the valley. Where the regional water-table slope is uniform, the entire upland is a recharge area. In hummocky terrain, numerous subbasins will be superimposed on the regional system.
3. A buried aquifer of significant permeability will act as a conduit that transmits water to principal discharge areas, and it will thus affect the magnitude of recharge and the position of the recharge area.
4. Stratigraphic pinchouts at depth can create recharge or discharge areas where they would not be anticipated on the basis of water-table configuration.
5. There is some depth in regions of reasonably horizontal sedimentation below which equipotential lines remain vertical.

Flow nets of groundwater basins may be usefully applied to gain an understanding of the interrelations between the processes of infiltration and recharge at topographically high parts of the basin, and of groundwater discharge via evapotranspiration and baseflow at topographically low points. Such studies permit a spatial evaluation of the flow of groundwater in the hydrologic cycle. For example, at least some of the water derived from precipitation which enters the ground in recharge areas will be transmitted to distant discharge points, and so cause a relative moisture deficiency in soils overlying recharge areas. Water which enters the ground in discharge areas cannot overcome the upward potential gradient, and therefore becomes subject to evapotranspiration in the vicinity of its point of entry. The hinge-line separating areas of upward and downward flow may thus serve as a boundary common to areas of relative soil-moisture surplus and deficiency (Toth, 1966a). In nonirrigated agricultural areas, this may be reflected by variations in crop yield. Further, the ramifications of man's activities in discharge areas are immediately apparent. Some of these include:

1. Water-logging problems associated with surface-water irrigation of lowlands
2. Water-logging problems associated with destruction of phreatophytes, or plants discharging shallow groundwater
3. Pollution of shallow groundwaters from gravity-operated sewage and waste-disposal systems located in valley bottoms in semiarid basins where surface water is inadequate for dilution.

The spatial distribution of flow systems will also influence intensity of natural groundwater discharge. From Fig. 6.3, the main stream of a basin may receive groundwater from the area immediately within the nearest topographic high, and possibly from more distant areas. If baseflow calculations are used as indicators of average recharge, significant error may be introduced in that baseflow may represent only a small part of the total discharge occurring down gradient from the hingeline. Hence, baseflow analysis based on lumped-parameter concepts may give numerical results that are of little practical use unless examined in the light of spatial flow characteristics.

Other interesting aspects resulting from such studies deal with determining the actual amount of groundwater that effectively participates in the hydrologic cycle. From Fig. 6.4, Toth (1963) calculated that about 90 percent of the recharge never penetrates deeper than 250 to 300 ft. Similar conclusions have been arrived at by tritium dating studies, which demonstrate a stratification of tritium and tritium-free waters (Carlston et al, 1960). It would appear that the use of theoretical flow analysis in conjunction with water-dating techniques holds considerable promise for some flow-system studies. The amount of water that effectively partakes in normal hydrologic cycle operations is also important to an understanding of chemical zonations of groundwater (Section 6.4).

The determination of flow quantities by use of flow nets is an application of Darcy's law. For this application, Darcy's law is expressed

$$Q = KiA \qquad (6.28)$$

The flow through any one of the flow channels Δq for a unit thickness of the flow system perpendicular to the plane of the flow diagram is

$$\Delta q = K \frac{\Delta h}{b} a \qquad (6.29)$$

where Δh is the head loss between any two equipotential lines, b is the distance between the equipotential lines, and a is the width of the flow channel. The hydraulic gradient between any two equipotential lines is then $\Delta h/b$. For isotropic, homogeneous conditions, the flow net is a system of squares, and a equals b. This leaves

$$\Delta q = K \Delta h \qquad (6.30)$$

The total discharge per unit thickness perpendicular to the plane of flow is then found by summing the quantities of flow in the individual flow channels.

Other than providing a conceptual framework for studies that are fundamentally hydrogeological in nature, flow-system concepts may be usefully applied to waste disposal and underground nuclear detonation problems. The safety aspects of disposal or nuclear detonation operations depend on delaying the forward movement of the products of such activities, which depends largely on the pattern of flow and the materials through which the flow takes place. Retention time may be increased by planning operations that take into account the locations where flow lines break surface. Other than providing information on the internal distribution of the transmissive properties, disposal studies should be equally concerned with the external boundaries of the flow systems which are to contain the wastes. Clearly, the scale of such a problem precludes detailed measurement of transmissive properties and flow rates at all points within the system, which suggests that regional approaches to flow-system delineation may provide the insight on which engineering judgment can be based. In all cases, the degree of accuracy with which performance can be predicted depends on the complexity of the unit, whereas performance itself depends on relatively few pertinent properties.

6.2 SURFACE FEATURES OF GROUNDWATER FLOW

Surface features of groundwater flow include all observations that can be used to ascertain the occurrence of groundwater, including springs, seeps phreatophytes, saline soils, and permanent or ephemeral streams, ponds, or bogs in hydraulic connection with underground water. The areal occurrence of these features is invariably restricted to areas of groundwater discharge, and their comprehension requires some knowledge of the nature of groundwater outcrops (Meyboom, 1966*b*). The significance of these features may best be understood by examining a few studies that focus on their interpretation.

THE PRAIRIE PROFILE

The prairie profile (Fig. 6.10), consisting of a central topographic high bounded on both sides by areas of major natural discharge, has been offered as a model of groundwater flow to which all observable groundwater phenomena in a prairie environment can be related (Meyboom, 1962; 1966*b*). Geologically, the profile consists of two layers, the uppermost being the least permeable, with a steady-state flow of groundwater toward the discharge areas. The unconfined-flow pattern has been substantiated by numerous borings for both small-scale systems, say, a typical knob and adjacent kettle common to rolling prairie topography, and for a scale of the magnitude of the prairie profile. Recharge and discharge areas have been delineated, and are characterized by a decrease in head with increased depth and by an increase in head with increased depth, respectively. The occurrence of flowing wells is noted in parts of the discharge area.

As it is apparent that most of the natural discharge occurs by evapotranspiration, considerable attention has been given to this phenomenon and to the surface features observed where it occurs. Included among these observations are (1) the occurrence of willow rings and the chemical character of water bodies centered within them, (2) the distribution and types of

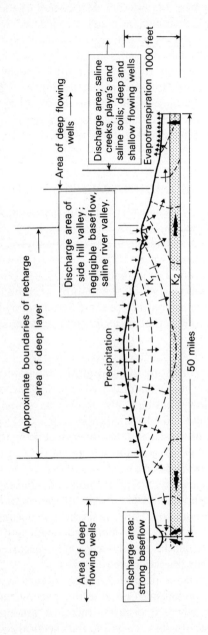

Fig. 6.10 The prairie profile. (*After Meyboom, 1966b.*)

REGIONAL GROUNDWATER FLOW

vegetation of saline soils with respect to the occurrence of local and regional flow systems, and (3) the location of ponds and bogs with respect to groundwater flow.

With regard to (1), Meyboom (1966b) noted that most of the willows are located in the higher areas, which suggests that the recharge area is covered by numerous discharging points, each of which is characterized by a willow ring. The explanation given is that willows have a very low tolerance for saline (alkali) conditions, so their occurrence is associated with waters that have not moved very far in the system. Although their occurrence in the higher watershed areas is widespread, they do occur elsewhere in the basin. These other occurrences are regarded as possible manifestations of local flow systems superimposed on the regional groundwater flow.

With regard to (2), extensive saline soil areas occur in areas of regional groundwater discharge, where a net upward movement of mineralized groundwater takes place. A consistent transition noted is from willow vegetation on the watershed areas to halophytic plant communities within the discharge areas (Fig. 6.11). A related transition is noted for local flow systems that receive water from highly saline formations. Where the local system is not replenished by saline water formations, the groundwater is relatively fresh, and saline zones fail to develop. In the latter case, fresh water phreatophytes occupy the area, and make possible its delineation as an end point of a local system.

In areas of hummocky terrain, ephemeral water bodies have been found to function as recharge points during spring and early summer, and discharge points during summer and fall. On the other hand, permanent lakes are areas of permanent groundwater discharge (Meyboom, 1966c; 1967). Four

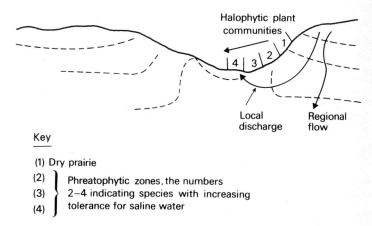

Fig. 6.11 Relation between groundwater movement and sequence of plant communities. (*After Meyboom*, 1966b.)

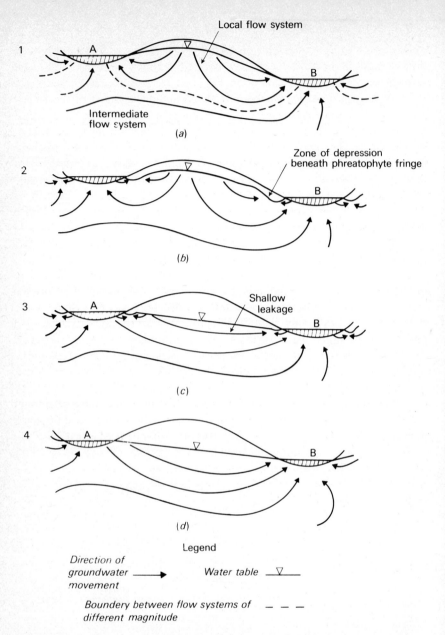

Fig. 6.12 Diagrams of flow conditions near permanent lakes with (*a*) a spring condition of discharge from local and intermediate systems, (*b*) a summer condition of seepage toward the phreatophyte fringe, (*c*) deterioration of local flow in the absence of recharge, and (*d*) a fall and winter condition for the deteriorated system, where shallow movement is superimposed on the intermediate system. (*After Meyboom, 1967. Used with permission from North Holland Publishing Company, Amsterdam.*)

typical flow conditions near permanent lakes are shown in Fig. 6.12, which demonstrate (1) a spring condition of discharge from local and intermediate flow systems; (2) a summer condition of seepage toward the phreatophyte fringe surrounding the lake; (3) a deterioration of local flow owing to insufficient recharge, which produces shallow movement from A to B; and (4) a fall and winter condition for the deteriorated system, where there is shallow movement from lake to lake superimposed on the intermediate flow system.

These studies demonstrate that lakes are dynamic bodies, and the movement of groundwater in their vicinity cannot be described in terms of static analysis. A number of potential measurements gives information about movement only at a particular moment in time. An identical attitude applies also to the chemical character of lakes, where a chemical analysis of a single water sample applies only to a specific set of circumstances (Livingstone, 1963; Garrels and Mackenzie, 1967).

Toth (1966a) reported the results of a related mapping method based on surface and other observations in the prairie environment of Alberta, Canada. Field observations are classified on the basis of whether they pertain to physical environment, such as climate, relief, and geology, or to the occurrence of water. The latter class was subdivided into features that (1) testify to the actual occurrence of water and its chemical properties, such as springs, seeps, groundwater levels in wells, and chemical analyses, and (2) are only associated with the occurrence of water, such as natural vegetation, salt precipitation, and moist and dry surface depressions. All observations were interpreted in terms of whether they represent a deficiency or a surplus of water relative to that available from surface sources. Figure 6.13 is the product of this mapping method, and is in good agreement with what is known from other methods about the distribution of recharge and discharge areas within the region. Note the correspondence of permanent and intermittent streams with areas of natural discharge. In addition, the observations support the midline hypothesis, which suggests that the downslope half of the hillside is the discharge area.

The mapping techniques discussed above are only of value in arid and semiarid regions, where surface water is not sufficiently abundant to mask or conceal the surface effects of groundwater flow.

GREAT BASIN STUDIES

In discussing the basins in the Great Basin and the lakes which occupied them during the Pleistocene epoch, Meinzer (1922) recognized three types: (1) those in which lakes still exist, (2) those which no longer have lakes and are discharging groundwater from the subterranean reservoirs into the atmosphere by evaporation and transpiration, and (3) those which do not have lakes and in which the water table is everywhere so deep that they do not discharge groundwater except by subterranean leakage out of the basin.

The basin of type (2) above is exemplified in early studies of Big Smokey Valley (Meinzer, 1917). Water enters the alluvial basin by influent seepage of streams on the alluvial fan areas, and at the contact between the surrounding mountain ranges and the alluvium. Groundwater is discharged by evaporation and transpiration in the valley lowlands. Studies of the Big Smokey Valley and similar environments provided the first understanding of

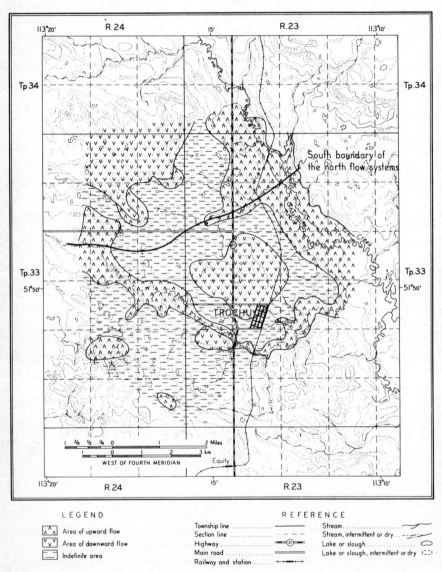

Fig. 6.13 Distribution of areas of upward and downward flow in a prairie environment. (*After Toth*, 1966a.)

the surficial manifestations of groundwater flow, including the distribution of soluble salts in discharge areas (Meinzer, 1927). The concentric arrangement of phreatophytes was first noted in these early studies, with salt grass occupying the inner belt where the water table is near land surface. Other species, such as greasewood in northern Nevada and mesquite in southern Nevada, occupy the outer belt, where the water table is further below land surface (Meinzer, 1927).

The basins of type (3) have been extensively investigated by Winograd (1962), Eakin (1966), and Mifflin (1968), which resulted in the delineation of two regional flow systems in carbonate terrain in southern and eastern Nevada. These systems are typified by large drainage areas encompassing several topographic basins, relatively long flow paths, and large spring areas of invariant discharge, where water temperature is several degrees higher than mean air temperature.

Eakin's (1966) study included the formulation of a water budget within individual basins in an attempt to demonstrate closure on a regional scale, and thereby to account for the magnitude and relative invariance of spring discharge (Fig. 6.14). Major spring areas include White River and the Pahranagat and Moapa Valley, the latter identified as the terminal point of the system. Closure of the water balance was accomplished within a 13-valley area, with 78 percent of the recharge estimated as occurring in the four northern valleys, and 62 percent of the discharge estimated to be from spring areas in Pahranagat and Moapa Valley.

Most of Winograd's (1962) conclusions are based on the results of an extensive drilling program within the Nevada test site and adjacent areas. Potential data suggests that groundwater moves from at least as far north as Yucca Flat to Ash Meadows, without regard to topographic divides (Fig. 6.15). The northern part of the system is a recharge area, receiving water from the overlying alluvial reservoirs. Discharge through spring areas occurs in Ash Meadows, the terminal point of the system. Boundaries to the system, at least to the south, appear to be controlled by thick sequences of clastic rock.

The examples cited above deal with discharge phenomena in conspicuous topographic lows. Such areas reflect the observable manifestations of the theories of groundwater flow discussed in Sections 4.2 and 6.1, and are logical starting places for study of the system of flow contributing to large spring areas in volcanic and carbonate terrain.

6.3 TEMPERATURE OF GROUNDWATER

The study of groundwater temperature within the earth has served many useful purposes. Some interests center on the relation between temperature of groundwater and geologic structures (Jones, 1967), and between temperature and water salinity (Garza, 1962). Theoretical studies have dealt with the simultaneous transfer of heat and water within the earth, and with the relation between this transfer and velocity of groundwater movement (Stallman, 1963; Bredehoeft and Papadopulos, 1965). Other studies have focused on the manner in which groundwater temperature is affected by

Fig. 6.14 General flow directions and recharge-discharge relations for the regional flow system in eastern Nevada. (*After Eakin*, 1966.)

REGIONAL GROUNDWATER FLOW

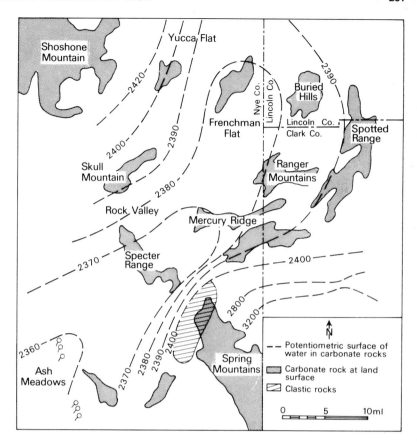

Fig. 6.15 Regional flow systems in rocks underlying the Nevada test site and adjoining area. (*After Winograd*, 1962.)

vegetative cover (Pluhowski and Kantrowitz, 1963), by infiltration of stream water (Rorabaugh, 1956; Schneider, 1962), and by variations in surface temperature (Stallman, 1965). Perhaps the most comprehensive studies deal with high-temperature groundwater and suggest that most, if not all, thermal water studied in detail is meteoric in origin, and therefore partakes in the normal hydrologic cycle (Craig et al., 1954; White, 1957a and 1957b; White et al., 1963; DeGrys, 1965).

Because of the insulating qualities of the earth, the extremes in seasonal air temperature are dampened out, which results in more or less uniform rock temperatures at depths greater than 30 ft. The temperature of groundwater within these units is therefore quite uniform and approximates the mean air temperature of the region. Below this depth, the primary control on temperature is the flow of heat from the earth's interior. The major factor contributing to the spatial redistribution of this heat is circulating groundwater. For this reason, groundwater temperature appears most useful in flow-system studies when considered a relic parameter of the environment through which the water has passed (Mifflin, 1968). Viewed in this manner, temperature depicts in an approximate way the depth of circulation, characterizing the vertical configuration of flow. Where temperature is uniform and approximates mean annual temperature, the depth of circulation is not great, as in local systems and near or within areas of groundwater recharge. Where temperature varies widely in range, both shallow and deep circulation may be present, as in intermediate systems. Where temperature is uniformly high, the groundwater is believed to be deeply circulated, as in regional systems. The accuracy and utility of these interpretations may be compromised by localized concentrations of thermal groundwater often associated with major structural features.

The cited relations between temperature distribution and groundwater movement have been noted during the course of several areal studies. Schneider (1964) found evidence for a very small vertical temperature gradient in carbonate rocks in regions characterized by large vertical components of flow. He suggested that the influx of meteoric water in recharge

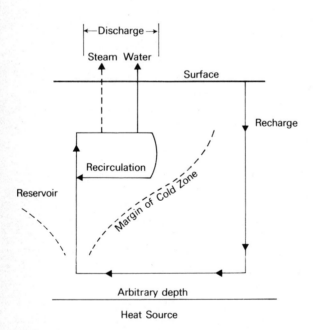

Fig. 6.16 Pipe model of a hydrothermal area. (*After Elder,* 1965.)

areas is sufficient to significantly alter the geothermal gradient. On a regional scale, the horizontal temperature distribution increases with increasing depth, which suggests a manifestation of the regional geothermal gradient. Mifflin (1968), working in the carbonate interbasin systems in Nevada, recognized a direct relation between direction and depth of groundwater flow and increasing temperature of the water. In some areas, thermal springs occur a few hundred feet from springs of normal temperature. Springs with large discharges are invariably characterized by high-temperature water; those with low discharge are of normal temperature. Extensive study of the Neogene deposits beneath the Gulf Coastal Plain revealed a marked increase in the geothermal gradient wherever abnormally high fluid pressures and saline water conditions were encountered (Jones, 1967).

Hydrothermal systems require special consideration, and have been defined as a heat transfer mechanism in the earth's crust. Such systems rely on the transport of water, but not necessarily the discharge of water at the earth's surface. These systems produce at the surface a so-called thermal area in which the heat flow is different from normal (Elder, 1965). The elements of such a system include a heat source, a recharge system, a recirculation system, and a surface-discharge system. Given these elements, a pipe model of a thermal area may be readily constructed (Fig. 6.16).

6.4 CHEMISTRY OF GROUNDWATER

The chemical evolution of groundwater with position in the groundwater flow has been studied from several viewpoints by various researchers, the most notable including Chebotarev (1955), Schoeller (1962), and Back (1961; 1966). Regardless of which viewpoint is examined, observations available to most students of chemical geohydrology relate to two fundamental premises: (1) The concentration of dissolved mineral matter is directly proportional to length of the flow path and to the residence time of the water, and (2) the chemical type of groundwater at any point in the system is a function both of the chemical composition of the rocks at that point and of the antecedent water quality. Beyond these accepted generalizations, interaction between groundwater and geologic environments under a multitude of possible influencing conditions does not permit the establishment of firm rules that hold for all conditions. However, a few concepts, methods, and empirical findings are available which may help relate the chemical character of the water to the geologic environment and prevailing flow pattern.

THE GEOCHEMICAL CYCLE

The primary features of the hydrologic cycle as it influences the geochemistry of natural waters includes the reactions between circulating water and the mineralogy of the environment, appropriately termed the geochemical cycle (Fig. 6.17). The concept was first proposed by Davis and others (1959) to describe the chemical characteristics of water in the San Joaquin Valley, California, but applies with slight modification to other areas. The geochemical cycle includes:

1. The transfer of dissolved mineral matter from the oceans via the vapor phase
2. The addition of nitrogen, oxygen, and carbon dioxide in the atmosphere
3. The consumption of oxygen and release of carbon dioxide, a result of biological organic decomposition, as water percolates through the soil
4. Chemical reactions between the water and rock fragments in the soil, aided by carbonic acid, liberating bicarbonates and carbonates which may be added to the percolating water
5. Numerous chemical and physiochemical reactions between water and rock in the groundwater environment

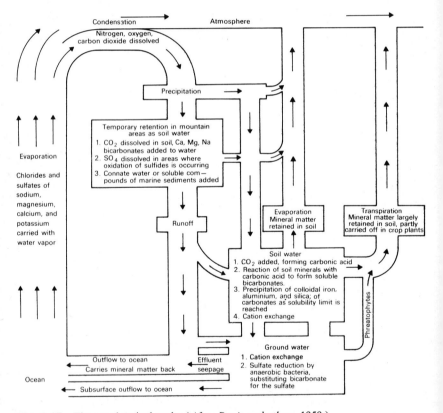

Fig. 6.17 The geochemical cycle. (*After Davis and others,* 1959.)

REGIONAL GROUNDWATER FLOW

6. Addition of other soluble compounds in the soil
7. The return of water to the atmosphere by evaporation or transpiration, leaving the mineral matter behind, or the return of water to the oceans as streamflow, carrying its mineral matter with it.

The geochemical cycle, like the hydrologic cycle within which it operates, can be divided into components with appropriate inputs and outputs. In this case, inputs and outputs no longer represent flows of water, but quantities and types of mineral matter in solution. The role of each component is that of chemical or physiochemical process and transformation, yielding an output that is chemically different from the input. This transformation process was described by Chebotarev (1955) as a "metamorphism of natural waters in the crust of weathering," and is controlled primarily by the movement of natural groundwaters and the mobility of chemical elements within rock units.

Mobility is a term used to describe the ease and tendency with which chemical elements pass into ionic solution. The process is the result of complex chemical and biological factors. Although a general relation exists between the mineral composition of groundwater and the mineral assemblage with which the water has been in contact, the relation may be simple or complex, depending largely on mobility of the constituent elements. For example, silica, aluminum, and iron are abundant in the earth's crust, but are relatively immobile, and are generally present in groundwater in small proportions. Calcium, sodium, and potassium are both abundant and mobile, and commonly form large proportions of the constituents in groundwater. On the other hand, chloride is scarce, but mobile, and is a common constituent of groundwater.

Chebotarev (1955) recognized the following constituents as chief formers of the soluble salts common to most groundwaters:

Cations
    Group A: Na, K, Ca, Mg, H
    Group B: NH_4, Al, Fe
Anions
    Group A: HCO_3, CO_3, Cl, SO_4
    Group B: NO_2, NO_3, SO_3, OH, SiO_3

Group A comprises ions which are constantly present in relatively large proportions, and group B represents ions of subordinate or inconsistent occurrence.

CHEMICAL CLASSIFICATIONS OF GROUNDWATER

Concentration of the common ions in groundwater are generally reported by weight in parts per million. One part per million defines one part by weight of the ion to a million parts by weight of water. For all practical purposes, parts per million and milligrams per liter are numerically equal. Total ionic concentration, or total dissolved solids, is also reported in this manner.

Given the results of several chemical analyses, it is generally simple to group waters together which have dissolved constituents that fall within selected ranges. These groupings provide the basis for most arbitrary classifications. Gorrell (1958), for example, based a simple classification system on ranges in total dissolved solids: (1) fresh water, 0 to 1,000 ppm; (2) brackish water, 1,000 to 10,000 ppm; (3) salty water, 10,000 to 100,000 ppm; (4) brine, greater than 100,000 ppm. Hem (1959), however, recognized that an immediate purpose of routine chemical quality data is to determine whether water is satisfactory for a proposed use. Whether a given quality is acceptable for a particular use depends on arbitrary standards of acceptability for that use. Three main uses are recognized: domestic (drinking water), agriculture (irrigation), and industrial. Drinking water standards are described by upper limits for bacterial, physical (turbidity, color, etc.), and mineralogical constituents. Agricultural standards are made contingent upon the effects of chemical constituents in water on soils and the plant life they support. Industrial standards vary with the particular industry. In all cases, emphasis is placed on the fact that the prescribed standards are not applicable to every situation.

Genetic classification systems have been proposed by White (1957*a*, 1957*b*), and encounter various problems concerning the chemical criteria which may be used to identify a given origin. Genetic terms have been defined as follows:

Meteoric water   Water that has recently been involved in atmospheric circulation
Marine water   Water that has recently invaded rocks along coastlines
Connate water   Water that was buried with sedimentation
Metamorphic water   Water associated with rocks during their metamorphism
Magmatic water   Water derived from a magma
Volcanic water   Water derived from a magma at shallow depth
Plutonic water   Water derived from a magma at considerable depth
Juvenile water   Water that has never been part of the hydrosphere

Meteoric water has an isotopic composition similar to surface water; marine water has an isotopic composition similar to sea water. Of the other types, connate, metamorphic, and magmatic waters are isotopically examined on the basis of their H^2/H^1 and O^{18}/O^{16} ratios.

For flow-system studies, the most useful chemical classification would attempt to relate water to rock types on the basis of dissolved mineral matter.

Hem (1959) discussed the main problems of a classification of this type. These include climate and other influences on weathering processes, which gives rise to different types of water from similar source rocks. For example, it is not totally unexpected to find that groundwater in silicate-rich igneous rocks is rich in the easily dissolvable components of silicate materials. On the other hand, sedimentary rocks vary markedly in chemical composition and the degree of alteration of the constituent minerals. For this reason, subdivisions which reflect these properties are required. Utilizing a classification system devised by Goldschmidt (1937), Hem (1959) proposed:

Resistates Rocks composed of residues not chemically decomposed in the weathering of parent rocks, such as sandstones, conglomerates, and other coarse clastics

Hydrolysates Rocks composed of insoluble products formed by chemical reactions in the weathering of the parent rock, such as shale

Precipitates Rocks formed by chemical precipitation of mineral matter from aqueous solutions, such as limestone

Evaporites Rocks deposited from highly concentrated natural brines consisting of readily soluble compounds

In that this classification is based on mineralogical composition and the degree of alteration of the mineral matter making up the rock, a wide range in the chemical composition of the water in a given class can be anticipated. For example, depending on the degree of weathering, resistates can vary in composition from quartz sands, as in many sandstones, to feldspars and other less resistant mineral species, as in arkose. Although the range in concentrations for each of the subclasses cannot be generalized, Hem (1959) described what may be expected under various conditions.

Classification systems based on rock types are also complicated by the modification of water due to reactions with the rock matrix. Ion exchange and sulfate reduction are often used as hypothetical processes for modifying the chemistry of water. One notable result of the former is the natural softening of water that comes into contact with natural zeolites (certain sodium and potassium aluminum silicates). The softening occurs when calcium and magnesium ions of the carbonates and bicarbonates are replaced by sodium and potassium ions from the zeolites. The process is described by the reaction

$$2Na\ Y + Ca \rightleftharpoons Ca\ Y_2 + 2Na$$

where Y represents a unit exchange capacity of the solid material. Sulfate reduction, on the other hand, not only accounts for diminishing quantities of sulfate, but can add to the alkalinity of the water. A possible reaction between waters containing sodium sulfate and methane, the latter a common constituent of natural gas, is

$$Na_2SO_4 + CH_4 \leftrightharpoons Na_2S + CO_2 + 2H_2O$$
$$Na_2S + 2CO_2 + 2H_2O \leftrightharpoons H_2S + Na_2(HCO_3)_2$$

HYDROCHEMICAL FACIES

The concept of hydrochemical facies offers a means for detection of regional relations between chemical character of groundwater, lithology, and regional flow patterns (Back, 1961). Although specific details concerning methodology vary somewhat, the facies are studied in much the same manner as lithofacies in geology, with Piper's (1944) trilinear diagram, or some slight modification, used to identify the various chemical types. The areal distribution of constituents is shown by fence diagrams, maps showing lines of equal concentrations, or concentration ratios of two constituents (Back and Hanshaw, 1965). The type of facies that develops is controlled largely by the mineralogy of the rocks, and its distribution is controlled by the flow pattern.

Constituents in solution in groundwater may be viewed as a chemical system, with cations and anions in equilibrium with each other. In discussing a graphic method of presentation for this system, Piper (1944) noted the abundance of the cations calcium, magnesium, and sodium; and the anions bicarbonate, sulfate, and chloride. He proposed that ionic solutions of groundwater be treated as though they contained three cation groups and three anion groups. The less abundant cation and anion constituents are summed with the major constituents in accordance with common physical properties, and are therefore accounted for in the plotting method. For example, barium may be combined with calcium; potassium with sodium; fluoride, nitrate, and nitrite with chloride; and carbonate with bicarbonate. In recent years, the sodium constituent has been referred to as the *sodium plus potassium constituent* because of the relative abundance of potassium in natural groundwaters.

To convert the results of a chemical analysis into a form suitable for graphic interpretation, parts per million are converted to equivalents per million (epm) by dividing the atomic weight of an ion by its valence, which gives the equivalent or combining weight, and dividing this value into parts per million. For example, calcium has an atomic weight of 40.08 and a valence of 2. Its equivalent weight is 20.04. The concentration in a water sample containing 80 ppm calcium is then 4 epm. For a complete analysis, the total equivalent weights of cations and anions in solution must be the same. To the extent that natural groundwater can be treated in terms of three cation constituents and three anion constituents, the chemical character can be shown graphically by a single point on a conventional trilinear diagram.

Figure 6.18 shows such a diagram, with various hydrochemical facies designated in the diamond-shape field. In the triangular field at the left, the proportion of each cation constituent is plotted as a single point. Thus, a sample with 3 epm calcium, 2 epm magnesium, and 1 epm sodium and potassium would be plotted in accordance with the percentages 50, 33.3, and 16.66, respectively. The same procedure is followed in the lower-right-hand field for the three anion groups. The central diamond-shape field is used to demonstrate the overall chemical character of the water by a third singular point, which is at the intersection of the rays projected from the plotting of cation and anion points.

The subareas of Fig. 6.18 serve as a basis for specific classification of

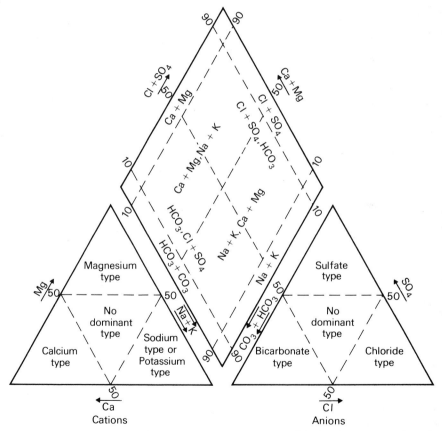

Fig. 6.18 Water analysis diagram showing hydrochemical facies, in percent of total equivalents per million. (*After Back*, 1966.)

Table 6.1 Classification of hydrochemical facies (*After Back*, 1966.)

	Percentage of constituents, epm			
	Ca + Mg	Na + K	$HCO_3 + CO_3$	Cl + SO_4
Cation facies:				
Calcium-magnesium	90–100	0 < 10		
Calcium-sodium	50–90	10 < 50		
Sodium-calcium	10–50	50 < 90		
Sodium-potassium	0–10	90–100		
Anion facies:				
Bicarbonate			90–100	0 < 10
Bicarbonate–chloride–sulfate			50–90	10 < 50
Chloride-sulfate–bicarbonate			10–50	50 < 90
Chloride-sulfate			0–10	90–100

hydrochemical facies, which may be studied in terms of anions, or cations, or both. A particularly useful classification is shown in Table 6.1. The nomenclature, although suggested by Fig. 6.18, is shown more clearly in Fig. 6.19.

Studies attempting to explain the genesis of the chemical character of groundwater make use of diagrams such as that in Fig. 6.18 in combination with knowledge of the mineralogy of the environment, the transmissivity of its component parts, and the distribution of piezometric head. The examples available for study suggest that no quantitative conclusions can be offered without full consideration of these controls. Back (1966), for example, identified a calcium-magnesium facies in recharge areas underlain by calcareous clays, and a sodium facies in discharge areas, the latter a result of ion exchange and salt-water intrusion. Bicarbonate content is low in recharge areas and high in discharge areas. The recharge area of the Englishtown formation in New Jersey is also characterized by a calcium-magnesium facies, followed by the sodium-calcium facies (Seaber, 1962). Toth (1966*b*), from investigations in Canada, described the anion facies development, and reported a general tendency for a shift from a pure bicarbonate facies in recharge areas to a sulfate facies in discharge areas. The explanation given is the utilization of carbon dioxide in the root zone to form bicarbonates in downward-moving groundwaters, and little addition of bicarbonates thereafter. With passage into the system, the sulfate ion, which is available in the rocks of the area, becomes the dominant constituent. In the Lake Bonneville area, hydrochemical data suggest a change from calcium magnesium bicarbonate to sodium bicarbonate with distance from the recharge area (Feth et al., 1966). Sulfate content decreases in the direction of flow, and bicarbonate increases, undoubtedly both occurrences being the result of sulfate reduction. Hence, no quantitative limits can be set that would apply to geochemical systems in general. However, a few general rules have been formulated by Chebotarev (1955), Schoeller (1959), and Back and Hanshaw (1965), more or less on the basis of their wide experiences:

1. There is generally an increase in total mineralization of water in proportion to the length of flow path and duration of content.
2. Within a given system, a major recharge area is characterized by water lower in dissolved solids than in an area where recharge is minor. Similarly, recharge areas may be leached of all soluble mineral matter, and the absence of chemical reactions would prevent the pH of water from becoming higher than that of rain water.
3. It is not uncommon for the ratio of sulfate to chloride to decrease in the direction of flow.

Included within the concept of hydrochemical facies are chemical zonations controlled by depth of circulation, sedimentological zones, and climatic factors (Schoeller, 1962). Chebotarev (1955), in particular, has dealt with vertical zonations as influenced by the dynamic aspects of natural

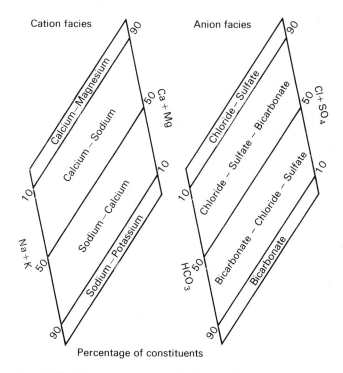

Fig. 6.19 Nomenclature for hydrochemical facies.

water, including velocity of movement, recharge-discharge relations, intensity of movement with depth, and residence time. Empirical evidence suggests an increase in salinity as movement deteriorates with depth, with distance from the recharge area, with nearness to the sea, and with the duration of contact. From this evidence, combined with the ideas on mobility and supported by almost 10,000 chemical analyses, Chebotarev developed his well-known sequence, which states that all groundwaters tend toward the composition of sea water,

$$HCO_3 \rightarrow HCO_3 + Cl \rightarrow Cl + HCO_3 \rightarrow Cl + SO_4 \text{ (or } SO_4 + Cl) \rightarrow Cl$$

This sequence, like many others in the geological sciences, must be viewed in terms of scale and with the normal provisions for interruption and incompletion.

In terms of facies development, Chebotarev's sequence can be viewed as three vertical zones:

1. The *uppermost zone* characterized by a high intensity of circulation through well-leached rocks. Water in this zone is of the bicarbonate type, and is low in mineralization.
2. An *intermediate zone* with less intensive circulation and higher total mineralization. Water in this zone is of the sulfate type.
3. A *lowermost zone* in a near stagnant condition, with virtually unleached rocks and highly mineralized water. Water in this zone is of the chloride type.

The chemical explanation for the sequence is given in Chebotarev's (1955) conclusions:

> While the least soluble salts are precipitated first and the most soluble salts last, at any given time and at any distance from the intake area, the chemical components of higher solubility will be found in water in greater relative abundance. The sodium chloride tends, therefore, to be in solution so long as the extremely high salinity concentration of water is achieved.

The development of a bicarbonate-sulfate-chloride sequence can be compared with the process of mineral formation by evaporation of surface-water bodies. With evaporation, concentration of the soluble salts occurs, and when super saturation with any salt is achieved, that salt is precipitated. The least soluble salts are precipitated first, and the most soluble last, with the order being calcite (bicarbonate), gypsum (sulfate), and halite (chloride). Halite remains in solution until its normal marine salinity of 35,000 ppm has increased to 337,000 ppm. Whereas evaporation is the mechanism of concentration in surface-water bodies, the relative solubility of the rocks in a dynamic flow system is the responsible factor in groundwater basins. With evaporation, a vertical zonation of evaporite deposits is anticipated; with groundwater flow, it is the chemical constituents in solution that reflect zonation.

Schoeller (1959) referred to the above cited sequence as the *Ignatovitch-Souline sequence* and, in agreement with Chebotarev, lists the same sequence of predominant ions.

The detection of certain parts of the Chebotarev sequence in large groundwater basins is inevitable, although the complete sequence may be rarely observed. Instead, different chemical types of water are more often associated with different types of rock or antecedent water quality. Hence, as a general rule, hydrochemical facies will be responsive to a particular physical environment, and no set limits can be advanced without study of the environment. The concept and various techniques associated with it can therefore be usefully applied to gain an understanding of the cause and effect relations between lithology, flow patterns, and chemical character of the water.

CHEMICAL DATA AND FLOW SYSTEM DELINEATION

Groundwater in motion is continually dissolving mineral matter in an attempt to reach equilibrium with its surroundings. Consequently, water chemistry, like temperature, may be considered a relic parameter of the environment, and may reflect the history of the flow path. Although some interpretations will necessarily be based on uncertain assumptions, much of the uncertainty may be removed when chemical data are examined in the context of flow system theory. The following examples are selected to demonstrate this point.

In spite of the fact that the artesian basin of the Dakotas has been studied more extensively than any other basin of comparable size, many questions dealing with observable anomalies remain unresolved. The basin is currently considered of the confined type described in Section 4.2, although this, too, may be challenged as more information becomes available. Pertinent to this discussion is the recent questioning of the widely held idea that recharge enters the equivalents of the Dakota sandstone where they are exposed on the flanks of the Black Hills (Swenson, 1968). According to Swenson's proposed theory, based largely on chemical data and stratigraphic relations, much of the water in the Dakota sandstone in eastern South Dakota enters the cavernous carbonate formations of the Madison group (Mississippian Age) on the flanks of the Black Hills. This water presumably travels eastward about 150 miles through carbonate rocks to move ultimately upward and enter the basal sandstones of the Dakota along the zone shown in Fig. 6.20a. In this area, pre-Dakota erosion has removed much of the

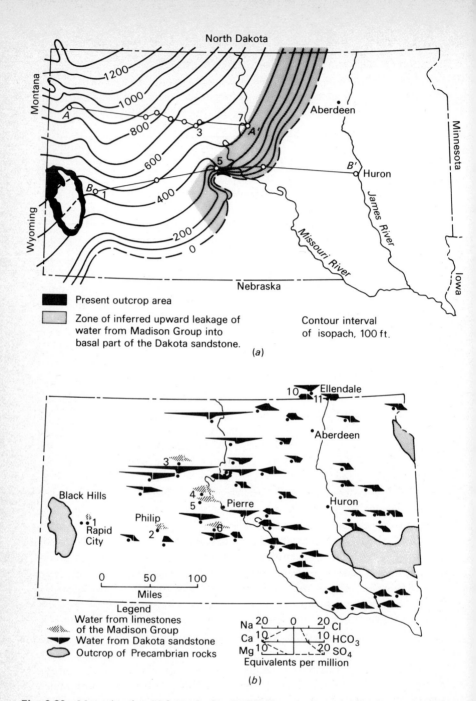

Fig. 6.20 Maps showing (a) isopachs for the Madison group and inferred zone of upward movement and (b) type of water in the Dakota sandstone and limestones of the Madison group. (*After Swenson*, 1968.)

intervening beds between the Mississippian and Cretaceous Dakota rocks. Although the general distribution of piezometric head makes such a hypothesis possible, the system has been extensively disturbed by pumping, which renders a direct use of piezometric data impossible.

All waters of the Madison group are typically of the calcium sulfate type. Water within the Dakota sandstone is of three distinct types:

1. A highly mineralized sodium chloride type east of the Black Hills and west of the hypothesized zone of upward movement. In the vicinity of the Black Hills, the water is less mineralized, and may contain local concentrations of sodium bicarbonate or sodium sulfate.
2. A calcium sulfate type first detectable in the vicinity of the hypothesized zone of upward movement, and extending eastward.
3. A sodium sulfate type east of Aberdeen and Huron, probably the result of natural softening of calcium sulfate waters by base exchange.

Figure 6.20b shows the chemical data by use of diagrams, the distinctive shapes of which facilitate comparison of the types of water. West of the inferred zone of upward flow, water from the Dakota sandstone is distinctively different from water in the Madison group. The calcium sulfate water in the vicinity of the zone is similar in most respects to the Madison water, from which it is believed to have originated. Hence, the relatively fresh water common to the Dakota sandstone in the eastern half of South Dakota is presumably explained by considering a system of flow rather than flow confined to an individual aquifer.

A method for characterizing a flow system by water chemistry of large springs in combination with other supporting data has been described for the regional carbonate systems in Nevada (Maxey and Mifflin, 1966). Water chemistry of springs believed to be associated with regional flow systems characteristically shows an increase in sodium, potassium, sulfate, and chloride with length of flow path (Fig. 6.21a). On the other hand, calcium and magnesium ions rapidly approach equilibrium with carbonate minerals, and achieve a relatively constant concentration (Fig. 6.21b). On the basis of these studies and supporting age determinations of water, Mifflin (1968) suggested that springs may be classified on the basis of their association with local or regional flow systems.

Several other examples of hydrochemical applications to flow-system delineation may be mentioned. Brown (1967) discussed some work in southern Manitoba, where a large supply of potable water is found in an area of highly saline water. Chemical data suggest the water has passed through underlying shales and is probably coming from bedrock highs outside the study area. Toth (1966b) related the chemical character of groundwater to areas of recharge and discharge in a region of southwestern Alberta. By chemical character, Henningsen (1962) distinguished between groundwaters

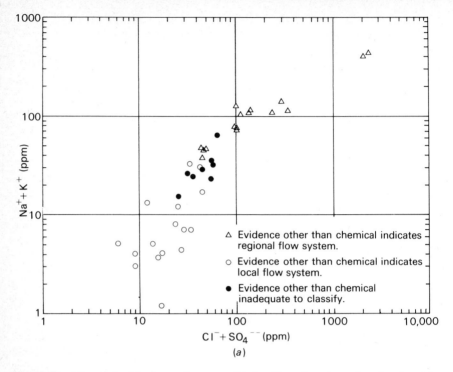

Fig. 6.21 Chemical plots for springs associated with carbonate rocks, showing (a) chloride plus sulfate versus sodium plus potassium and (b) calcium plus magnesium versus sodium plus potassium. (*After Maxey and Mifflin*, 1966.)

in two recharge areas in the Trinity aquifers, Texas. Where the waters intermingle farther down gradient, a third type of water reflects the characteristics of the primary sources, and is readily distinguishable.

6.5 EXTENSIONS FOR ENVIRONMENTAL, ENGINEERING, AND GEOLOGICAL APPLICATIONS

This chapter has thus far been oriented to the flow system as visualized by the groundwater hydrologist—a potential field of well-defined movements, capable of supporting plant growth and surface-water bodies; of dissolving, transporting, and depositing mineral constituents; and of redistributing the heat flux derived from the earth's interior and atmospheric sources. There has been no consideration given to these systems as they relate to man's activities, to their role in the accumulation of valuable minerals, nor to the important geologic work which they might accomplish. It is hardly possible to remedy these omissions completely in a few pages of a single chapter.

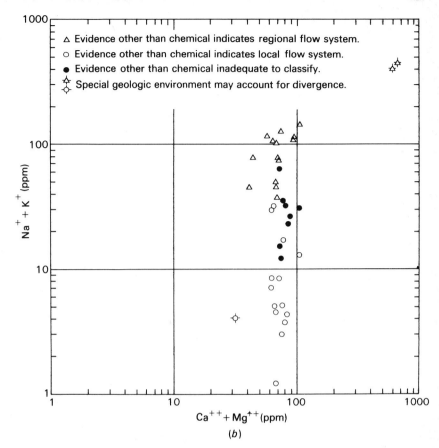

(b)

However, their inclusion, no matter how abbreviated, is in the spirit of the flow-system point of view adopted thus far.

THE IMPACT OF MAN'S ACTIVITIES ON GROUNDWATER FLOW SYSTEMS

As complex societies expand, both in area and population, their impact on the environment expands at an accelerating rate. This impact on the geologic environment is first received by the groundwater system, the subsystem of the lithologic cycle which is most mobile. Ignorance of preexisting flow systems or no anticipation of probable effects of construction can result in environmental modifications that are unnecessarily expensive, inefficient, or potentially destructive.

There are four basic types of projects which can or will modify groundwater flow systems: (1) use of groundwater as a resource, (2) use of the deep subsurface environment to protect the surface environment, (3) allocation of surface and near-surface space for disposal of solid waste products, and (4) engineering construction in which the desired effect is the modification of a surface-water configuration. With items (1) and (2), the modification of groundwater flow is a direct consequence of the major objectives of the project. With items (3) and (4), the modifications take the form of incidental effects not related to major goals. The treatment of item (1) is reserved for the next chapter.

Subsurface waste disposal is the introduction of a liquid at one or more points in a flow system. Because we live in virtual isolation from such environments, their use as a disposal site or buffer from lethal liquids, both chemical and radioactive, is naturally attractive. The goal of such disposal schemes is that of confinement, or containment, of the material to be disposed of. Confinement is the restriction of movement, whereas containment is the preclusion of movement (Drescher, 1965). Fortunately, containment is necessary only in relatively few cases where toxic materials cannot be expected to break down within a reasonable period of time. Such materials are as noxious after several thousand years as they are when disposed of. At present, disposal through wells is not considered for such high-level wastes.

Confinement is a more reasonable level of achievement for a wide variety of liquids, including many acids and bases and low- and intermediate-level radioactive wastes. Acids and bases may be expected to react with mineral species in the host rock, and so be neutralized. Reaction with natural waters in the receiving horizon is not a great problem (Bernard, 1957; Warner, 1969). Radioactive wastes, on the other hand, decay with time, and need only be removed from the surface environment until their activities have declined to acceptable limits. A good example of "time in transient" deactivation is provided by the disposal techniques used at Hanford atomic energy plants, Washington (Schwendiman and others, 1959). For higher-level wastes, flow systems with greater residence times are required, with the materials introduced at a point within the flow that is at a greater distance from the surface.

As most deep saline formations are in interior structural basins or beneath coastal plains, the influence of geology and topography on the flow patterns is already familiar to the reader. Although the injection of wastes may cause a significant increase in hydraulic potential in the injection zone, the regional movement of such wastes will be controlled by the regional flow, and will not greatly exceed the rate of movement of the native fluids. If the waste is a high-density fluid, there will, of course, be a tendency for it to migrate toward the bottom of the formation, and vice versa. Indeed, the relative propensity for the waste to move toward surface sinks is reflected through its potential, which depends, in part, upon its density.

Several other aspects of the disposal problem can be cited briefly. If the medium is anisotropic, the actual flow paths will differ somewhat from the inferred regional flow of the native waters. Dispersion of wastes due to anisotropicity, however, may be beneficial by increasing sorption and dilution (Drescher, 1965). Further, a multitude of geochemical effects are possible, the most notable including solution and depositional mechanisms that can change permeability, and reactions that result in the formation of gas. Temperature changes are of some importance because they affect viscosity, and thereby affect velocity of movement. Temperature changes in the ambient groundwater can also increase the solution and deposition of mineral matter, which can establish openings through which rapid movement may take place (Birch, 1958).

In general, deep sandstones or, more rarely, vuggy limestones used for deep-well disposal act elastically, and increase in porosity to accommodate wastes pumped or injected into them. Hence, whenever liquid waste, particularly waste under pressure, is injected into an aquifer, the proportion of load supported by the liquid increases, which decreases the load on the rocks themselves (Section 5.5). Tremors and shifts may result. Recent, well-publicized experiences in the Denver, Colorado, area demonstrate that such earth movements need not be confined to known-fault or active-earthquake regions.

Gas storage structures are much like deep-well waste disposal projects. The gas is usually placed in some type of structural trap caused by faulting, or in a domal structure in the rock. The gas is placed in sandstone or vuggy limestone and dolomite, since these rocks have sufficient permeability and porosity to accept and hold the gas. An impermeable cap rock is mandatory. Gas is injected to force the water from the pores in the rock. This water is usually high in mineral content, and therefore not potable.

An understanding of the configuration of the groundwater flow in potential gas storage sites is essential. Prevention of pollution of the aquifers above the gas storage reservoir is the major problem. The low density of the gas will cause it to rise in the reservoir to the impermeable (to gas) cap rock. Some migration laterally is inevitable, and discontinuities in the cap rock can be disastrous.

One of the major types of projects which can have secondary effects on groundwater is the surface disposal of solid wastes in sanitary landfills. The sanitary landfill is a means of controlled dumping where, ideally, sites are given a preliminary investigation to determine direction of water movement, surface points of discharge and recharge, depth to the water table, and local and regional aquifers which might be affected by the operation. Hence, the efficient construction and operation of landfills are based on simple hydrological principles. In arid climates, it is relatively easy for landfills to be placed above the regional water table. Since rainfall is low, no serious effects are anticipated. In more humid climates, a leachate may be generated when water moves through the refuse.

Hughes (1967) described leaching as the movement of water, present either initially in the fill or from percolating rain or groundwater, that leaches the soluble materials in the refuse. Carbon dioxide, produced as the refuse decomposes, dissolves in the water, and forms a weak acid. The acid facilitates the solution and mobilization of some potential contaminants. Once water or leachate reaches the zone of saturation, it moves with the local or regional flow. This leachate transmits bacterial pollutants, and is a potential pollution hazard. Hughes and others (1969) concluded that leachate quality improves both with distance from, and depth below, the landfill, and with age of the landfill. This suggests that the leachate is rapidly dispersed, or that favorable chemical reactions take place, or both.

Flow-system study is essential in choosing a site for a new waste-disposal facility, or for reexamination of existing sites. Important considerations include knowledge of where the present system will carry the leachate, of the chemical reactions that are possible or likely, and of the possible changes in the systems and leachate paths brought about by the landfill itself (Hughes and others, 1969). Frequently, if refuse disposal creates a low relief mound, springs of leachate may be generated at its perimeter. In essence, this reflects the establishment of a man-made local flow system. Also, the placement of a landfill in either a recharge or discharge area will have different effects. If leachate is produced in a recharge area, its movement will not present a serious problem provided it does not reach a usable aquifer or surface-water resource, or it becomes sufficiently decontaminated before reaching such resources. The idea here is quite analogous to that of confinement discussed in deep-well disposal. Leachate produced in discharge areas cannot seriously contaminate aquifers, but may affect surface-water quality. In Illinois, a low-permeability layer ranging in thickness from 30 to 50 ft is recommended between the base of the landfill and the shallowest aquifer. Cartwright and Sherman (1970) present several criteria that may be used for selecting landfill locations.

Another important modification of groundwater flow occurs in response to engineering construction. Although many construction projects can affect flow systems locally, dams and canals have the greatest potential for affecting large areas. Indeed, in any hydraulic project where an alteration of the surface-water configuration is a desired objective, a corresponding alteration of the groundwater configuration must be expected. The degree of alteration, however, is not a simple function of the linear dimensions of the project. Raising the water level of a stream by 1 ft over a 100-mile reach will not materially affect the configuration of groundwater flow. However, a deep reservoir created by a high dam may have widespread effects on local and even regional flow systems.

The interrelations between groundwater flow and engineering construction can be examined from several points of view. The stability of a slope, for example, is greatly influenced by groundwater flow, and the success of methods employed for stabilization depends on knowledge of the flow. As topographically low positions often function as groundwater sinks, the slopes surrounding regional topographic lowlands are likely to be part of a groundwater discharge area. The presence of low-permeability units overlying high-permeability units within slopes in discharge areas suggests a condition that is particularly susceptible to development of high fluid pressures. The slopes of the Bow valley near the city of Calgary, Canada, have such geologic conditions, and the valley is noted for its landslide problem (Meyboom, 1961). The operating mechanism, of course, is the reduction in resistance to sheer with increase in fluid pressure [Eq. (5.87)]. During periods of heavy rainfall, the landslide problem becomes particularly acute owing to the development of small local systems on the slopes themselves.

A thorough understanding of groundwater flow is essential to the success of large water-storage projects. If a reservoir is constructed in a discharge area (as is usually the case), and if the water level in the reservoir is above the regional groundwater level, the reservoir will act as a new point of recharge. In the vicinity of the dam, the new gradients will be very steep, and result in active movements through natural materials on either side of the dam, or through deposits under the dam. The spacing of the flow lines and equipotential lines in deposits under the dam will be further affected by the increase in flux and by the detail of the subsurface, which will cause potential stability problems (Terzaghi, 1960). At downstream points, the added flux may go into groundwater storage, and thus raise the water table with possible effects on agricultural lands (Cady, 1941). Eventually, the total discharge from the system will be increased by the amount of added recharge.

If a high-level reservoir will act as a new source of recharge, man-made navigable waterways excavated below the water table will establish new lines of discharge. The sea-level canal is a case in point. By their very nature, sea-level canals are built in coastal areas of low topographic relief, such as the Florida and Panama peninsulas. The natural equilibrium position of the freshwater-saltwater interface is already at a fairly high elevation, although still below sea level. Excavation below the water table permits the canal to act as a natural drain, and so will allow the interface to rise. In a preliminary investigation of a sea-level canal across Florida, Paige (1938a) contended that these effects would be minimal, and would occur only in the vicinity of the canal. His opponents, however, contended that potentials would be lowered for several miles, and salt water would rise to unacceptable levels (Paige, 1938b). Although Hubbert (1940) demonstrated that Paige was essentially correct in his analysis, none of the principals involved anticipated the amount of groundwater development that would take place in future years. Observations by Brown and Parker (1945) along a completed portion of the canal showed that saltwater encroachment into surface alluvial aquifers had been facilitated by hydraulic gradients created by pumping wells.

Most, if not all, navigable waterways are in environments of groundwater discharge, and the groundwater resources are protected from pollutant spills because hydraulic gradients are toward surface water bodies. This assumes, of course, a natural flow undisturbed by pumping. On the other hand, canals excavated above the water table can be a direct source of groundwater pollution. Here, again, the destination of these pollutants is controlled largely by the regional flow pattern.

MIGRATION AND ACCUMULATION OF HYDROCARBONS

Commercial accumulations of oil and gas often occur in clusters, or with well-defined trends. Outside of these trends, numerous dry holes attest to the lack of commercial hydrocarbons in otherwise productive rocks. As pointed out by Hodgson and others (1964), hydrocarbon accumulation from widely dispersed hydrocarbons in sedimentary environments takes place through an interplay between the hydrologic cycle and a corresponding hydrocarbon cycle. In that the processes leading to the formation of a hydrocarbon are reasonably well known, attention will be focused on their migration and accumulation in water-saturated environments.

In order to use some of the ideas on flow systems in hydrocarbon exploration, three significant features must be understood. First, as mentioned above, it is important to distinguish between primary migration from the source rock, in which the hydrocarbon presumably originated (largely shales), and the movement and entrapment in reservoir rocks themselves (permeable media). The forces acting on oil in source rocks are largely capillary in nature, with a resulting impelling force exerted in the direction of decreasing capillary pressure, that is, in the direction of the steepest rate of increase in the grain size of the rock. Hence, a sand-shale boundary may be considered an impermeable barrier to oil present in the sand, but not to water. Oil or gas can flow across the boundary in the direction of the sand, but cannot flow from the sand to the shale unless its pressure is greater than the opposing capillary pressure of oil in shale.

Second, the versatility of the concept of potential should not be overlooked in multiple-phase flow systems. In particular, the quantity known by the term *potential* is determinable not only in the space occupied by a given fluid, but also at any point capable of being occupied by the fluid (Hubbert, 1953). This has already been demonstrated in Section 4.4, where the potential of salt water was expressed in terms of the potential of fresh water. Since groundwater acts as a transport medium for a hydrocarbon, a logical procedure would require the determination of the energy field for oil or gas in terms of the energy field of the groundwater.

REGIONAL GROUNDWATER FLOW

Third, the term *hydrodynamic* refers broadly to the motion of fluids, and at least four mechanisms of fluid motion may be postulated (Chapter 4): thermal osmosis, electroosmosis, chemical osmosis, and gravitational flow. In addition, a fifth mechanism, through the dissipation of excess-fluid pressures, is possible (Sections 7.7 and 7.8). Of these mechanisms, very little is known about oil migration due to the dissipation of fluid pressures, and thermo- and electroosmosis are considered negligible. Hence, only chemical osmosis and gravitational flow are of sufficient significance to warrant discussion. Whichever of these mechanisms is operative, it is recognized that groundwater moves from areas where its energy is high to areas where it is low. As the groundwater acts as a transport medium for the hydrocarbon, the hydrocarbon in its dispersed state will come to rest in any environment in which it is completely surrounded by higher energy levels, or jointly by impermeable boundaries and higher energy levels (Hubbert, 1953).

Formation waters in many deep aquifers have been mapped as flowing inward from topographic highs from all directions (Hill and others, 1961; Hitchon, 1969b). In general, the low-potential area toward which water is moving is given in two dimensions by a closed equipotential line (Fig. 6.22). Along with anomalous salinity and low-pressure zones, the closure of equipotential lows suggests that water is flowing cross-formationally to other

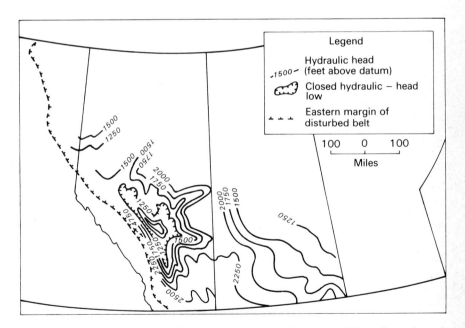

Fig. 6.22 Hydraulic-head distribution in the Lower Cretaceous Viking formation of Western Canada. (*After Hitchon*, 1969b.)

hydrologic units. The flow, of course, is attributed to osmotic withdrawal. Experimental evidence suggests that, for highly compacted shale membranes having an abundance of clay minerals, osmotic pressures across the shale can be approximately 12 to 15 lb/in.² for each 1,000 ppm difference in water salinity (Hill and others, 1961). Of importance to this discussion is that a closed equipotential line in a low-pressure osmotic region may be a stable, minimum-energy position for hydrocarbon accumulation for the following reasons:

1. The transport medium is directed toward such lows.
2. The low-potential region is bounded in two dimensions by higher-energy barriers, and in a third direction by an impermeable (to oil) shale membrane.

Hill and others (1961) discuss in detail some potentiometric surfaces that appear to be modified by osmotic withdrawals. The large Mesa Verde gas accumulation in the Point Lookout sandstone of the San Juan basin, New Mexico, is described as a hydrodynamically entrapped hydrocarbon.

The second mechanism of migration and accumulation is founded on the concepts of gravitational flow. At least two approaches to the exploration problem may be examined here. The first has been cited by Hodgson and Hitchon (1966), who suggested that the diminution of hydrocarbons in a dynamic water system and the entry of the water into a quasistagnant region are of first-rank importance in the accumulation of hydrocarbons. The problem, then, reduces to evaluating the relations of recharge, discharge, and topography, and to ascertaining those areas in the flow system that are of a quasistagnant nature. It is recognized, of course, that as long as the water table is not a flat surface, each molecule of water has the potential to move toward a system sink, and stagnant regions cannot exist. However, regions where two or more flow systems meet are characterized by slow movement, and approach a quasistagnant condition. Such regions may be determined by a combination of fluid-potential slice maps and cross sections (Hitchon, 1969a; 1969b).

The second approach to this problem is based on an expanded concept of gravitational flow as presented in detail by Hubbert (1953). The more salient features of this paper are best discussed within a framework of scalar potentials and impelling forces. First, it is clear that an element of petroleum in a state of migration in reservoir rocks will possess a potential Φ_o, which is a function of position and environment

$$\Phi_o = gz + \frac{P}{\rho_o} \tag{6.31}$$

where ρ_o is density. This element will be acted on by an impelling force $\mathbf{E}_o$ per unit fluid mass, or

$$\mathbf{E}_o = -\text{grad } \Phi_o = \mathbf{g} - \frac{1}{\rho_o}\text{grad } P \tag{6.32}$$

Equation (6.32) merely states that at every point in space in which the fluid potential is not constant, a force of intensity $\mathbf{E}$ will tend to act on the fluid, driving it in the direction of decreasing potential. The force $\mathbf{E}$ is the vector sum of two independent forces, gravity and the negative gradient of the pressure divided by the density of the fluid.

Similarly, equations may be stated for the ambient groundwater, which are

$$\Phi_w = gz + \frac{P}{\rho_w} \tag{6.33}$$

and

$$\mathbf{E}_w = -\text{grad } \Phi_w = \mathbf{g} - \frac{1}{\rho_w}\text{grad } P \tag{6.34}$$

respectively. Solving Eq. (6.33) for the pressure P and substituting this result in Eq. (6.31) gives

$$\Phi_o = \frac{\rho_w}{\rho_o}\Phi_w - \frac{\rho_w - \rho_o}{\rho_o}gz \tag{6.35}$$

Solving Eq. (6.34) for $-\text{grad } P$ and substituting the result in Eq. (6.32) gives

$$\mathbf{E}_o = \mathbf{g} + \frac{\rho_w}{\rho_o}(\mathbf{E}_w - \mathbf{g}) \tag{6.36}$$

Equation (6.35) expresses the potential of oil at every point in terms of the potential of water and the elevation z; Eq. (6.36) expresses the impelling force of oil in terms of the impelling force of water and of gravity. In a similar manner, the impelling force of gas may be expressed

$$\mathbf{E}_g = \mathbf{g} + \frac{\rho_w}{\rho_g}(\mathbf{E}_w - \mathbf{g}) \tag{6.37}$$

Here, the subscript g stands for the gaseous state.

Arriving at these relations, Hubbert (1953) pointed out two possible lines of attack. By utilizing Eq. (6.35), one can map the family of oil equipotential surfaces whose negative gradient will give the force vector $\mathbf{E}_o$. Or, one can utilize Eq. (6.36) to obtain $\mathbf{E}_o$ directly. The first suggestion is appropriate in field applications, whereas the second is important for an understanding of the mechanics of petroleum migration.

306 THE GROUNDWATER BASIN AS A DISTRIBUTED-PARAMETER SYSTEM

If Hubbert's reasoning is followed for the second line of attack discussed above, the vector terms of Eqs. (6.34), (6.36), and (6.37) are plotted graphically from a common origin (Fig. 6.23). If $\mathbf{E}_w$ is either zero (hydrostatics) or vertical, $\mathbf{E}_o$ and $\mathbf{E}_g$ will be vertical. For the hydrostatic case, this result is in agreement with the familiar gravitational theory of oil and gas accumulation with a conformable horizontal interface (Section 4.4). On the other hand, if $\mathbf{E}_w$ is neither zero nor vertical, $\mathbf{E}_o$ and $\mathbf{E}_g$ will be tilted away from the vertical. If the density of oil or gas is made to approach the density of water, $\mathbf{E}_o$ and $\mathbf{E}_g$ will tend to coincide with $\mathbf{E}_w$, and the hydrocarbons will migrate in the direction of the groundwater. As the density of the gas approaches the density of the oil, the hydrocarbons tend to migrate together. As the densities diverge greatly in value, so do the flow directions. For water flowing directly down a steeply dipping reservoir bed, it is possible for the gas to migrate updip, and the oil downdip, with an ultimate stable position in two different traps.

A more direct approach is possible by recognizing that

$$\Phi_o = gh_o \qquad \Phi_w = gh_w \tag{6.38}$$

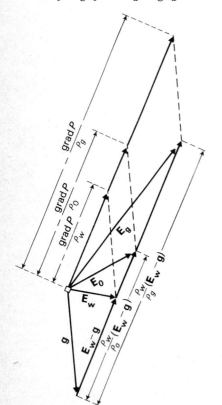

Fig. 6.23 Impelling forces on water, oil, and gas in a hydrodynamic environment. (*After Hubbert, 1953. Used with permission of The American Association of Petroleum Geologists.*)

REGIONAL GROUNDWATER FLOW

Substituting these results in Eq. (6.35) gives

$$h_o = \frac{\rho_w}{\rho_o} h_w - \frac{\rho_w - \rho_o}{\rho_o} z \qquad (6.39)$$

with every term expressed in units of length. For this case, z may be taken as a point on a structure contour map of a reservoir, and h_w as the height above datum to which water would stand in a well terminated at the point z, that is, a point on the potentiometric surface. Each of these surfaces can be contoured after appropriate modification by the density amplification factors of Eq. (6.39). Subtraction of one surface from the other gives the family of curves h_o = a constant. For reasons similar to those discussed in the osmotic case, a hydrodynamic trap is present wherever low values of h_o close completely upon themselves.

PALEOHYDROLOGY

Paleohydrology is based on a concept of uniformitarianism, which means, hydrologically, that ancestral flow systems must have operated throughout geologic time in much the same manner as those observed in modern times. In many respects, paleohydrology may be considered an extension of the methodology introduced in the section on surface manifestations of flow (Section 6.2), with the exception that the features observed are presumed to be the result of ancestral flow systems. In general, one observes certain features in the rock and makes inferences on the nature of the flow that must have existed to cause them.

In many instances, more questions are raised than are answered. The question of the formation of redbeds is a case in point. If one accepts the position of Walker (1967) that the redistribution of iron is best explained by variations in the redox potential and pH of the interstitial fluids, the question of ancestral flow becomes somewhat indeterminate. Where an environment is such that the interstitial water lies in the stability field for ferrous ions (Fig. 4.28), iron in solution will remain in solution and migrate with the flow. Where the interstitial environment lies in the field of $Fe(OH)_3$, the iron will precipitate. Reversals, however, are possible whenever Eh/pH conditions change. Hence, sediments may become depleted and enriched several times throughout their history. As these reversals are not registered in the rock, very little can be said about the paleoflow systems. What can be done, however, is to examine Eh/pH relations in modern flow systems.

Another difficult question arises in speculating about the role that groundwater plays in the formation of nonmarine evaporites. The evidence presented by the prairie profile and Great Basin studies suggests that salt accumulation in discharge areas is a rule rather than an exception in closed basins in arid to semiarid regions. Indeed, Langbein (1961) has stated that saline residues are so common in discharge areas of closed basins that the absence of such residues may be taken to indicate that the basin is not closed hydrologically. The importance of deflation in removing the salts is uncertain in that transport of silt and sand is observed at the present time, but areas with a salt crust appear to yield little airborne material. In the mudflats of the Great Salt Lake basin, Utah, the rate of upward discharge is calculated by actually measuring the rate of salt crust accumulation (Feth and Brown, 1962). However, it is also common for surface waters derived from occasional bursts of rainfall to accumulate in low-discharge areas. These waters are usually heavy in brines. Hence, whether some extensive nonmarine evaporite deposits may be attributed to ancestral groundwater flow systems remains speculative.

An interesting line of evidence concerning greater flux in ancestral flow systems is provided by geologic manifestations of groundwater discharge during the pluvial periods in the Pleistocene. The term *pluvial* refers to intervals of geologic time when, because of climatic variations, more moisture was available for precipitation and recharge. In several of the discharge valleys of the great carbonate flow system in eastern Nevada (Fig. 6.14), deposits of calcareous, fine-grained sediments extend well beyond the limits of modern groundwater discharge (Mifflin, 1968). These deposits are very similar to those forming in modern discharge areas, but their distribution places them many feet above the present elevation of groundwater discharge. These deposits give indirect evidence of greater flux in Pleistocene flow systems.

In addition to such theories of mineralization, which depend largely on some knowledge of paleoclimates in unglaciated regions, a mechanism of mineralization requiring a flow-system reversal has been described (McGinnis, 1968). Consolidation of basin shales due to the weight of a continental ice sheet and the hydraulic connection between the base of an ice sheet and water in bedrock are presumed capable of providing a temporary head that could cause reversals of a preglacial flow system. Sedimentary basin brines could then be discharged in high regions near the margins of continental ice sheets. Owing to the transport of heat by water moving out of deep basins at high rates, the bedrock at the point of discharge is capable of reaching high temperatures. This mechanism has been described by McGinnis (1968) to account for zinc-lead deposits of the Mississippi Valley that are found on broad uplifts which are unrelated to igneous activity.

Perhaps the most definitive paleohydrologic studies are those treating of the variations in permeability of a limestone terrain. As stated by Stringfield and LeGrand (1966), the question becomes: "What features of limestone hydrology are related to present hydrogeologic conditions, and what features have been retained from earlier times?" The authors have set about answering these questions by examining the present circulation system in the southeastern states, and the geological history following deposition.

The distribution of the cavities and their general decrease in size and numbers with respect to present recharge and discharge areas indicate that the present pattern of circulation occurred when sea level was below its present position during the Pleistocene. The geological history is reflected in the paleocirculation systems themselves (Stringfield and LeGrand, 1966): (1) If the deposits overlying the limestone are of low permeability and are thick, and if the limestone was never elevated into a groundwater circulation system, little secondary permeability will develop; (2) a lack of overlying deposits and early elevation into a groundwater circulation system will lead to development of secondary permeability, and to partial or complete removal of the limestone.

6.6 CONCLUDING STATEMENT

The whole burden of flow system theory seems to consist of this: from the concepts of potential theory, to investigate the regional movement of groundwater, and then from this movement, to demonstrate the other phenomena. Quantitative measurements or observations include those on the distribution of potential, chemical characteristics, temperature, and surface manifestations of flow. The potential distribution may be measured directly, or may be generated from models of geologic basins. If the latter course is chosen, field investigations are required to ascertain the boundaries of the system and the permeability distribution, and to verify the accuracy of the model.

Chemistry and temperature are not quantified in the sense that they may be "generated" from theoretical models, and they are generally determined in the field. Along with the surface manifestations of flow, they constitute the "other phenomena" that must be viewed within the concept of a single system of flow, rather than independent measurements or observations. Logical extensions include consideration of flow systems in relation to man's activities, their role in the accumulation of valuable minerals, and the geologic work which they accomplish.

PROBLEMS AND DISCUSSION QUESTIONS

6.1 List the methods that can be used to obtain regional flow patterns. Briefly discuss each method.

6.2 What is the main difference between analytic and finite-difference flow models? Between continuous- and lumped-element electrical analog models? Discuss or point out further analogies between lumped-element analogs and models based on finite-difference mathematics. Between continuous analogs and models based on analytic mathematics.

6.3 Briefly explain the physical meaning of the boundary conditions in Fig. 6.2. What boundary conditions are included in Fig. 6.5 that are not in Fig. 6.2? Why?

6.4 Briefly discuss the conditions under which local, intermediate, and regional flow systems may develop in homogeneous terrain.

6.5 Briefly discuss how the following geologic factors can contribute to interbasin flow:
   a. Rolling topography superimposed on a regional slope
   b. A lenticular high-permeability unit, extending in the subsurface over two or more basins
   c. A regionally extending high-permeability unit at considerable depth

6.6 What is meant by surficial features of groundwater flow? Why are these features important in reconnaissance and mapping studies in regions where water wells are scarce?

6.7 What factors are responsible for a diminishing sulfate/chloride ratio in the direction of groundwater flow?

6.8 Explain how equilibrium concepts (Section 4.5) in combination with base exchange and sulfate reduction reactions are completely adequate to account for the development of a bicarbonate, sulfate, chloride sequence in groundwater flow.

6.9 Examine Fig. 6.21, and answer the following questions:
   a. Is there any reason to expect that any of these ions in solution are in chemical equilibrium with mineral matter of the rock? Explain.
   b. What is the evidence for or against the process of base exchange taking place?
   c. What is a logical explanation for the increase of K and Na with distance in the flow system?

6.10 Describe in some detail the methodology you would apply in a field-oriented approach to flow-system delineation to satisfy the stated objectives in the following environments. Familiarity with any given region can be obtained from published material not included in the text.
   a. A region where lakes, ponds, and bogs are relatively abundant, such as Minnesota, for which it is desired to determine which surface water bodies are points of recharge and which are points of discharge. The overall objective is to determine which of the surface water bodies are permanent, and which may diminish significantly in the event of long-term drought.
   b. Saucerlike structural basins, such as the Illinois basin, which have outcrops along parts of the periphery and numerous deep and shallow observation wells, but little or no surface evidence of groundwater discharge. The main objective is to identify and delineate the areas of natural discharge.
   c. A volcanic terrain, such as the Columbia River plateau basalts, where the main objective is to delineate the interbasin flow regions contributing to large spring discharges.

REFERENCES

Allen, D. N. G.: "Relaxation Methods," McGraw-Hill Book Company, New York, 1954.

Back, W.: Techniques for mapping of hydrochemical facies, *U.S. Geol. Surv., Profess. Papers,* 424-D, pp. 380–382, 1961.

———: Hydrochemical facies and groundwater flow patterns in northern part of Atlantic Coastal Plain, *U.S. Geol. Surv., Profess. Papers,* 498-A, 1966.

——— and B. Hanshaw: Chemical geohydrology, in V. T. Chow (ed.), "Advances in Hydroscience," Academic Press, Inc., New York, vol. 2, pp. 49–109, 1965.

Bernard, G. G.: Effects of reactions between interstitial and injected waters on permeability of reservoir rocks, *Illinois State Geol. Surv., Bull.* 80, pp. 86-98, 1957.
Birch, F.: Thermal considerations in deep disposal of radioactive waste, *Natl. Acad. Sci. Natl. Res. Council, Publ.* 588, Washington, D.C., 1958.
Bredehoeft, J. D., and I. S. Papadopulos: Rates of vertical groundwater movement estimated from the earth's thermal profile, *Water Resources Res.*, vol. 1, no. 2, pp. 325-328, 1965.
Brown, I. C.: Introduction chap. 1, in "Groundwater in Canada," *Can. Dept. Mines Tech. Surv., Geol. Surv. Can., Econ. Geol. Ser.* 24, pp. 1-30, 1967.
Brown, R. H., and G. G. Parker: Salt-water encroachment in limestone of Silver Bluff, Miami, Florida, *Econ. Geol.*, vol. 40, pp. 235-262, 1945.
Cady, R. C.: Effect upon groundwater levels of proposed surface-water storage in Flatland Lake, Montana, *U.S. Geol. Surv., Water Supply Papers,* 849b, pp. 51-81, 1941.
Carlston, C. W., L. L. Thatcher, and E. C. Rhodehamel: Tritium as a hydrologic tool, the Wharton Tract study, *Intern. Assoc. Sci. Hydrol.*, Publ. no. 52, pp. 503-512, 1960.
Cartwright, K., and F. B. Sherman, Groundwater and engineering geology in setting of sanitary landfills: AIME, preprint no. 70-I-57, 1970.
Chebotarev, I. I.: Metamorphism of natural waters in the crust of weathering, *Geochim. Cosmochim. Acta,* vol. 8, pp. 22-48, 137-170, and 198-212, 1955.
Craig, H., G. Boato, and D. E. White: Isotopic geochemistry of thermal waters (abstr.), *Bull. Geol. Soc. Amer.*, vol. 65, p. 1243, 1954.
Davis, G. H., and others: Groundwater conditions and storage capacity in the San Joaquin Valley, California, *U.S. Geol. Surv., Water Supply Papers*, 1496, 1959.
DeGrys, A.: Some observations on the hot springs of central Chile, *Water Resources Res.*, vol. 1, no. 3, pp. 415-428, 1965.
Drescher, W. J.: Hydrology of deep-well disposal of radioactive liquid wastes, in "Fluids in Subsurface Environments," Amer. Assoc. Petrol. Geologists, Tulsa, Oklahoma, pp. 399-406, 1965.
Dusinberre, G. M.: "Numerical Analysis of Heat Flow," McGraw-Hill Book Company, New York, 1949.
Eakin, T. A.: A regional interbasin groundwater system in the White River area, southeastern Nevada, *Water Resources Res.*, vol. 2, no. 2, pp. 251-271, 1966.
Elder, J. W.: Physical processes in geothermal areas, in "Terrestrial Heat Flow," *Geophys. Monogr.* no. 8, Amer. Geophys. Union, pp. 211-239, 1965.
Fayers, F. J., and J. W. Sheldon: The use of a high speed digital computer in the study of the hydrodynamics of geologic basins, *J. Geophys. Res.*, vol. 67, no. 6, pp. 2421-2431, 1962.
Feth, J. H., and R. J. Brown: Method of measuring upward leakage from artesian aquifers, using rate of salt-crust accumulation, *U.S. Geol. Surv., Profess. Papers*, 450-B, pp. 100-101, 1962.
―――― and others: Lake Bonneville: Geology and hydrology of the Weber Delta district, including Ogden, Utah, *U.S. Geol. Surv., Profess. Papers*, 518, 1966.
Freeze, R. A.: Theoretical analysis of regional groundwater flow, Ph.D. thesis for hydraulic engineering, University of California, 1966.
―――― and P. A. Witherspoon: Theoretical analysis of regional groundwater flow, I: Analytical and numerical solutions to the mathematical model, *Water Resources Res.*, vol. 2, pp. 641-656, 1966.
―――― and ――――: Theoretical analysis of regional groundwater flow, II: Effect of water table configuration and subsurface permeability variations, *Water Resources Res.*, vol. 3, no. 2, pp. 623-634, 1967.

Garrels, R. M., and F. T. Mackenzie: Origin of the chemical compositions of some springs and lakes, in "Equilibrium Concepts in Natural Water Systems" (Werner Stumm, Symp. chm.), *Advan. Chem., Amer. Chem. Soc., Ser.* 67, pp. 222–242, 1967.

Garza, S.: Recharge, discharge, and changes in groundwater storage in the Edwards and associated limestones, San Antonio area, Texas: A progress report on studies, 1955–1959, *Texas Board Water Engrs., Bull.* 6201, 1962.

Goldschmidt, V. M.: The principles of the distribution of the chemical elements in minerals and rocks, *J. Chem. Soc. London*, pp. 655–673, 1937.

Gorrell, H. A.: Classification of formation waters based on sodium chloride content, *Bull. Amer. Assoc. Petrol. Geologists*, vol. 42, p. 2513, 1958.

Harr, M. E.: "Groundwater and Seepage," McGraw-Hill Book Company, New York, 1962.

Hem, J. D.: Study and interpretation of the chemical characteristics of natural water, *U.S. Geol. Surv., Water Supply Papers*, 1473, 1959.

Henningsen, E. R.: Water diagenesis in Lower Cretaceous Trinity aquifers of central Texas, *Baylor Geol. Studies, Bull.* 3, 1962.

Hill, G. A., W. A. Colburn, and J. W. Knight: Reducing oil finding costs by use of hydrodynamic evaluations, in "Economics of Petroleum Exploration, Development and Property Evaluation," Prentice-Hall, Inc., Englewood Cliffs, N.J., pp. 38–69, 1961.

Hitchon, B.: Fluid flow in the western Canada sedimentary basin, Pt. I: Effect of topography, *Water Resources Res.*, vol. 5, no. 1, pp. 186–195, 1969a.

———, Fluid flow in the western Canada sedimentary basin, Pt. II: Effect of geology, *Water Resources Res.*, vol. 5, no. 2, pp. 460–469, 1969b.

Hodgson, G. W., B. Hitchon, and K. Taguchi, The water and hydrocarbon cycles in the formation of oil accumulations, in "Recent Researches in the Fields of Hydrosphere, Atmosphere, and Nuclear Geochemistry," Maruzen Co. Ltd., Tokyo, Japan, pp. 217–242, 1964.

——— and ———: Research trends in petroleum genesis, *Commonwealth Mining and Met. Congr., 8th, Australia and New Zealand*, vol. 5, pp. 9–19, 1966.

Hubbert, M. K.: The theory of groundwater motion, *J. Geol.*, vol. 48, pp. 785–944, 1940.

———: Entrapment of petroleum under hydrodynamic conditions, *Bull. Amer. Assoc. Petrol. Geologists*, vol. 37, pp. 1954–2026, 1953.

Hughes, G.: Selection of refuse disposal sites in northeastern Illinois, *Illinois State Geol. Surv., Envir. Geol., Note* 17, 1967.

———, R. Farvolden, and R. Landon: Hydrogeology and waste quality at a solid waste disposal site, in *Eng. Geol. and Soils Eng. Symp., Moscow, Idaho*, pp. 116–130, 1969.

Jones, P. H.: Hydrology of Neogene deposits in the northern Gulf of Mexico basin, *Proc. Symp. Abnorm. Subsurface Pressure, 1st, Louisiana State Univ.*, pp. 91–207, Apr. 28, 1967.

Karplus, W. J.: "Analog Simulation," McGraw-Hill Book Company, New York, 1958.

Langbein, W. B.: Salinity and hydrology of closed lakes, *U.S. Geol. Surv., Profess. Papers*, 412, 1961.

Livingstone, D. A.: Chemical composition of rivers and lakes, *U.S. Geol. Surv., Profess. Papers*, 440-G, 1963.

Maxey, G. B., and M. D. Mifflin: Occurrence and movement of groundwater in carbonate rocks of Nevada, *Natl. Speleol. Soc. Bull.*, vol. 28, no. 3, pp. 141–157, 1966.

McCracken, D. D., and W. S. Dorn: "Numerical Methods and Fortran Programming with Applications in Engineering and Science," John Wiley & Sons, Inc., New York, 1964.

McGinnis, L. D.: Glaciation as a possible cause of mineral deposition, *Econ. Geol.*, vol. 63, pp. 390–400, 1968.
Meinzer, O. E.: Geology and water resources of Big Smokey, Clayton, and Alkali Spring Valleys, Nevada, *U.S. Geol. Surv., Water Supply Papers*, 423, 1917.
———, Map of the Pleistocene lakes of the basin and range province and its significance, *Bull. Geol. Soc. Amer.*, vol. 33, pp. 541–552, 1922.
———: Plants as indicators of groundwater, *U.S. Geol. Surv., Water Supply Papers*, 577, 1927.
Meyboom, P.: Groundwater resources of the city of Calgary and vicinity, *Res. Council of Alberta (Can.), Geol. Div., Bull.* 8, Edmonton, Alberta, 1961.
———: Patterns of groundwater flow in the prairie profile, *Can. Hydrol. Symp.*, 3d, pp. 5–33, 1962.
———: Current trends in hydrogeology, *Earth Sci. Revs.*, vol. 2, no. 4, pp. 345–364, 1966a.
———: Groundwater studies in the Assiniboine River drainage basin, Part I: The evaluation of a flow system in south-central Saskatchewan, *Can. Dept. Mines Tech. Surv. Geol. Surv. Can., Bull.* 139, 1966b.
———: Unsteady groundwater flow near a willow ring in hummocky moraine, *J. Hydrol.*, vol. 4, no. 1, pp. 38–62, 1966c.
———: Mass transfer studies to determine the groundwater regime of permanent lakes in hummocky moraine of western Canada, *J. Hydrol.*, vol. 5, no. 2, pp. 117–142, 1967.
Mifflin, M. D.: Delineation of groundwater flow systems in Nevada, *Desert Res. Inst., Tech. Rept. Ser.* H-W, no. 4, Reno, Nev., 1968.
Muskat, M.: "The Flow of Homogeneous Fluids through Porous Media," McGraw-Hill Book Company, New York, 1937.
Paige, S.: Effect of a sea-level canal on the groundwater level of Florida, *Econ. Geol.*, vol. 31, pp. 537–570, 1938a.
———: Effect of sea-level canal on the groundwater level of Florida, a reply, *Econ. Geol.*, vol. 3, pp. 647–665, 1938b.
Piper, A. M.: A graphic procedure in the geochemical interpretation of water analyses, *Trans. Amer. Geophys. Union*, vol. 25, pp. 914–923, 1944.
Pluhowski, E. J., and I. H. Kantrowitz: Influence of land surface conditions on groundwater temperatures in southwestern Suffolk County, Long Island, N.Y., *U.S. Geol. Surv., Profess. Papers*, 475-B, pp. 186–188, 1963.
Polubarinova-Kochina, P. Ya.: "Theory of Groundwater Movement" (trans. by R. J. M. DeWeist), Princeton University Press, Princeton, N.J., 1962.
Rorabaugh, M. I.: Groundwater in northeastern Louisville, Kentucky, with reference to induced infiltration, *U.S. Geol. Surv., Water Supply Papers*, 1360-B, pp. 101–169, 1956.
Schneider, R.: An application of thermometry to the study of groundwater, *U.S. Geol. Surv., Water Supply Papers*, 1544-B, 1962.
———: Relation of temperature distribution to groundwater movement in carbonate rocks of central Israel, *Bull. Geol. Soc. Amer.*, vol. 75, pp. 209–216, 1964.
Schoeller, H.: Arid zone hydrology, recent developments, *UNESCO Rev. Reicardi* 12, 1959.
———: "Les Eaux souterraines," Mason et Cie, Paris, 1962.
Schwendiman, L. C., and others: Disposal of industrial radioactive waste waters at Hanford, *Amer. Soc. Testing Mater., Spec. Tech. Publ.*, 273, pp. 3–19, 1959.
Scott, R. F.: "Principles of Soil Mechanics," Addison-Wesley Publishing Company, Inc., Reading, Mass., 1963.

Seaber, P. B.: Cation hydrochemical facies of groundwater in the Englishtown formation, New Jersey, *U.S. Geol. Surv., Profess. Papers,* 450-B, pp. 124–126, 1962.

Southwell, R. V.: "Relaxation Methods in Theoretical Physics," Oxford University Press, London, 1946.

Stallman, R. W.: Computation of groundwater velocity from temperature data, in "Methods of collecting and interpreting groundwater data," *U.S. Geol. Surv., Water Supply Papers,* 1544-H, pp. 36–46, 1963.

———: Steady one-dimensional fluid flow in a semiinfinite porous medium with sinusoidal surface temperature, *J. Geophys. Res.,* vol. 70, no. 12, pp. 2821–2827, 1965.

Stringfield, V. T., and H. E. LeGrand: Hydrology of limestone terranes in the Coastal Plain of the southeastern United States, *Geol. Soc. Amer., Spec. Paper,* 93, 1966.

Swenson, F. A.: New theory of recharge to the artesian basin of the Dakotas, *Bull. Geol. Soc. Amer.,* vol. 79, pp. 163–182, 1968.

Terzaghi, K.: Effect of minor geologic details on the safety of dams, in L. Bjerrum and others, "From Theory to Practice in Soil Mechanics," John Wiley and Sons, Inc., New York, pp. 119–132, 1960.

Thomas, H. E., and L. B. Leopold: Groundwater in North America, *Science,* vol. 143, pp. 1001–1006, 1964.

Toth, J.: A theory of groundwater motion in small drainage basins in central Alberta, Canada, *J. Geophys. Res.,* vol. 67, no. 11, pp. 4375–4387, 1962.

———: A theoretical analysis of groundwater flow in small drainage basins, *J. Geophys. Res.,* vol. 68, no. 16, pp. 4795–4812, 1963.

———: Mapping and interpretation of field phenomena for groundwater reconnaissance in a prairie environment, Alberta, Canada, *Intern. Assoc. Sci. Hydrol.,* 11 Année, no. 2, pp. 1–49, 1966a.

———: Groundwater geology, movement chemistry, and resources near Olds, Alberta, *Res. Council Alberta (Can.), Geol. Div., Bull.,* 17, 1966b.

———: A hydrogeological study of the Three Hills area, Alberta, *Res. Council Alberta (Can.), Geol. Div., Bull.,* 24, 1968.

Walker, T. R.: Formation of redbeds in modern and ancient deserts, *Bull. Geol. Soc. Amer.,* vol. 78, pp. 353–368, 1967.

Warner, D. L.: Deep well waste injection—reaction with aquifer water, *J. Sanit. Eng. Div., Amer. Soc. Civil Engrs.,* vol. 92, pp. 45–69, 1969.

White, D. E.: Thermal waters of volcanic origin, *Bull. Geol. Soc. Amer.,* vol. 68, no. 12, pp. 1637–1657, 1957a.

———: Magmatic, connate, and metamorphic waters, *Bull. Geol. Soc. Amer.,* vol. 68, no. 12, pp. 1659–1682, 1957b.

———, J. D. Hem, and G. A. Waring: Chemical composition of subsurface waters, *U.S. Geol. Surv., Profess. Papers,* 440-F, 1963.

Winograd, I. H.: Interbasin movement of groundwater at the Nevada test site, Nevada, *U.S. Geol. Surv., Profess. Papers,* 450-C, pp. 108–111, 1962.

7
Theoretical Models of the Unsteady State: Well Flow, Consolidation Theory, Simulation

The steady-flow regime discussed in the previous chapter is well suited for study of natural flow in groundwater basins when small transient perturbations are relatively unimportant. In these applications, distinctions between high- and low-permeability materials are important only in the sense that their influence on the natural flow pattern is important. With this approach, aquifers and confining layers are assigned little or no significance in the conventional water-supply sense, in that movement of water to pumping wells under man-induced gradients is ignored.

Consideration now of movement under hydraulic gradients far in excess of those characteristic of the natural flow regime requires that the idea of a steady-state flux in an ideal groundwater basin be abandoned in favor of a transient response in one or more ideal units within the basin. The large gradients under consideration will be due exclusively to some external disturbance (as, for example, pumping wells), and the system response may be in a transient state for years, or even decades. Therefore, for the cases to be discussed in this chapter, any steady-state assumptions are useful only in establisning initial (pretransient) or terminal (posttransient) conditions, but are inadequate to deal with intermediate states.

It may be recalled that the unsteady-flow equation was derived with the aid of a differential element by equating the difference between influx and efflux with the time rate of change of fluid mass within the element. With some additional manipulations, one is led, ultimately, to the diffusion equation. This equation, like Laplace's equation, provides a solution in the form of two families of intersecting curves for prescribed initial and boundary conditions. However, any given configuration represents an instantaneous condition; that is, the magnitude of the potential now varies not only with space, but with time. It follows that unsteady flow is merely a continuous succession of steady states, each of which is "steady" for very short periods of time. The succession of one instantaneously steady condition to that of another, but different, steady condition is referred to as *transient* (passing through), or unsteady.

This chapter is a continuation of the preceding chapter, with the main exception that a new dimension, time, is incorporated in the system models. The contents are described in three parts: well flow, consolidation theory, and simulation. As in all preceding chapters, concentration will be on ideal models that are considerably less complex than the real systems they represent, but they may satisfactorily explain some aspect of real behavior.

A. WELL FLOW

That aspect of groundwater hydrology generally included under the heading "well flow" deals chiefly with the steady- or transient-response surface in the vicinity of pumping wells in ideal aquifers. As a quantitative field, well-flow investigations had their early beginnings shortly following Darcy's publication of the empirical law that currently bears his name. Related steady-state developments extended into the early 1900s, after which the theory appeared to have taken a prolonged and comfortable rest, until 1935, when Theis invoked the now classic heat-flow analogy. Remarkably, Theis did not set out to establish a theory of transient flow to pumping wells, but was concerned at the time with a related problem; namely, to determine the manner in which an extensive groundwater basin in an arid zone responds to pumping when the area of pumping is far removed from the area of natural discharge (Section 1.2). The analogy invoked dealt with the withdrawal of heat from an infinite, homogeneous slab, and the time-space consequences of such withdrawal.

Following Theis' publication in 1935, mathematically inclined hydrologists became so absorbed in transient-flow theory that three decades later, almost every meaningful problem in well hydraulics had been solved in one form or another. It is not intended here to discuss every meaningful

7.1 THE HEAT–FLOW ANALOGY: RADIAL FLOW TO A WELL IN AN INFINITE AQUIFER

The mathematical expression for a removal of heat at a constant rate from a homogeneous, infinite slab has provided a useful analogy for study of groundwater flow to a pumping well. It is assumed, initially, that the slab is at some uniform temperature. An infinitesimal rod of lower temperature parallel to the z axis is then allowed to draw off the heat (Fig. 7.1a). In mathematical terminology, the rod represents a continuous line sink. The temperature change at some distance r from the rod is a function of the rate at which heat is withdrawn, the properties of the slab, and time. By analogy, the infinite, homogeneous slab is replaced by an extensive, homogeneous aquifer, and the rod by a well of infinitesimal diameter (Fig. 7.1b). Similarly, the rate of pumping is analogous to the rate of heat withdrawal, and water-level change at any distance r from the pumping well is a function of the pumping rate, the properties of the aquifer, and time.

A solution to the heat-flow problem stated above is given by Carslaw and Jaeger (1959), and is of the form

$$T_e = \frac{\lambda}{4\Pi\kappa} \int_{r^2/4\kappa t}^{\infty} \frac{e^{-z}}{z} dz \qquad (7.1)$$

where T_e is temperature at any distance r from the source or sink at some time t after heat is added or liberated; λ is the strength of the sink, or the amount of heat removed divided by the specific heat per unit volume; κ is

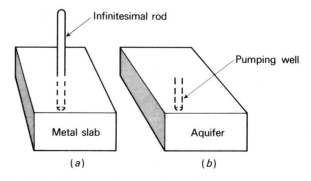

Fig. 7.1 Schematic diagram of (a) infinite slab from heat flow theory and (b) the well-flow analogy.

the diffusivity; and the exponential integral is a well-known tabulated function, the value of which is given by the infinite series

$$\int_u^\infty \frac{e^{-z}}{z} dz = -0.577216 - \ln u + u - \frac{u^2}{2.2!} + \frac{u^3}{3.3!} - \frac{u^4}{4.4!} + \cdots \tag{7.2}$$

where

$$u = \frac{r^2}{4\kappa t} \tag{7.3}$$

Equation (7.1) is a solution of the diffusion equation

$$\frac{\partial^2 T_e}{\partial x^2} + \frac{\partial^2 T_e}{\partial y^2} = \frac{1}{\kappa} \frac{\partial T_e}{\partial t} \tag{7.4}$$

or, in polar coordinates,

$$\frac{\partial^2 T_e}{\partial r^2} + \frac{1}{r} \frac{\partial T_e}{\partial r} = \frac{1}{\kappa} \frac{\partial T_e}{\partial t} \tag{7.5}$$

for the initial and boundary conditions

$$T_e(r,0) = T_{e_0}$$
$$T_e(\infty,t) = T_{e_0}$$
$$\lim_{r \to 0} \left(r \frac{\partial T_e}{\partial r} \right) = \frac{\lambda}{2\Pi\kappa} \quad \text{for } t > 0$$

The first two of these are read: The temperature at some radius r from the rod at time zero equals the initial temperature T_{e_0}, and the temperature at an infinite distance at any time t equals the initial temperature T_{e_0}. The first condition is self-explanatory. The second condition stipulates that heat withdrawal via the rod may affect distant parts of the slab, but the exterior boundary is never encountered (as expected for a slab of infinite extent). The third condition provides for a constant withdrawal rate at r_r, which is the radius of the rod, which, in turn, is taken as infinitesimal.

Theis (1935) used the analogy between heat and groundwater flow to investigate the flow of water to a pumped well, the results of which stand as a classic example of application of knowledge from one branch of physics to problems of another. Analogous quantities include temperature and head, and the strength of the sink to the ratio of discharge to storativity. With recognition, further, that diffusivity is the ratio of thermal conductivity (analogous to transmissivity) to the product of the average density and specific heat (analogous to storativity), κ is replaced by T/S. Equation

THEORETICAL MODELS OF THE UNSTEADY STATE

(7.1) is then transformed into one describing the effect of pumping from an ideal infinite aquifer that possesses all the properties assumed for the slab,

$$h_o - h = \frac{Q/S}{4\Pi T/S} \int_{r^2/(4Tt/S)}^{\infty} \frac{e^{-z}}{z} dz \tag{7.6}$$

where h_o is the original head at any distance r from a fully penetrating well at time t equals zero, h is the head at some later time t, Q is a steady pumping rate, T is the transmissivity, and S is the storativity. For a given condition, r is fixed in space, and Q, S, T, and h_o are constants. As the situation described is one of transient flow, the difference between h_o and h at a fixed distance r gets larger with increasing time. Designating this difference as the drawdown s,

$$s = \frac{Q}{4\Pi T} \int_{r^2S/4Tt}^{\infty} \frac{e^{-z}}{z} dz \tag{7.7}$$

It follows that Eq. (7.7) is a solution to the polar-coordinate form of the diffusion equation, or

$$\frac{\partial^2 h}{\partial r^2} + \frac{1}{r}\frac{\partial h}{\partial r} = \frac{S}{T}\frac{\partial h}{\partial t} \tag{7.8}$$

for the initial and boundary conditions

$$h(r,0) = h_o$$
$$h(\infty,t) = h_o$$
$$\lim_{r \to 0}\left(r\frac{\partial h}{\partial r}\right) = \frac{Q}{2\Pi T} \quad \text{for } t > 0$$

THE CONE OF DEPRESSION IN AN IDEAL AQUIFER

A verbal interpretation of the parameters incorporated in the so-called nonequilibrium equation will provide insight into the shape and growth to be expected of a cone of depression caused by a pumping well in an ideal aquifer. It is first noted that the exponential integral of Eq. (7.7) is a function only of the lower limit of integration, so that the equation can be written

$$s = \frac{Q}{4\Pi T} W(u) \tag{7.9}$$

where

$$u = \frac{r^2 S}{4Tt} \tag{7.10}$$

The term $W(u)$ is termed *well function of u*, which indicates its dependence on values of u expressed in Eq. (7.10). It follows that the value of the exponential integral may be easily ascertained and tabulated for each of several values of u. Such a tabulation is shown in Table 7.1 for values of u ranging from 10^{-15} to 9.0. If u equals 4.0×10^{-10}, then $W(u)$ equals 21.06, which is the value of the series of Eq. (7.2) for this particular value.

Of the variables comprising u, the storativity and transmissivity may be considered constant for a given set of conditions, and distance and time considered variables. Values of $W(u)$ may be plotted against values of u or $1/u$ in such a way that, for any given time, the plotted curve reveals the exact profile of a cone of depression as a function of distance r^2 from a pumping well (Fig. 7.2a); or such that for any fixed distance r, the plotted curve reveals the exact shape of the drawdown curve as a function of time t (Fig. 7.2b). Actual values for drawdown depend on the pumping rate and the hydraulic properties [Eq. (7.9)] and (1) the time of observation, in Fig. 7.2a; (2) the distance at which the observations are made, in Fig. 7.2b. Curves showing the relation between $W(u)$ and u are termed "type" curves.

In that the value of u depends on time, distance, transmissivity, and storativity, u determines the radius of a cone of depression (Theis, 1940). Hence, the radius not only increases with increasing time, but for a given time, is larger for decreasing values of storativity and increasing values of transmissivity. By examining the complete statement for drawdown [Eq. (7.9)], drawdown at any point for a given time is proportional to discharge and inversely proportional to transmissivity. The lateral extent of a cone of depression at any given time, and its rate of growth, are independent of the pumping rate.

Table 7.1 Values of $W(u)$ for values of u (After Wenzel, 1942.)

u	1.0	2.0	3.0	4.0	5.0	6.0	7.0	8.0	9.0
$\times 1$	0.219	0.049	0.013	0.0038	0.0011	0.00036	0.00012	0.000038	0.000012
$\times 10^{-1}$	1.82	1.22	0.91	0.70	0.56	0.45	0.37	0.31	0.26
$\times 10^{-2}$	4.04	3.35	2.96	2.68	2.47	2.30	2.15	2.03	1.92
$\times 10^{-3}$	6.33	5.64	5.23	4.95	4.73	4.54	4.39	4.26	4.14
$\times 10^{-4}$	8.63	7.94	7.53	7.25	7.02	6.84	6.69	6.55	6.44
$\times 10^{-5}$	10.94	10.24	9.84	9.55	9.33	9.14	8.99	8.86	8.74
$\times 10^{-6}$	13.24	12.55	12.14	11.85	11.63	11.45	11.29	11.16	11.04
$\times 10^{-7}$	15.54	14.85	14.44	14.15	13.93	13.75	13.60	13.46	13.34
$\times 10^{-8}$	17.84	17.15	16.74	16.46	16.23	16.05	15.90	15.76	15.65
$\times 10^{-9}$	20.15	19.45	19.05	18.76	18.54	18.35	18.20	18.07	17.95
$\times 10^{-10}$	22.45	21.76	21.35	21.06	20.84	20.66	20.50	20.37	20.25
$\times 10^{-11}$	24.75	24.06	23.65	23.36	23.14	22.96	22.81	22.67	22.55
$\times 10^{-12}$	27.05	26.36	25.96	25.67	25.44	25.26	25.11	24.97	24.86
$\times 10^{-13}$	29.36	28.66	28.26	27.97	27.75	27.56	27.41	27.28	27.16
$\times 10^{-14}$	31.66	30.97	30.56	30.27	30.05	29.87	29.71	29.58	29.46
$\times 10^{-15}$	33.96	33.27	32.86	32.58	32.35	32.17	32.02	31.88	31.76

THEORETICAL MODELS OF THE UNSTEADY STATE

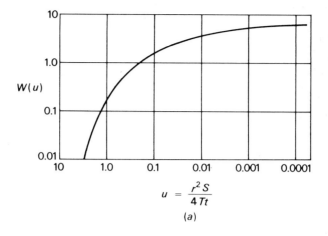

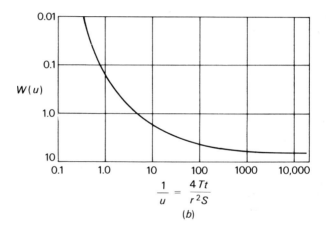

Fig. 7.2 Values of $W(u)$ plotted against (a) values of u and (b) values of $1/u$.

NONEQUILIBRIUM PUMPING TEST METHOD

The Theis (or nonequilibrium) equation is used extensively to determine the hydraulic properties of aquifers. Once these values are determined, it is possible to determine:

1. The theoretical drawdown at any distance r from a single well pumping at a given rate Q for any time t
2. The theoretical drawdown at any point in the aquifer for any time t in response to a battery of wells pumping at various rates

To carry out these applications, it will be recalled that for a fixed distance r, drawdown versus time is of the form of the curve in Fig. 7.2b.

Hence, if drawdown s can be measured for one value of distance r and several values of time t, and if the discharge is steady and is known, the coefficients of transmissivity and storativity can be determined by a graphical method of superposition. Field data composed of drawdown versus time collected at a nonpumping observation well at a known distance r from a pumped well are plotted on logarithmic paper of the same scale as the type curve (Fig. 7.3a). The field curve is superimposed on the type curve, with the coordinate axes of the two curves kept parallel while matching field data to the type curve (Fig. 7.3b). Any point on the overlapping sheets is selected arbitrarily

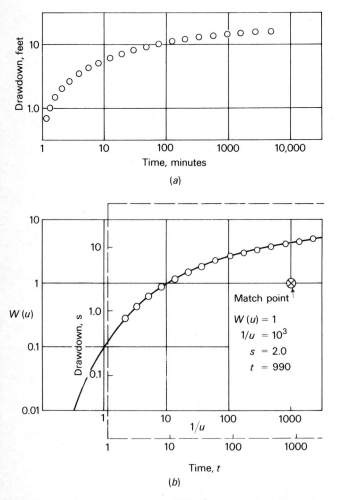

Fig. 7.3 Graphs of (a) a plot of field data and (b) superposition of the field data on the type curve to obtain the formation parameters.

THEORETICAL MODELS OF THE UNSTEADY STATE 323

(the point need not be on the matched curves). The selected point is defined by four coordinate values: $W(u)$ and s, and $1/u$ and t (Fig. 7.3b). The unique relation between these coordinate values can be examined by expressing Eqs. (7.9) and (7.10) in logarithmic form,

$$\log s = \log \left[\frac{Q}{4\Pi T}\right] + \log W(u) \tag{7.11}$$

and

$$\log t = \log \left[\frac{r^2 S}{4T}\right] + \log \frac{1}{u} \tag{7.12}$$

The bracketed parts of the equations are constant for a given pumping rate, and the unbracketed parts are variables. It follows that the variable $W(u)$ is related to $1/u$ (see type curve) in the same manner as the variable s is to t (see field-data curve). Equation (7.9) can be solved for transmissivity by using the match-point coordinates s and $W(u)$ and the discharge Q. Equation (7.10) can be solved for storativity by using the match-point coordinates $1/u$ and t, the distance r from the pumped well to the observation well, and the value of transmissivity determined above.

Example 7.1 Suppose that the data of Fig. 7.3 were collected at a distance of 500 ft from a well pumped at a rate of 1,000 gal/min (about 192×10^3 ft^3/day). From Fig. 7.3,

$W(u) = 1 \qquad s = 2 \text{ ft}$

$\dfrac{1}{u} = 10^3 \qquad t = 990 \text{ min (about 0.7 day)}$

Transmissivity is determined from Eq. (7.9),

$$T = \frac{Q}{4\Pi s} W(u) = \frac{192 \times 10^3 \text{ ft}^3/\text{day} \times 1}{4 \times 3.12 \times 2 \text{ ft}} = 7.7 \times 10^3 \text{ ft}^2/\text{day}$$

Storativity is determined from Eq. (7.10),

$$S = \frac{4uTt}{r^2} = \frac{4 \times 1 \times 10^{-3} \times 7.7 \times 10^3 \text{ ft}^2/\text{day} \times 7 \times 10^{-1} \text{ day}}{250 \times 10^3 \text{ ft}^2} = 8.6 \times 10^{-5}$$

For the so-called American practical hydrology units, transmissivity is expressed in gallons per day per foot, and the storativity remains a dimensionless constant. This requires a mixed system of volume measurement: gallons for transmissivity and, inherently, cubic feet for storativity. This mixed system is likely the result of the early introduction of arbitrary units of measurement, such as the Meinzer unit (gallons per day per square foot), and has led to several dimensional versions of the nondimensional Eqs. (7.9) and

(7.10). Hence, transmissivity calculated above is easily converted to gallons per day per foot,

$$7.7 \times 10^3 \text{ ft}^2/\text{day} \times 7.48 \text{ gal/ft}^3 = 57{,}600 \text{ gpd/ft}$$

or it may be computed directly in American practical hydrologic units by dimensional versions of Eqs. (7.9) and (7.10),

$$s = \frac{114.6Q}{T} W(u) \tag{7.13}$$

and

$$u = \frac{2{,}693 r^2 S}{Tt} \tag{7.14}$$

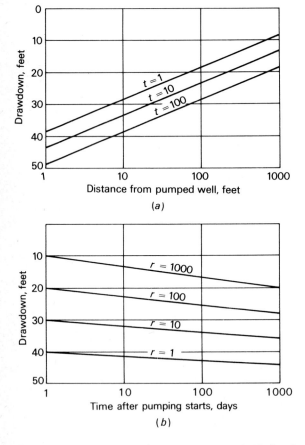

Fig. 7.4 Plots of (*a*) distance/drawdown and (*b*) time/drawdown, computed from the data of Table 7.2.

THEORETICAL MODELS OF THE UNSTEADY STATE 325

where s = drawdown, ft
Q = well discharge, gpm
T = transmissivity, gpd/ft
r = distance from pumping well to observation well, ft
S = the storativity, dimensionless
t = time, min since pumping started

These equations allow direct use of data obtained from pumping tests with little or no preliminary conversions. It should be clear, however, that the conversion factors 114.6 and 2,693 have hidden units to make the equations dimensionally correct. Equations (7.13) and (7.14) are given here only because of their widespread use in the United States In the interests of a consistent system of units, these equations should be abandoned.

Given the hydraulic properties T and S and the assumption they are constant in space, it is possible to calculate a theoretical drawdown at any point for various times and pumping rates. For the example cited above, drawdown s is expressed in terms of $W(u)$,

$$s = \frac{Q}{4\Pi T} W(u) = \frac{192 \times 10^3 \text{ ft}^3/\text{day}}{4 \times 3.2 \times 7.7 \times 10^3 \text{ ft}^2/\text{day}} W(u) = 2.0 W(u)$$

with the value of $W(u)$ provided in Table 7.2. Figure 7.4 is easily constructed from these data, and shows the theoretical effects of pumping from this ideal aquifer. As the relation between drawdown and pumping rate is one to one [Eq. (7.9)], drawdowns for other pumping rates can be easily determined.

The effect of two or more wells pumping from the idealized aquifer can be analyzed with the graph of Fig. 7.4.

Example 7.2 Two wells are pumping for 10 days at rates of 384×10^3 ft^3/day (about 2,000 gpm) and 96×10^3 ft^3/day (about 500 gpm), respectively. Five days after these wells start pumping, a third well starts pumping at a rate of 144×10^3 ft^3/day (about 750 gpm). The pumping wells are at a distance of 1,000, 100, and 10 ft, respectively, from an observation well. At the end of a 10-day period, the drawdown at the point of observation caused by each of the wells is determined from Fig. 7.4 for the postulated pumping rates

$s_1 = 30.48$ ft $s_2 = 12.22$ ft $s_3 = 24$ ft

For linear differential equations, the superposition principle permits calculation of the total drawdown, or 67 ft. Similar calculations at other points in time and space allow a two-dimensional (plan-view) representation of water-level change.

ASSUMPTIONS AND INTERPRETATIONS

In that the nonequilibrium equation was derived by consideration of a heat-flow situation, there is need to compare the properties of the slab, and the assumptions concerning it, with the assumptions concerning an aquifer. It

Table 7.2 Tabulated values for the drawdown curves of Fig. 7.4

Time = 1 day

$$u = \frac{r^2 S}{4Tt} = \frac{r^2 \times 8.6 \times 10^{-5}}{4 \times 7.7 \times 10^3 \times 1} = 2.74 \times 10^{-9} r^2$$

r, ft	u	W(u)	s, ft
1	2.74×10^{-9}	19.3	38.26
10	2.74×10^{-7}	14.53	29.06
100	2.74×10^{-5}	9.92	19.84
1,000	2.74×10^{-3}	5.32	10.64

Time = 10 days

$$u = \frac{r^2 S}{4Tt} = \frac{r^2 \times 8.6 \times 10^{-5}}{4 \times 7.7 \times 10^3 \times 10} = 2.74 \times 10^{-10} r^2$$

1	2.74×10^{-10}	21.43	42.86
10	2.74×10^{-8}	16.83	33.66
100	2.74×10^{-6}	12.22	24.44
1,000	2.74×10^{-4}	7.62	15.24

Time = 100 days

$$u = \frac{r^2 S}{4Tt} = \frac{r^2 \times 8.6 \times 10^{-5}}{4 \times 7.7 \times 10^3 \times 100} = 2.74 \times 10^{-11} r^2$$

1	2.74×10^{-11}	23.74	46.48
10	2.74×10^{-9}	19.13	38.26
100	2.74×10^{-7}	14.53	29.06
1,000	2.74×10^{-5}	9.92	19.84

should be recognized that the operational hydraulic equations were derived for an ideal model aquifer, whereas time/drawdown data represents real aquifer response. The procedure called for requires a matching of real response data with the response expected under ideal conditions. Deviations from the ideal behavior are to be expected, especially for long periods of pumping, and they represent a measure of how much the real aquifer departs from the ideal (Ferris and others, 1962).

The most conspicuous assumption is that the aquifer is infinite in areal extent. This means the following:

1. The cone of depression will never intersect a boundary to the system.
2. An infinite amount of water is stored in the aquifer.

Condition 1 is likely to be met, if it is met at all, for very short times, or for longer times in extensive aquifers of uniform material. In that the cone of depression is expanding with time, any geologic boundary within a short distance from the pumping well will immediately disrupt the postulated behavior. According to assumption 2, water levels will eventually return to their prepumping level once the wells are shut down.

In addition, it is assumed that the well is of infinitesimal diameter and fully penetrates the aquifer. This means that storage in the well can be ignored, and that the well receives water from the entire thickness of the aquifer. Methods of treatment for deviations from full penetration have been presented in a number of papers (Muskat, 1937; Hantush and Jacob, 1955; Hantush, 1957). A solution for drawdown in large-diameter wells which takes into consideration the storage within the well has been presented by Papadopulos and Cooper (1967).

As removal of water from the aquifer is taken to be analogous to removal of heat from the slab, it is assumed that water is released instantaneously with decline in head. This assumption may be reasonably met when water is released by compression of the aquifer and expansion of the water, but fails to describe adequately the gravity flow system, characteristic of the unconfined case. The delay of storage release has been treated by Boulton (1954; 1963) and Prickett (1965).

Homogeneity and isotropicity are major assumptions in the development of the descriptive differential equation of which the well-flow equation is a solution. This means that the transmissivity and storativity are assumed to be constants, both in space and in time. Hence, the geologic medium is assumed to be the simplest type conceivable.

At first glance, it appears that the assumptions are rather demanding, which renders application of this model of little practical value. However, for short periods of pumping in extensive uniform aquifers (a few days to a few weeks), the model, as developed, may work reasonably well. For non-uniform aquifers or aquifers of restricted extent, the model may or may not work, depending on how long one wishes to examine the system response.

7.2 THE STEADY RESPONSE

If time is very large, the lower limit of Eq. (7.1) approaches zero, and the heat flow expression reduces to

$$T_e = \frac{\lambda}{4\Pi\kappa} \ln \frac{4\kappa t}{r^2} - \frac{0.577216\lambda}{4\Pi\kappa} \qquad (7.15)$$

approximately (Carslaw and Jaeger, 1959). By considering the temperature $T_{e_o} - T_{e_w}$ at two points r_o and r_w, with $r_o > r_w$,

$$T_{e_o} - T_{e_w} = \left(\frac{\lambda}{4\Pi\kappa} \ln \frac{4\kappa t}{r_o^2} - \frac{0.577\lambda}{4\Pi\kappa}\right) - \left(\frac{\lambda}{4\Pi\kappa} \ln \frac{4\kappa t}{r_w^2} - \frac{0.577\lambda}{4\Pi\kappa}\right) \qquad (7.16)$$

or

$$T_{e_o} - T_{e_w} = \frac{\lambda}{4\Pi\kappa}\left(\ln\frac{4\kappa t}{r_o^2} - \ln\frac{4\kappa t}{r_w^2}\right) \qquad (7.17)$$

With recognition that

$$\ln\frac{1}{x} - \ln\frac{1}{y} = \ln\frac{y}{x} \qquad (7.18)$$

Eq. (7.17) can be expressed

$$T_{e_o} - T_{e_w} = \frac{\lambda}{4\Pi\kappa}\left(\ln\frac{4\kappa t/r_w^2}{4\kappa t/r_o^2}\right) \qquad (7.19)$$

or

$$T_{e_o} - T_{e_w} = \frac{\lambda}{4\Pi\kappa}\ln\frac{r_o^2}{r_w^2} \qquad (7.20)$$

This can be expressed

$$T_{e_o} - T_{e_w} = \frac{\lambda}{4\Pi\kappa}\left(\ln\frac{1}{r_w^2} - \ln\frac{1}{r_o^2}\right) \qquad (7.21)$$

In that

$$\ln\frac{1}{r^2} = 2\ln\frac{1}{r} \qquad (7.22)$$

Eq. (7.21) may be stated

$$T_{e_o} - T_{e_w} = \frac{\lambda}{2\Pi\kappa}\left(\ln\frac{r_o}{r_w}\right) \qquad (7.23)$$

Substituting the hydrologic entities head for temperature, the ratio of discharge to storativity for the strength of the sink, and the ratio of transmissivity to storativity for the diffusivity

$$h_o - h_w = \frac{Q}{2\Pi T}\ln\frac{r_o}{r_w} \qquad (7.24)$$

This equation describes the steady-state response of water levels in the vicinity of a pumping well.

It follows that Eq. (7.24) should be easily obtained as a direct solution to Laplace's equation, which can be expressed in one-dimensional radial form

$$\frac{1}{r}\frac{\partial}{\partial r}\left(r\frac{\partial h}{\partial r}\right) = 0 \qquad (7.25)$$

with the axis of the well at the origin. By integrating twice, the general solution for head distribution is of the form

$$h = a \ln \frac{r}{b} \tag{7.26}$$

where a and b are constants to be determined from the boundary conditions.

In plan view, the radial flow system is bounded by two concentric circles. In Fig. 7.5a, term r_w corresponds to the radius of the pumped well, and h_w to the head in the formation adjacent to the well; r_o represents the radius of the outer circle, and h_o the total head at r equals r_o. The boundary conditions depicted can be visualized as those inherent in a well of radius r_w discharging water from the center of a circular island of radius r_o.

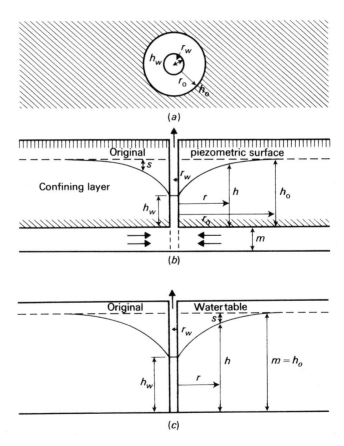

Fig. 7.5 Diagrams of (a) concentric circular boundary conditions, (b) steady radial flow to a well in a confined aquifer, and (c) steady radial flow to a well in an unconfined aquifer.

STEADY CONFINED FLOW

For the confined state of Fig. 7.5b, the gradient of flow toward the well is

$$i = \frac{dh}{dr} \tag{7.27}$$

and the cylindrical area of flow is

$$A = 2\Pi rm \tag{7.28}$$

The flow rate through the cylinder of radius r and height m is expressed by Darcy's law

$$Q = K\frac{dh}{dr} 2\Pi rm \tag{7.29}$$

Rearranging terms and integrating for the boundary conditions of Fig. 7.5b,

$$\int_{h_w}^{h_o} dh = \frac{Q}{2\Pi Km} \int_{r_w}^{r_o} \frac{dr}{r} \tag{7.30}$$

which gives

$$h_o - h_w = \frac{Q}{2\Pi T} \ln \frac{r_o}{r_w} \tag{7.31}$$

which is of the form promised by Eq. (7.26), and is identical with Eq. (7.24) arrived at through the temperature analog. The difference between h_o and h_w is termed the *steady drawdown* at the pumping well, again designated as s.

Equation (7.31) is known as the *equilibrium*, or *Thiem equation* (Thiem, 1906), and is sometimes used to obtain field values of transmissivity. Jacob (1950) has shown that drawdown varies with the logarithm of distance from a pumped well so that, for a given discharge and various steady drawdowns at different distances from a pumping well, a plot of drawdown on an arithmetic scale versus distance on logarithmic scale will give a straight line. Hence, any two points will define the drawdown curve, and the recommended field procedure requires one pumping well and two observation wells. This eliminates the use of drawdown data obtained in the pumping well, which are generally affected by head losses caused by flow through the casing or screen openings, and by the upward movement of water in the well to the pump intake.

In field applications, the well is pumped at a steady rate until a steady state is approximated, that is, until additional drawdown with time in a pair of observation wells is negligible. (It will be demonstrated later that this near steady condition is not necessary.) The transient response is not used in the method, only the approximate steady-state drawdowns being used. Designating s_1 as the steady drawdown in the closest observation well at a distance r_1 from the pumping well, and s_2 as the steady drawdown in the well at a distance r_2, Eq. (7.31) is solved for transmissivity

$$T = \frac{Q}{2\Pi(s_1 - s_2)} \ln \frac{r_2}{r_1} \tag{7.32}$$

This equation is valid for any consistent system of units. Since ln x equals 2.3 log x,

$$s = \frac{2.3Q}{2\Pi T} \log \frac{r_2}{r_1} \tag{7.33}$$

STEADY UNCONFINED FLOW

For the unconfined condition, the gradient of flow toward the well of Fig. 7.5c is

$$i = \frac{dh}{dr} \tag{7.34}$$

The cylindrical area of flow is $2\Pi r(m - s)$, where s is the dewatered drawdown, and $(m - s)$ equals h. The area of flow is expressed

$$A = 2\Pi rh \tag{7.35}$$

Applying Darcy's law,

$$Q = K\frac{dh}{dr} 2\Pi rh \tag{7.36}$$

Rearranging terms and integrating for the boundary conditions of Fig. 7.5c,

$$\int_{h_w}^{h_o} h \, dh = \frac{Q}{2\Pi K} \int_{r_w}^{r_o} \frac{dr}{r} \tag{7.37}$$

This gives

$$h_o^2 - h_w^2 = \frac{Q}{\Pi K} \ln \frac{r_o}{r_w} \tag{7.38}$$

As in the application for the confined state, any pair of distances may be used.

ASSUMPTIONS

The following assumptions are generally cited as necessary for the application of the steady-flow equations:

1. The aquifer is homogeneous, isotropic, and infinite in areal extent.
2. The well fully penetrates the aquifer.

3. The tangent to the angle of inclination of the water table is equal to the sine.
4. Flow is uniform and horizontal with depth.

Assumptions 1 and 2 have already been encountered in the unsteady state. Mansur and Kaufman (1962) give a good discussion of partial penetration for steady-flow methods. Assumptions 3 and 4 are serious limitations on the unconfined state. They require that the hydraulic gradient be constant with depth and equal to the slope of the drawdown curve at any point. These assumptions are commonly referred to as the *Dupuit-Forchheimer assumptions*. Such assumptions are not generally justified in the vicinity of pumping wells, where hydraulic gradients are steep, and where large vertical components of flow occur.

As to other so-called equilibrium equations, such as developed by Slichter (1899), Turneaure and Russel (1901), and Israelson (1950), they may be omitted on the grounds that they are merely modified forms of the Thiem equation (Ferris and others, 1962).

7.3 EXTENSIONS AND MODIFICATIONS OF THE NONEQUILIBRIUM EQUATION

At least three important modifications of the nonequilibrium equation can be traced to a very simple observation made by Cooper and Jacob (1946), namely, that the sum of the series of Eq. (7.2) beyond $\ln u$ becomes negligible when u becomes small. This occurs for large values of time t or small value of distance r. By neglecting the series beyond $\ln u$, Eq. (7.7) can be expressed

$$s = \frac{Q}{4\Pi T}\left(-0.5772 - \ln \frac{r^2 S}{4Tt}\right) \tag{7.39}$$

or

$$s = \frac{Q}{4\Pi T}\left(\ln \frac{4Tt}{r^2 S} - 0.5772\right) \tag{7.40}$$

As 0.5772 equals $\ln 1.78$, Eq. (7.40) becomes

$$s = \frac{Q}{4\Pi T} \ln \frac{2.25Tt}{r^2 S} \tag{7.41}$$

Further, as $\ln x$ equals $2.3 \log x$, Eq. (7.41) becomes

$$s = \frac{2.3Q}{4\Pi T} \log \frac{2.25Tt}{r^2 S} \tag{7.42}$$

which is presented as the modified nonequilibrium equation. This equation may be applied to:

THEORETICAL MODELS OF THE UNSTEADY STATE

1. Drawdown/time observations made in a single observation well, as in the original nonequilibrium method (modified nonequilibrium method)
2. Recovery-time observations made in a single observation well after a pumping well is shut down (recovery method)
3. Drawdown observations made in different wells at the same time (distance-drawdown method)

MODIFIED NONEQUILIBRIUM PUMPING TEST METHOD

If drawdown observations are made in a single well for various times, a plot of drawdown versus the logarithm of time will yield a straight line (Fig. 7.6). At time t_1, the drawdown s_1 is expressed

$$s_1 = \frac{2.30Q}{4\Pi T} \log \frac{2.25Tt_1}{r^2 S} \tag{7.43}$$

and at time t_2, the drawdown s_2 will be

$$s_2 = \frac{2.3Q}{4\Pi T} \log \frac{2.25Tt_2}{r^2 S} \tag{7.44}$$

It follows that

$$s_2 - s_1 = \frac{2.3Q}{4\Pi T} \log \frac{t_2}{t_1} \tag{7.45}$$

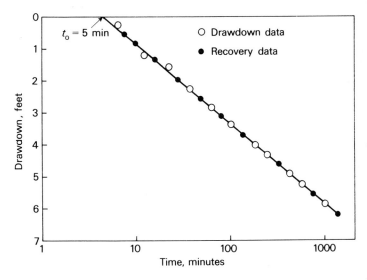

Fig. 7.6 Semilogarithmic plot of drawdown and recovery versus time in an observation well.

If t_1 and t_2 are selected one log cycle apart,

$$\log \frac{t_2}{t_1} = 1 \tag{7.46}$$

Equation (7.45) then becomes

$$\Delta s = \frac{2.3Q}{4\Pi T} \tag{7.47}$$

where Δs is the drawdown per log cycle. Thus, the value of T can be obtained from Eq. (7.47).

The storativity can be obtained by selecting any drawdown in Fig. 7.6 for a given time and substituting this value in Eq. (7.41). For convenience, s equals zero is selected, so that Eq. (7.41) becomes

$$s = 0 = \frac{Q}{4\Pi T} \ln \frac{2.25 T t_0}{r^2 S} \tag{7.48}$$

This requires that

$$\frac{2.25 T t_0}{r^2 S} = 1 \tag{7.49}$$

and

$$S = \frac{2.25 T t_0}{r^2} \tag{7.50}$$

where t_0 is the time intercept where the drawdown line intercepts the zero drawdown axis.

Example 7.3 Suppose that the data of Fig. 7.6 were collected at an observation well 1,000 ft from a well pumping at a rate of 192×10^3 ft^3/day (approximately 1,000 gpm). Drawdown per log cycle is determined from the graph to be 2.4 ft. Solving for T and S,

$$T = \frac{2.3Q}{4\Pi \Delta s} = \frac{2.3(192 \times 10^3) \text{ ft}^3/\text{day}}{4 \times 3.12 \times 2.4 \text{ ft}} = 14.7 \times 10^3 \text{ ft}^2/\text{day}$$

$$S = \frac{2.25 T t_0}{r^2} = \frac{2.25(14.7 \times 10^3 \text{ ft}^2/\text{day})(5 \text{ min})}{(1 \times 10^6 \text{ ft}^2)(1{,}440 \text{ min}/\text{day})} = 1.1 \times 10^{-4}$$

The recovery curve of Fig. 7.6 should plot on the same line as the drawdown curve if all assumptions of the nonequilibrium equation are satisfied. Generally, the coincidence of the two curves becomes a reality only after the well has been recovering for some time.

In American practical hydrologic units, Eqs. (7.47) and (7.50) become

$$\Delta s = \frac{264 Q}{T} \tag{7.51}$$

THEORETICAL MODELS OF THE UNSTEADY STATE

and

$$S = \frac{Tt_0}{4{,}790 r^2} \tag{7.52}$$

respectively, where Δs is in feet, Q is in gallons per minute, T is in gallons per day per foot, t_0 is in minutes, and r is in feet.

RECOVERY METHOD

As the heat-flow analogy is equally valid for both a sink and source, the nonequilibrium development is suitable for analysis of recovery of water levels in a well. If a well is pumped for a given period of time t and then shut down, the residual drawdown (the original prepumping water level minus the water level at any time after shut down) can be approximated as the numerical difference between the drawdown in the well if the discharge had continued and the recovery of the well in response to an imaginary recharge well, of the same flow rate, superimposed on the discharging well at the time it is shut down. Designating original head as h_0 and the recovered head at any time as h', residual drawdown is expressed

$$h_0 - h' = \frac{Q}{4\Pi T}\left(\int_{r^2 S/4Tt}^{\infty} \frac{e^{-z}}{z}\,dz - \int_{r^2 S/4Tt'}^{\infty} \frac{e^{-z}}{z}\,dz\right) \tag{7.53}$$

where t is the time since pumping started, and t' is the time since pumping stopped. Employing the Cooper-Jacob assumptions of the modified nonequilibrium method,

$$\Delta s' = \frac{2.3 Q}{4\Pi T}\left(\log \frac{2.25 Tt}{r^2 S} - \log \frac{2.25 Tt'}{r^2 S}\right) \tag{7.54}$$

where $\Delta s'$ is residual drawdown. This equation reduces to

$$\Delta s' = \frac{2.3 Q}{4\Pi T} \log \frac{t}{t'} \tag{7.55}$$

Field procedure requires a drawdown measurement at the end of pumping (at time t) and recovery measurements during the recovery period (starting at time t'). The graphic procedure is to plot residual drawdown on the arithmetic scale, and the value of t/t' on logarithmic scale. If calculations are made over one log cycle of t/t',

$$\Delta s' = \frac{2.3 Q}{4\Pi T} \tag{7.56}$$

where $\Delta s'$ is the residual drawdown per log cycle.

Figure 7.7 shows a typical time/drawdown and recovery curve to which this equation may be applied. The storativity is not determined directly with this method.

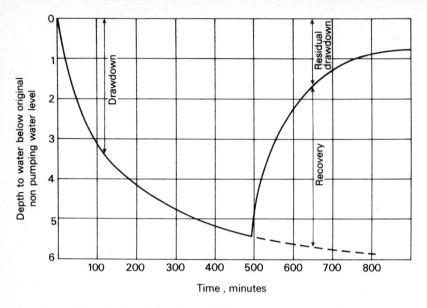

Fig. 7.7 Arithmetic plot of drawdown and recovery curve versus time.

Example 7.4 It is possible to use the mathematical models of Fig. 7.4 to determine residual drawdown for various periods of pumping and quiescence. From this figure, the drawdown 10 ft from a pumped well after 100 days of discharge would be about 38 ft, while after 10 days it would be 34 ft. The residual drawdown after 90 days of pumping and 10 days of quiescence at this point would be about 4 ft, or the difference between these amounts. The drawdown 100 ft from the pumped well after 100 days of pumping would be 29 ft, and after 10 days would be 24 ft. The residual drawdown at this point after 90 days of pumping and 10 days of nonpumping would be 5 ft.

DISTANCE-DRAWDOWN METHOD

From Fig. 7.4, a plot of drawdown versus distance on semilogarithmic paper gives a series of straight lines. As each line represents a series of points on the cone of depression at a given instant in time, and as unsteady flow is merely a continuous succession of steady states, it is possible to arrive at the hydraulic properties by examining drawdown at two or more points at one instant in time. At time t, the drawdown s_1 at a distance r_1 is [Eq. (7.42)]

$$s_1 = \frac{2.3Q}{4\Pi T} \log \frac{2.25Tt}{r_1^2 S} \qquad (7.57)$$

and the drawdown s_2 at a distance r_2 is

$$s_2 = \frac{2.3Q}{4\Pi T} \log \frac{2.25Tt}{r_2^2 S} \qquad (7.58)$$

THEORETICAL MODELS OF THE UNSTEADY STATE

It follows that

$$s_1 - s_2 = \frac{2.3Q}{4\Pi T} \log \frac{r_2{}^2}{r_1{}^2} \tag{7.59}$$

Equation (7.59) can be expressed

$$s_1 - s_2 = \frac{2.3Q}{4\Pi T} \left(\log \frac{1}{r_1{}^2} - \log \frac{1}{r_2{}^2} \right) \tag{7.60}$$

On recognition that

$$\ln \frac{1}{r^2} = 2 \ln \frac{1}{r} \tag{7.61}$$

Eq. (7.60) becomes

$$s_1 - s_2 = \frac{2.3Q}{2\Pi T} \log \frac{r_2}{r_1} \tag{7.62}$$

which is identical with the Thiem equation for steady confined flow [Eq. (7.33)]. Hence, the steady-state drawdown requirement generally cited for the application of the Thiem method is not necessary.

The graphic procedure calls for the plotting of drawdowns observed at the end of a particular pumping period in two or more observation wells at different distances from the pumped well against the logarithms of the respective distances (Fig. 7.8). By considering drawdown per log cycle, Eq. (7.62) becomes

$$\Delta s = \frac{2.3Q}{2\Pi T} \tag{7.63}$$

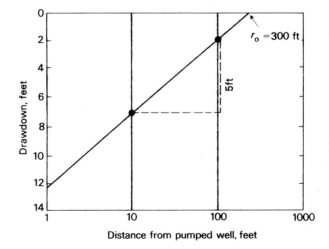

Fig. 7.8 Semilogarithmic plot of drawdown versus distance.

By extrapolating the distance-drawdown curve to its intersection with the zero drawdown axis, the storativity can be determined in the same manner as described for the modification nonequilibrium method,

$$S = \frac{2.25Tt}{r_0^2} \tag{7.64}$$

where r_0 is the intersection of the straight-line slope with the zero drawdown axis.

Example 7.5 Consider the following data for Fig. 7.8. Observation wells are placed 10 ft and 100 ft, respectively, from a pumping well. The well is pumped at a rate of 192×10^3 ft^3/day (1,000 gpm). At the end of 200 min (1.39×10^{-1} days), the drawdowns in the observations wells are 7 ft and 2 ft, respectively. From Eqs. (7.63) and (7.64),

$$T = \frac{2.3Q}{2\Pi \, \Delta s} = \frac{2.3(192 \times 10^3) \text{ft}^3/\text{day}}{2(3.14)(5 \text{ ft})} = 14 \times 10^3 \text{ ft}^2/\text{day}$$

$$S = \frac{2.25Tt}{r_0^2} = \frac{2.25(14 \times 10^3 \text{ ft}^2/\text{day})(1.39 \times 10^{-1} \text{ day})}{9 \times 10^4 \text{ ft}^2} = 5 \times 10^{-2}$$

SPECIALIZED SOLUTIONS

The modified forms of the nonequilibrium equation given above are valid only when the sum of the series of Eq. (7.2) beyond ln u becomes negligible. They do not by any means exhaust the specialized techniques available for determining the formation constants. Solutions are available for a varying discharge and constant drawdown (Jacob and Lohman, 1952), for bailed wells (Skibitzke, 1963), for cyclic pumping (Theis, 1963; Brown, 1963), and for wells subject to an instantaneous charge of water (Cooper and others, 1967).

7.4 A MORE GENERAL THEORY OF RADIAL FLOW

The material to be discussed in this section was developed by Hantush (1960; 1964) and constitutes what is probably one of the most significant contributions to the field of well hydraulics, not so much because it adds significantly to new methods, but because it unifies concepts developed over a period of several years.

Figure 7.9 is a schematic representation of three flow conditions. The aquifer in each case is a layer of thickness m, and is assumed to be homogeneous and isotropic. The layers of thickness m' and m'' are also homogeneous and isotropic with respect to permeability, but of different lithology. Each of the layers is characterized by its own permeability k, k', k'' and specific storage S_s, S_s', S_s''. It is specified further that k greatly exceeds k' and k'' to the extent that the low-permeability units serve as confining, semipervious

THEORETICAL MODELS OF THE UNSTEADY STATE

layers. The water table in conditions 1 and 3 is assumed to be a constant head boundary.

If the aquifer in any of the cases discussed is capable of development by wells, the water pumped will be accompanied by a reduction in storage in the three-layered system. With such pumping, water levels in each of the layers will decline. The time required for full development of a cone of depression in any layer depends to a great extent on the ratio of specific storage to permeability. From a practical point of view, the main concern is the response of the aquifer itself, although this, in fact, is controlled by the combined flow (or lack of flow) in the three-layered system. The purpose here is to investigate expressions for drawdown in the aquifer under conditions stipulated for the boundary layers.

It is assumed that flow toward a fully penetrating well in the aquifer is radial. Flow in the boundary layers, if it occurs at all, is vertical. At the interface between the boundary layers and the aquifer, the flow is considered totally refracted because of permeability differences. The semipervious layers are assumed to be elastic.

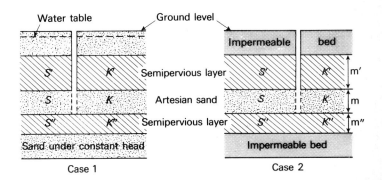

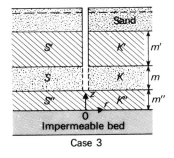

Fig. 7.9 Schematic representation of three flow systems. (*After Hantush*, 1960.)

For the three conditions to be discussed, Hantush developed asymptotic solutions for small and large values of time. If time is small, the drawdown for all three states can be expressed

$$s = \frac{Q}{4\Pi T} H(u,\beta) \tag{7.65}$$

where

$$u = \frac{r^2 S}{4Tt} \tag{7.66}$$

$$\beta = \tfrac{1}{4} r \lambda \tag{7.67}$$

$$\lambda = \left(\frac{k'/m'}{T} \frac{S'}{S}\right)^{1/2} + \left(\frac{k''/m''}{T} \frac{S''}{S}\right)^{1/2} \tag{7.68}$$

and $H(u,\beta)$ is an infinite integral whose value for short times is approximated

$$H(u,\beta) \approx W(u) - \frac{4\beta}{\sqrt{\Pi u}} \left(0.2577 + 0.6931 \exp \frac{-u}{2}\right) \tag{7.69}$$

where $W(u)$ is the well function of u. Some typical type curves of the function $H(u,\beta)$ are shown in Fig. 7.10.

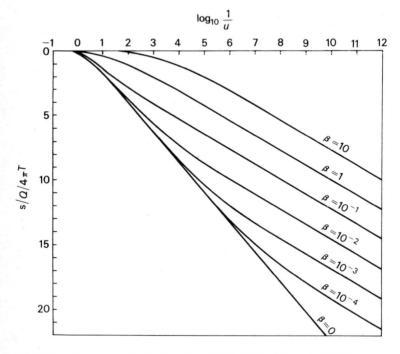

Fig. 7.10 Type curves for the function $H(u,\beta)$. (*After Hantush*, 1960.)

THEORETICAL MODELS OF THE UNSTEADY STATE   341

For large values of time, the drawdown in each of the states in Fig. 7.9 requires a separate expression.

For case 1,

$$s = \frac{Q}{4\Pi T} W(u\delta_1, \alpha) \tag{7.70}$$

where

$$\delta_1 = 1 + \frac{(S' + S'')}{3S} \tag{7.71}$$

$$\alpha = r\left(\frac{k'/m'}{T} + \frac{k''/m''}{T}\right)^{\frac{1}{2}} \tag{7.72}$$

and $W(u\delta_1, \alpha)$ is a well function to be considered later.

For case 2,

$$s = \frac{Q}{4\Pi T} W(u\delta_2) \tag{7.73}$$

where

$$\delta_2 = 1 + \frac{(S' + S'')}{S} \tag{7.74}$$

and $W(u\delta_2)$ is another well function to be considered shortly.

For case 3,

$$s = \frac{Q}{4\Pi T} W\left(u\delta_3, r\left(\frac{k'/m'}{T}\right)^{\frac{1}{2}}\right) \tag{7.75}$$

where

$$\delta_3 = 1 + \frac{(S'' + S'/3)}{S} \tag{7.76}$$

INCOMPRESSIBLE AND IMPERMEABLE CONFINING LAYERS

It is now possible to examine the effect of some modifications of the pertinent parameters given in Fig. 7.9. If the semipermeable layers are considered relatively incompressible and flow through them is retarded to the extent that it is negligible, S_s' and S_s'' and k' and k'' tend to zero. Accordingly, for very short times, the parameter β approaches zero [Eqs. (7.67) and (7.68)]. Equation (7.69) then becomes

$$H(u, \beta) \approx W(u) \tag{7.77}$$

The short-term drawdown expressed by Eq. (7.65) reduces to

$$s = \frac{Q}{4\Pi T} W(u) \tag{7.78}$$

which is the nonequilibrium equation of Theis.

Examining, now, cases 1, 2, and 3 for long times under the assumption of impermeable and incompressible confining layers:

Case 1: $\quad \delta_1 = 1 \quad\quad \alpha = 0 \quad\quad W(u\delta_1, \alpha) = W(u)$
Case 2: $\quad \delta_2 = 1 \quad\quad W(u\delta_2) = W(u)$
Case 3: $\quad \delta_3 = 1 \quad\quad r\left(\frac{k'/m'}{T}\right)^{\frac{1}{2}} = 0 \quad\quad W\left(u\delta_3, r\left(\frac{k'/m'}{T}\right)^{\frac{1}{2}}\right) = W(u)$

Hence, for long-term drawdown in all three cases,

$$s = \frac{Q}{4\Pi T} W(u) \tag{7.79}$$

In other words, when the confining layers are incompressible and impermeable, the drawdown for both short and long periods of pumping is described by the nonequilibrium equation of Theis. This describes the conditions for which application of the Theis equation is best suited.

INCOMPRESSIBLE CONFINING LAYERS WITH FINITE PERMEABILITY

The condition that the lower semipervious layer is incompressible and impervious (S_s'' and k'' equal zero) and the upper semipervious layer is incompressible (S_s' equals zero) but pervious ($k' \neq 0$) will now be examined. Interpreted literally, the upper semipervious layer cannot yield its stored water to the pumped aquifer, but is pervious to the vertical passage of water through it. The contributing water originates in the upper layer for cases 1 and 3, but does not exist for case 2 (Fig. 7.9).

For very short times, β approaches zero in cases 1 and 3 [Eqs. (7.67) and (7.68)]. Equation (7.69) then becomes

$$H(u, \beta) \approx W(u) \tag{7.80}$$

The short-time drawdown expressed by Eq. (7.65) then reduces to

$$s = \frac{Q}{4\Pi T} W(u) \tag{7.81}$$

which is the nonequilibrium development. For large values of time:

Case 1 $\quad \delta_1 = 1 \quad\quad \alpha = r\left(\frac{k'/m'}{T}\right)^{\frac{1}{2}} \quad\quad W(u\delta_1, \alpha) = W\left(u, r\left(\frac{k'/m'}{T}\right)^{\frac{1}{2}}\right)$

Case 3 $\quad \delta_3 = 1 \quad\quad \alpha = r\left(\frac{k'/m'}{T}\right)^{\frac{1}{2}} \quad\quad W(u\delta_3, \alpha) = W\left(u, r\left(\frac{k'/m'}{T}\right)^{\frac{1}{2}}\right)$

By letting

$$\frac{1}{B} = \left(\frac{k'/m'}{T}\right)^{\frac{1}{2}} \qquad (7.82)$$

the long-term drawdown for cases 1 and 3 is expressed

$$s = \frac{Q}{4\Pi T} W\left(u, \frac{r}{B}\right) \qquad (7.83)$$

which is the Hantush-Jacob nonsteady-radial-flow equation for an infinite leaky aquifer (Hantush and Jacob, 1955; Hantush, 1956). The term $W(u,r/B)$ is the tabulated well function for leaky aquifers.

It follows that short-term drawdowns for leaky aquifers are described by the Theis nonequilibrium equation, whereas long-term drawdowns are described by the well function $W(u,r/B)$. Values of $W(u,r/B)$ are given in Table 7.3, and a type curve is presented as Fig. 7.11. Note that the family of curves converges on the nonleaky curve for small values of $1/u$ (which correspond to small values of time).

The graphic procedure for solving for the formation constants is similar to that described previously. Early drawdown data from the field curve are matched with the nonleaky part of the curve, but they soon deviate and follow one of the leaky r/B curves (Fig. 7.12). The match point yields values of $W(u,r/B)$, $1/u$, t, and s. In addition, the r/B curve followed by the field data is noted. Transmissivity and storativity are readily determined from

$$T = \frac{Q}{4\Pi s} W\left(u, \frac{r}{B}\right) \qquad (7.84)$$

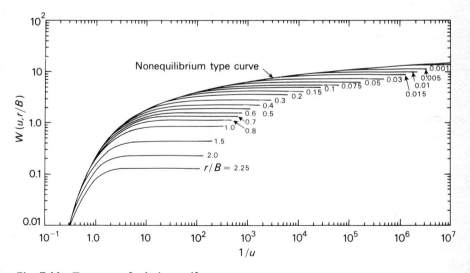

Fig. 7.11 Type curve for leaky aquifers.

Table 7.3 Values of $W(u, r/B)$ (*After Hantush*, 1956.)

r/B \ u	0.01	0.015	0.03	0.05	0.075	0.10	0.15	0.2	0.3	0.4
0.000001										
0.000005	9.4413									
0.00001	9.4176	8.6313								
0.00005	8.8827	8.4533	7.2450							
0.0001	8.3983	8.1414	7.2122	6.2282	5.4228					
0.0005	6.9750	6.9152	6.6219	6.0821	5.4062	4.8530				
0.001	6.3069	6.2765	6.1202	5.7965	5.3078	4.8292	4.0595	3.5054		
0.005	4.7212	4.7152	4.6829	4.6084	4.4713	4.2960	3.8821	3.4567	2.7428	2.2290
0.01	4.0356	4.0326	4.0167	3.9795	3.9091	3.8150	3.5725	3.2875	2.7104	2.2253
0.05	2.4675	2.4670	2.4642	2.4576	2.4448	2.4271	2.3776	2.3110	1.9283	1.7075
0.1	1.8227	1.8225	1.8213	1.8184	1.8128	1.8050	1.7829	1.7527	1.6704	1.5644
0.5	0.5598	0.5597	0.5596	0.5594	0.5588	0.5581	0.5561	0.5532	0.5453	0.5344
1.0	0.2194	0.2194	0.2193	0.2193	0.2191	0.2190	0.2186	0.2179	0.2161	0.2135
5.0	0.0011	0.0011	0.0011	0.0011	0.0011	0.0011	0.0011	0.0011	0.0011	0.0011

r/B \ u	0.5	0.6	0.7	0.8	0.9	1.0	1.5	2.0	2.5
0.000001									
0.000005									
0.00001									
0.00005									
0.0001									
0.0005									
0.001									
0.005									
0.01	1.8486	1.5550	1.3210	1.1307					
0.05	1.4927	1.2955	1.2955	1.1210	0.9700	0.8409			
0.1	1.4422	1.3115	1.1791	1.0505	0.9297	0.8190	0.4271	0.2278	
0.5	0.5206	0.5044	0.4860	0.4658	0.4440	0.4210	0.3007	0.1944	0.1174
1.0	0.2103	0.2065	0.2020	0.1970	0.1914	0.1855	0.1509	0.1139	0.0803
5.0	0.0011	0.0011	0.0011	0.0011	0.0011	0.0011	0.0010	0.0010	0.0009

and

$$S = \frac{4uTt}{r^2 S} \tag{7.85}$$

In addition, the coefficient of vertical permeability can be determined from Eq. (7.82). For convenience, both sides are multiplied by the distance r

$$\frac{r}{B} = r\left(\frac{k'/m'}{T}\right)^{\frac{1}{2}} \tag{7.86}$$

so that r/B is a dimensionless quantity, and

$$\left(\frac{r}{B}\right)^2 = r^2\left(\frac{k'/m'}{T}\right) \tag{7.87}$$

THEORETICAL MODELS OF THE UNSTEADY STATE   345

Solving for vertical permeability,

$$k' = \frac{Tm'(r/B)^2}{r^2} \tag{7.88}$$

Example 7.6 The data in Fig. 7.12 are similar to those in Fig. 7.3 except for the leakage. Assuming the data were obtained from an observation well located 500 ft from a well pumped at a rate of 192×10^3 ft^3/day (about 1,000 gpm), the coefficient of transmissivity is calculated

$$T = \frac{Q}{4\Pi s} W\left(u, \frac{r}{B}\right) = \frac{(192 \times 10^3 \text{ ft}^3/\text{day})(1)}{4(3.12)(2 \text{ ft})} = 7.7 \times 10^3 \text{ ft}^2/\text{day}$$

The coefficient of storage is

$$S = \frac{4uTt}{r^2} = \frac{4(1 \times 10^{-3})(7.7 \times 10^3 \text{ ft}^2/\text{day})(7 \times 10^{-1} \text{ days})}{250 \times 10^3 \text{ ft}^2} = 8.6 \times 10^{-5}$$

Both results are identical with the numerical values obtained for Fig. 7.3. Assuming the semipervious layer is 10 ft in thickness.

$$k' = \frac{Tm'(r/B)^2}{r^2} = \frac{(7.7 \times 10^3 \text{ ft}^2/\text{day})(10 \text{ ft})(5.6 \times 10^{-3})}{250 \times 10^3 \text{ ft}^2} = 15.7 \times 10^{-4} \text{ ft/day}$$

From the preceding developments, the leaky- and nonleaky-radial-flow equations are special cases of the general flow equations developed by Hantush (1960; 1964). Solutions for the general condition are rarely used, although Hantush (1960) tabulated a few values of the function $H(u,\beta)$ which

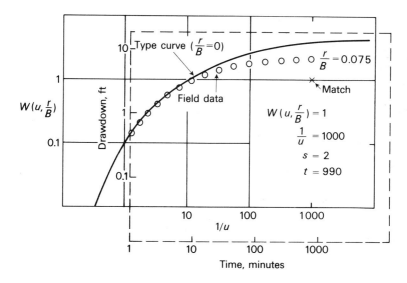

Fig. 7.12 Graphic procedure for solving for formation constants for leaky aquifers.

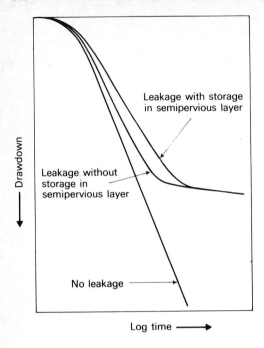

Fig. 7.13 Comparison of time/drawdown variations for leaky and nonleaky aquifers. (*After Hantush*, 1960.)

may be adequate for application. The procedure for analysis is as follows:

1. A family of type curves for the function $H(u,\beta)$ and the early data of a pumping test will produce a solution for T [Eq. (7.65)], S [Eq. (7.66)], and β (β is a parameter of the type curve in the same manner that r/B is a parameter for the leaky curve).
2. The usual procedures for leaky aquifers, using Eq. (7.70) for case 1 for large values of time will produce a solution for T, α, and $\delta_1 S$.
3. From S and $\delta_1 S$, the storativity of the semipervious layer is obtained. From α and r, the vertical permeability of the semipermeable layer is obtained.
4. Steps 1, 2, and 3, are (by using the pertinent parameters) applicable to case 3.
5. The usual procedure for nonleaky aquifers, using Eq. (7.73) for case 2 and the data collected for large values of time, will produce a solution for T and $\delta_2 S$. From S of step 1, and $\delta_2 S$, the storativity of the semipervious layer is determined.

Figure 7.13 shows a comparison of time/drawdown variations for the cases discussed.

7.5 RADIAL FLOW TO A WELL IN A FINITE AQUIFER

An assumption employed throughout this chapter is that the aquifer is infinite in areal extent. Geologic boundaries limit the extent of real aquifers, and serve to distort the calculated cones of depression forming around pumping wells. The method of images, which plays an important role in the mathematical theory of electricity and is employed in the solution of some geophysical problems, aids in the evaluation of the influence of aquifer boundaries on well flow. This theory, first developed in published form by Ferris (1959), permits treatment of the aquifer limited in one or more directions. However, the additional assumption of straight-line boundaries has been added. This gives aquifers of rather simple geometric form.

The image well theory can be explained as follows (Stallman, 1952): Formation A is bounded by the relatively impermeable formation B, the boundary between the two located at a variable distance r from a pumping well (Fig. 7.14a). As formation B is relatively impermeable, no flow can occur from it toward the pumping well, and the boundary is referred to as a *no-flow (barrier) boundary*. The effect of a barrier boundary is to increase the drawdown in a well. The problem now is to duplicate this physical situation by substituting a hydraulic entity which serves this purpose. As no flow occurs across a groundwater divide, the barrier boundary is simulated

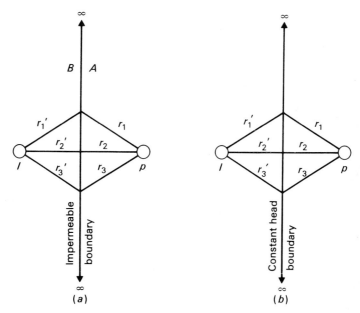

Fig. 7.14 Simple two-well system for an aquifer bounded by (*a*) a no-flow boundary and (*b*) a constant head boundary. (*After Stallman, 1952.*)

by the supposition that formation A is infinite in areal extent, and that an imaginary well is located across the real boundary, on a line at right angles thereto, and at the same distance from the boundary as the real pumping well. If the imaginary well is assumed to start pumping at the same time and at the same rate as the real well, the boundary will be transformed into a groundwater divide.

Similarly, if the aquifer is bounded by a stream that provides recharge to the aquifer, the effect is to decrease the drawdown in a well. A zero-drawdown (constant-head) boundary can be simulated by an imaginary well, located as above, with the exception that the imaginary well must recharge water at the same rate as the pumping well.

The drawdown at any point in the real aquifer or at the boundary for the simple two-well system is the sum of the effects of the real and imaginary wells operating simultaneously,

$$s = s_p \pm s_i = \frac{Q}{4\Pi T}[W(u)_p \pm W(u)_i] = \frac{Q}{4\Pi T}\sum W(u) \qquad (7.89)$$

where s is the observed drawdown at any point, consisting of the sum of the effects of the real well s_p and the imaginary well s_i; the well function for the real well is $W(u)_p$, and $W(u)_i$ is the well function for the imaginary well. If the imaginary well is a recharging well, the negative sign is used.

By similar reasoning,

$$u_p = \frac{r_p^2 S}{4Tt} \qquad u_i = \frac{r_i^2 S}{4Tt} \qquad (7.90)$$

where r_p is the distance from the pumping well to the observation point, and r_i is the distance from the imaginary well to the observation point. Following the procedure of Kazmann (1946),

$$u_i = \left(\frac{r_i}{r_p}\right)^2 u_p \qquad (7.91)$$

or

$$u_i = K^2 u_p \qquad (7.92)$$

where K equals the constant r_i/r_p. For any point on the boundary, r_p equals r_i (k equals 1), and u_i equals u_p. It follows that $W(u)_i$ equals $W(u)_p$ [Eq. (7.89)], and the drawdown is either zero (constant-head boundary) or twice the effect of one well pumping in an infinite aquifer.

To demonstrate that the boundary is actually a groundwater divide if there are barriers to flow, consider the following: At any point in the real aquifer, $r_i > r_p$, $K^2 > 1$, and $W(u)_p > W(u)_i$. For a corresponding point across the boundary, $r_i < r_p$, $K^2 < 1$, and $W(u)_p < W(u)_i$. As mentioned above, $W(u)_p$ equals $W(u)_i$ at the boundary.

When a well in a bounded aquifer is pumped, water levels in observation wells will initially decline under the influence of the pumping well only. When the cone of depression reaches an exterior boundary, deviations from the ideal response will be noted in the observation well. Under these conditions, the early part of the time/drawdown curve unaffected by the boundary behaves as if the aquifer were infinite in areal extent, and this part of the curve can be used to determine the hydraulic properties.

Figure 7.15 shows the effects of pumping near a barrier and recharge boundary.

MULTIPLE-BOUNDARY AQUIFERS

Aquifers are often bounded by two or more boundaries, and time/drawdown data will respond accordingly. As mentioned above, the resulting geometry must be rather simple in order to apply the image theory. Hence, two converging boundaries delineate a wedge-shaped aquifer, two parallel boundaries intersected at right angles by a third boundary forms a semiinfinite strip, and four intersecting right-angle boundaries form a rectangular aquifer.

A number of image wells associated with each pumping well characterizes a multiple-boundary system. Clearly, primary image wells placed across a boundary will balance the effect of a pumping well at that boundary, but will cause an unbalanced effect at the opposite boundary in violation of the no-flow or constant-head requirement. It is then necessary to add secondary image wells at appropriate distances to satisfy the conditions of no flow or zero drawdown. In examining this problem, Ferris and others (1962) recommended the following:

> The simplest way to analyze any multiple-boundary problem is to consider each boundary separately and determine how best to meet the condition of no flow or no drawdown, as the case may be, at that boundary. After the positions of the primary image wells have been established, the boundary positions should be re-examined to see if the net drawdown effects of the primary image wells satisfy all stipulated conditions of no flow or no drawdown. For each primary image causing an unbalance at a boundary position, or extension thereof, it is necessary to place a secondary image well at the same distance from the boundary, but on the opposite side, both wells occupying a common line perpendicular to the boundary. When the combined drawdown (or buildup) effects of all image wells are found to produce the desired effect at this boundary, the same procedure is executed with respect to the second boundary. Thus, the inspection and balancing process is repeated around the system until everything is in balance and all boundary conditions are satisfied, or until the effects of additional image wells are negligible compared to the total effect.

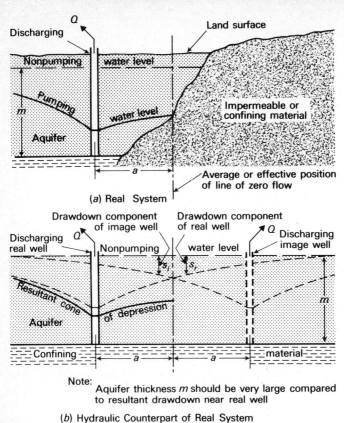

Fig. 7.15 Diagrams of the effect of pumping near barrier and constant-head boundaries and appropriate hydraulic counterparts for image well theory. (*After Ferris and others*, 1962.)

Figure 7.16 shows some plan views of image well systems for some simple aquifers.

Once the boundaries of a finite aquifer have been simulated by means of image wells, analysis of drawdown effects can proceed as if the aquifer were infinite in areal extent. In other words, graphic models such as given in Fig. 7.4 can be used to find the drawdown in response to one or several wells pumping simultaneously. Once the distance from real and imaginary wells to the point of interest has been determined, drawdown (or recovery) may be obtained from the graphs. For example, the drawdown at any point in the real aquifer of Fig. 7.16a is the algebraic sum of the effect of one real well, one discharging imaginary well, and two recharging imaginary wells. For the infinite array (Fig. 7.16c), image wells are added until the most remote pair has negligible influence on water-level response.

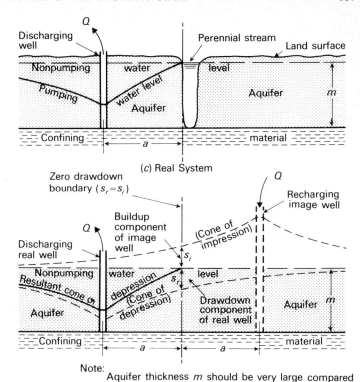

(c) Real System

(d) Hydraulic Counterpart of Real System

Note: Aquifer thickness m should be very large compared to resultant drawdown near real well

7.6 WATER–LEVEL RESPONSE AT THE PUMPED WELL

Well capacity, or discharge per unit time, commonly expressed in gallons per minute, is often used as a measure of well yield. For comparative purposes, it is better to express well capacity in relation to some arbitrary, but constant, standard. The accepted standard of comparison is the unit drawdown, giving gallons per minute per foot of drawdown, or "specific" capacity. Theis and others (1963) demonstrated that the theoretical specific capacity of a well can be determined from the abbreviated nonequilibrium equation

$$T = \frac{Q}{4\Pi s}\left(-0.5772 - \ln\frac{r^2 S}{4Tt}\right) \tag{7.93}$$

or, solving for Q/s in practical American hydrologic units,

$$\frac{Q}{s} = \frac{T}{264 \log (Tt/1.87 r_w^2 S) - 65.5} \tag{7.94}$$

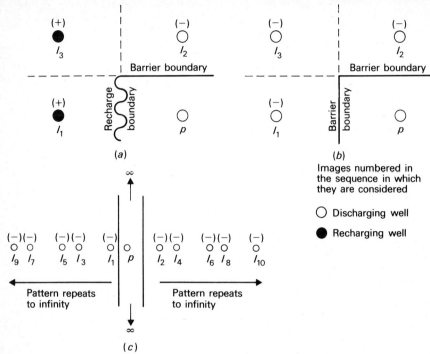

Fig. 7.16 Image well systems for (a) wedge-shaped aquifer with both barrier and recharge boundaries, (b) wedge-shaped aquifer with barrier boundaries, and (c) infinite strip.

where t = pumping period, days

r_w = effective radius of the well, ft

Q/s = theoretical specific capacity, gpm/ft of drawdown.

The following assumptions were made: (1) Well loss is negligible; (2) the effective radius of the well has not been affected by drilling and well development, and is equal to the nominal radius of the well.

From this equation, theoretical specific capacity of a well is directly proportional to T, and inversely proportional to $\log t$, $\log 1/r_w^2$, and $\log 1/S$. Hence, large changes in T cause correspondingly large changes in specific capacity. Large changes in t, r_w, and S cause comparatively small changes in specific capacity. Given the actual specific capacity of a pumped well, the radius of the well, and the duration of pumping, and with the assumption of a value for S, the transmissivity can be estimated with Eq. (7.94).

The effective radius of a well is seldom equal to actual well radius. Because specific capacity gets large as r_w gets large, well development techniques employed by drillers are aimed at increasing effective radius. These include gravel packing, overpumping, backwashing, and surging (Johnson, Inc., 1959).

WELL LOSS

Drawdown in a pumping well not subject to interference from other pumping wells is composed of two parts: (1) drawdown due to laminar flow of water in the aquifer toward the well, referred to as *formation loss s* and calculated with the well-flow equations discussed in this chapter; and (2) drawdown resulting from the turbulent flow of water in the immediate vicinity of the well, through the well screen or casing openings and in the well casing, referred to as *well loss s_w*. Jacob (1950) stated that well loss is proportional to some power of the discharge exceeding the first power and approaching the second. As an approximation,

$$s_w = CQ^2 \tag{7.95}$$

where s_w = well loss (L)
C = well-loss constant (T^2L^{-5})
Q = pumping rate (L^3T^{-1})

Total drawdown in a pumping well, s_t, is then expressed

$$s_t = s + s_w = \frac{Q}{4\Pi T} W(u) + CQ^2 \tag{7.96}$$

For a given time and well radius, $W(u)/4\Pi T$ is a constant, and Eq. (7.96) becomes

$$s_t = BQ + CQ^2 \tag{7.97}$$

where B is a constant. The development of Eq. (7.94) assumed C is zero.

Procedures for solving for the well-loss constant C have been discussed by Jacob (1950), Bruin and Hudson (1958), and Walton (1962). A step-drawdown test is generally required, in which a well is operated during successive periods at constant fractions of its full capacity (Fig. 7.17). Bruin and Hudson described a graphic procedure based on Jacob's original work, where s_t/Q versus Q is plotted on arithmetic paper. The value of B is the intercept of the line, and C is the slope. Walton calculates C directly from Jacob's equations

$$C = \frac{\Delta s_i/\Delta Q_i - \Delta s_{i-1}/\Delta Q_{i-1}}{\Delta Q_{i-1} + \Delta Q_i} \tag{7.98}$$

or, for steps 1 and 2 of Fig. 7.17,

$$C = \frac{\Delta s_2/\Delta Q_2 - \Delta s_1/\Delta Q_1}{\Delta Q_1 + \Delta Q_2} \tag{7.99}$$

and for steps 2 and 3,

$$C = \frac{\Delta s_3/\Delta Q_3 - \Delta s_2/\Delta Q_2}{\Delta Q_2 + \Delta Q_3} \tag{7.100}$$

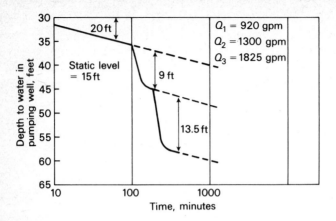

Fig. 7.17 Semilogarithmic plot of time/drawdown curve obtained during a step-drawdown test.

Given the results of a step-drawdown test, it is possible to construct drawdown/yield curves for a pumping well (Fig. 7.18). For any pumping rate, well efficiency is expressed as the ratio of drawdown due to formation loss to total drawdown, or

$$E = \frac{s}{s_t} \times 100 \tag{7.101}$$

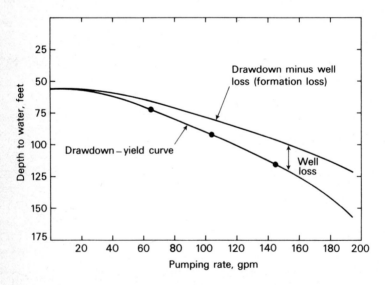

Fig. 7.18 Arithmetic plot of drawdown versus yield.

GEOLOGIC CONTROLS ON SPECIFIC CAPACITY

Many wells are furnished water from more than one aquifer. To study the relative yields of individual aquifers, specific capacity per foot of well penetration can be tabulated as to order of magnitude, and frequencies computed with the following formula

$$F = \frac{m_0}{n_w + 1} 100 \qquad (7.102)$$

where m_0 is the order number of the well, n_w is the total number of wells, and F is the percentage of wells whose specific capacities are equal to, or greater than, the specific capacity of the order number (Walton, 1962). Values of specific capacity per foot of penetration are then plotted against percentage of wells on logarithmic probability paper.

For the three categories in Fig. 7.19, specific capacity per foot of penetration decreases with increased well depth, which indicates that where they overlie one another, the Niagaran and Alexandrian series are more

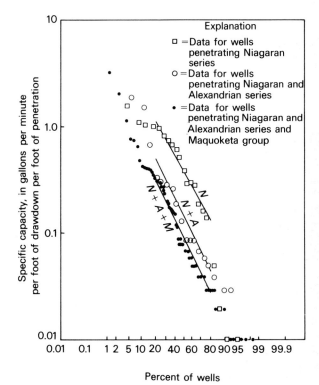

Fig. 7.19 Specific capacity/frequency graphs. (*After Zeizel and others*, 1962.)

productive than the Maquoketa formation, and the Niagaran series is more productive than the Alexandrian series. Other studies by Walton and Neill (1963) indicated that significant correlations exist between specific capacity and thickness of the Niagaran series, and specific capacity and thickness of reef deposits. Csallany and Walton (1963) demonstrated that specific capacity values are influenced by anticlinal or synclinal location of a well, and thickness of glacial drift overlying a dolomite aquifer. LeGrand (1954) has made similar studies in an area underlain by igneous and metamorphic rock.

B. CONSOLIDATION THEORY

The previous sections in this chapter have treated the problem of unsteady flow to a pumping well in a saturated porous medium. Most of the solutions discussed represent solutions of the diffusion equation for a prescribed set of initial and boundary conditions. It will be recalled that this equation was derived by equating the net inward flux of a differential element with the rate of change of fluid mass within the element. Storage occurs in such an element if the fluid, or the medium, or both, are compressible. It is now appropriate to consider other problems that may be described by the diffusion equation, such as the one-dimensional flow of an incompressible fluid in a compressible medium. Several interesting conditions may be described by such an analysis, including the transient flow in low-permeability units in response to changes in hydraulic head in boundary aquifers, in response to rates of sedimentation in water-saturated environments, or in response to other forms of loading. The development that follows will be quite general, with emphasis on the fact that one descriptive differential equation may be employed to describe these phenomena.

7.7 THEORETICAL DEVELOPMENT FOR ONE-DIMENSIONAL CONSOLIDATION

The slow compression of saturated clays was demonstrated as presenting a problem related to the hydraulic rather than the strength properties of the material (Section 5.4). It will be recalled that when a saturated elemental volume is confined laterally and is subject to vertical pressure, volume reduction can come about only if some of the pore water is extruded. The volume reduction at any time is totally accounted for by the decrease in void volume. The process of volume reduction in response to an extrusion of pore water is termed *consolidation*. Total volume change is a function of the compressibility (storage) properties of the medium, whereas the time rate of volume change is a function of both compressibility and permeability.

Soft clays have a relatively high specific storage and low permeability,

THEORETICAL MODELS OF THE UNSTEADY STATE 357

so that their contained water is expelled slowly. During the deposition of such a sediment, it is expected that a small element of the clay deposit is characterized by a high void ratio. This element may gradually be squeezed down by the weight of additional material during the process of sedimentation. In order to examine the internal pore-water characteristics within the element, it is necessary to reexamine the concept of effective stress. Restated for the condition of excess fluid pressures,

$$\sigma = \bar{\sigma} + (P_s + \bar{P}) \tag{7.103}$$

where the total pore-water pressure consists of two parts,

$$P = P_s + \bar{P} \tag{7.104}$$

where P_s is the steady-state pressure, and $\bar{P}$ is a transient pore-water pressure in excess of the steady-state pressure. For the problem under consideration, continued deposition generates an excess component of fluid pressure, which causes some of the water to move slowly to regions of less rapid sedimentation. It is understood that this flow takes place laterally as well as vertically. However, because the areal extent of deposition is large with respect to sediment thickness, lateral deformation of any element may be negligible, and it is assumed that the flow process and attendant volume reduction occurs in a vertical direction only. With complete dissipation of the excess pressure, the water pressure in the element is reduced to its steady-state value.

In that the decay of pore-water pressure increases effective stress, its effect on void ratio is clearly shown by the void-ratio–effective-stress curve obtained from laboratory testing of clay specimens (Fig. 7.20). The slope of this curve represents the coefficient of compressibility

$$\frac{\partial e}{\partial \bar{\sigma}} = -a_v \tag{7.105}$$

the negative sign indicating that the coefficient of compressibility decreases as the pressure increases. The time rate of compression may be examined from Eq. (7.103),

$$\frac{\partial \sigma}{\partial t} = \frac{\partial \bar{\sigma}}{\partial t} + \frac{\partial P_s}{\partial t} + \frac{\partial \bar{P}}{\partial t} \tag{7.106}$$

As the excess pressure is responsible for the flow and attendant volume changes, the steady-state pressure P_s is assumed to be invariant with time, and vanishes. Substituting Eq. (7.105) in Eq. (7.106), and assuming $\partial P_s/\partial t$ equals zero, gives, upon rearranging terms,

$$\frac{1}{a_v}\frac{\partial e}{\partial t} = \frac{\partial \bar{P}}{\partial t} - \frac{\partial \sigma}{\partial t} \tag{7.107}$$

From developments described in Section 5.4,

$$a_v = \frac{1 + e}{E_c} \tag{7.108}$$

Substituting this result in Eq. (1.107),

$$\frac{E_c}{1 + e}\frac{\partial e}{\partial t} = \frac{\partial \bar{P}}{\partial t} - \frac{\partial \sigma}{\partial t} \tag{7.109}$$

A verbal interpretation of this equation is as follows: The time rate of change of water expulsion from a compressible element (left-hand side) equals the difference between the time rate of change of excess-pressure development and the time rate of change of the external load causing the compression (right-hand side).

Given this intuitive insight into the nature of Eq. (7.109), it is possible now to examine some special cases. Under conditions of very slow sedimentation in a water-saturated environment, pore-water expulsion may be able to maintain itself at the same rate as deposition, which renders the excess-pressure buildup essentially zero. For this condition, the first term on the right-hand side of Eq. (7.109) vanishes, and

$$\frac{E_c}{1 + e}\frac{\partial e}{\partial t} = \frac{-\partial \sigma}{\partial t} \tag{7.110}$$

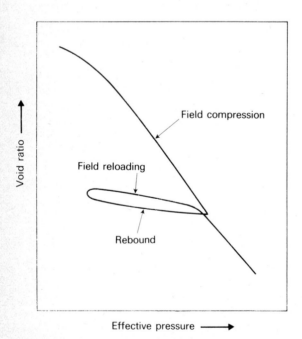

Fig. 7.20 Void ratio versus effective pressure as obtained from laboratory consolidation.

Interpreted literally, the rate of water expulsion is equal to the rate of sediment deposition. It follows that the water pressure in the clay deposit following deposition will be "normal" for its depth, described approximately by a hydrostatic law. For continued deposition, volume change is adequately described by the void-ratio–effective stress curve, with instantaneous compression accompanying the deposition.

If the geologic basin is uplifted slightly so that deposition is halted and erosion takes place, the effect is the same as a decrease in effective stress. This gives rise to rebound. Because of irreversible changes in the compressible matrix, the rebound curve is flat (Fig. 7.20). Continued basin deposition some time after rebound promotes the formation of a hysteresis loop, which is merely the result of the application of a stress to a given level, removal of that stress, and application of another cycle of loading.

If the stress causing the excess pressure is applied rapidly, and then held constant, the second term on the right-hand side of Eq. (7.109) vanishes, and

$$\frac{E_c}{1+e}\frac{\partial e}{\partial t} = \frac{\partial \bar{P}}{\partial t} \qquad (7.111)$$

That is, the rate of water expulsion equals the rate of decay of excess pressure. A solution to the differential equation equivalent of Eq. (7.111) is the Terzaghi (1925) consolidation equation. This equation may be applied to situations where the loading is applied instantaneously, and thereafter held constant. This is generally the case in construction of structures on soils overlaying compressible clay units, or for the rapid, one-time decrease in hydraulic head by a given amount in aquifers bounding compressible clay units (Domenico and Mifflin, 1965).

Another example is clay deposition in a water-saturated environment, where the clay acts as a hermetically sealed container, allowing no excess-pressure decay. Here, the increase in stress caused by continued deposition is carried exclusively by the pore water in the clay. As the rate of increase in excess fluid pressure is equal to the rate of sedimentation,

$$\frac{E_c}{1+e}\frac{\partial e}{\partial t} = 0 \qquad (7.112)$$

and the volume of the element remains unchanged. Stated in another way, no water is expelled in the absence of compression. If, upon completion of deposition, the excess pressure is assumed to commence its dissipation process, Eq. (7.111) applies. These assumptions have been employed to study the decay of excess pressure in thick sequences of clay (Bredehoeft and Hanshaw, 1968), and are implicit in studies relating excess fluid pressure to overthrust faulting (Rubey and Hubbert, 1959). A more reasonable approach would take into account the maintenance of excess fluid pressures during sedimentation, where $\partial\sigma/\partial t$ of Eq. (7.109) equals the rate of sedimentation.

7.8 MATHEMATICAL ANALYSIS

The analysis presented in the preceding section is sufficient in most circumstances for an intuitive insight into the consolidation process, but is not sufficiently descriptive for a quantitative description of the actual flow process. Clearly, partial differential equations incorporating both space and time variables are required. Most of the effort in obtaining such descriptive equations has already been made in Chapter 5.

It will be recalled that the transient change in fluid-mass equation was originally expressed in terms of time rates of change of effective and fluid pressure

$$\frac{\partial(\Delta M_w)}{\partial t} = \left[-\rho_w n\alpha \, \Delta z \, \frac{\partial \bar{\sigma}}{\partial t} - \rho_w \, \Delta z \, \alpha(1-n)\frac{\partial \bar{\sigma}}{\partial t} + \rho_w \, \Delta z \, n\beta \, \frac{\partial \bar{P}}{\partial t}\right] \Delta x \, \Delta y \tag{7.113}$$

Following this development, it was assumed that a change in fluid pressure was accompanied by an instantaneous change in effective stress, which allowed Eq. (7.113) to be expressed in terms of the time rate of change of fluid pressure and, ultimately, the time rate of change of total head. Faced now with the possibility that stresses may not always be applied instantaneously, or they may not be accompanied instantaneously by strains, one is required to consider the complete statement of effective stress [Eq. (7.106)],

$$\frac{\partial \bar{\sigma}}{\partial t} = \frac{\partial \sigma}{\partial t} - \frac{\partial \bar{P}}{\partial t} \tag{7.114}$$

where the first term on the right-hand side accounts for the rate of stress application, and only excess pressures are considered. Substituting this result in Eq. (7.113) and assuming the fluid is incompressible gives, upon clearing terms,

$$\frac{\partial(\Delta M_w)}{\partial t} = \rho_w \alpha \left(\frac{\partial \bar{P}}{\partial t} - \frac{\partial \sigma}{\partial t}\right) \Delta x \, \Delta y \, \Delta z \tag{7.115}$$

Combining this equation with the one-dimensional version of the conservation-of-mass principle [Eq. (5.91)],

$$-\left(\frac{\partial}{\partial z}\rho_w q_z\right) = \rho_w \alpha \left(\frac{\partial \bar{P}}{\partial t} - \frac{\partial \sigma}{\partial t}\right) \tag{7.116}$$

gives the statement of continuity required.

THEORETICAL MODELS OF THE UNSTEADY STATE   361

As water flows in response to a gradient in total head, Darcy's law may be restated to incorporate three components of head

$$q = -k' \frac{\partial h}{\partial z} = -k' \frac{\partial}{\partial z}\left(z + \frac{P_s}{\rho_w g} + \frac{\bar{P}}{\rho_w g}\right) \qquad (7.117)$$

where k' is vertical permeability. However, the flow process in consolidation theory is a transient one, and occurs in response to a gradient in excess pressure, not total head. By considering both the position and steady-state pressure head to be invariant throughout the process, they are conveniently eliminated, which gives

$$q = \frac{-k'}{\rho_w g}\frac{\partial \bar{P}}{\partial z} \qquad (7.118)$$

Substituting Eq. (7.118) in Eq. (7.116) gives, upon rearranging terms,

$$\frac{k'}{\rho_w g \alpha}\frac{\partial^2 \bar{P}}{\partial z^2} = \frac{\partial \bar{P}}{\partial t} - \frac{\partial \sigma}{\partial t} \qquad (7.119)$$

It follows that $k'/\rho_w g \alpha$ is the ratio of permeability to specific storage for the case where water is incompressible. This equation is equivalent to Eq. (7.109), and may be given similar interpretation. In essence, it states that the volume of water expelled from the voids of an element of clay per unit area per unit time (left-hand side) equals the difference between the time rate of change of excess-fluid pressure and the time rate of change of application of external load causing the compression (right-hand side). The larger the specific storage, or the smaller the permeability, the longer the time required to approach a terminal steady state. It follows further that the expulsion process is not independent of the rate of loading. At one extreme, the rate of increase of water pressure would be equal to the rate of loading if no dissipation took place. At the other extreme, a slow rate of loading (say deposition in a water-saturated environment) accompanied by "instantaneous" water-pressure dissipation may render the excess-pressure buildup essentially zero, which gives the special case equivalent of Eq. (7.110)

$$\frac{k'}{S'_s}\frac{\partial^2 \bar{P}}{\partial z^2} = -\frac{\partial \sigma}{\partial t} \qquad (7.120)$$

Here, the rate of effective pressure development equals the rate of sediment deposition, which gives the condition of normal water pressure.

If the loading, or stress, causing the excess-pressure buildup is applied rapidly and then held constant, the equivalent of Eq. (7.111) arises,

$$\frac{k'}{S'_s}\frac{\partial^2 \bar{P}}{\partial z^2} = \frac{\partial \bar{P}}{\partial t} \qquad (7.121)$$

which is the Terzaghi (1925) consolidation equation. This equation may be derived directly from the equation of transient change in fluid mass by employing Jacob's (1950) assumption of an instantaneous stress transfer from fluid pressure to intergranular pressure.

BOUNDARY CONDITIONS, DIMENSIONLESS VARIABLES, AND FINAL SOLUTIONS

The linear differential equations cited above merely represent a physical statement of the general problem under consideration. To obtain specific solutions, some assumptions must be made which pertain to initial and boundary conditions. As mentioned in Section 5.4, a variety of initial and boundary conditions may be employed.

As an example of the above discussion, consider the following problem treating the decay of excess pressure in a thick sequence of clays. The clay body is initially at some uniform excess pressure, whereas the pressure at its upper surface is somewhat lower. The problem is similar to that encountered in heat-flow theory, where the initial temperature distribution in a semiinfinite medium is uniform and suddenly is subjected to a lower temperature at the surface (or, identically, where the temperature throughout the medium is raised instantaneously to a constant value at all points, except at the upper surface). These assumptions are clearly demonstrated in the boundary conditions (Bredehoeft and Hanshaw, 1968)

$$h'(z,0) = h_i \quad \text{for } t = 0$$
$$h'(0,t) = 0 \quad \text{for } t > 0$$
$$h'(\infty,t) = h_i \quad \text{for } t > 0$$

where h' is excess head, or $\bar{P}/\rho_w g$, and h_i is initial head, or $\bar{P}_i/\rho_w g$. The first (initial) condition stipulates a uniform excess pressure with depth z. The second condition is required to maintain a pressure gradient across the upper surface for all times. The third condition stipulates that initial excess head shall be maintained at an infinite depth for all times. The differential equation is of the form of Eq. (7.121), and the solution is (Carslaw and Jaeger, 1959)

$$h' = h_i \operatorname{erf} \frac{z}{2(k't/S_s')^{\frac{1}{2}}} \tag{7.122}$$

where erf is the tabulated error function.

It is important to notice in this case that the result expressed in Eq. (7.122) depends on the single dimensionless parameter

$$\frac{z}{2(k't/S_s')^{\frac{1}{2}}}$$

THEORETICAL MODELS OF THE UNSTEADY STATE   363

It follows that, in any material, the time required for any point to reach a given pressure is proportional to the square of its distance from the surface. Further, the time required to attain a given pressure varies directly with specific storage and inversely with permeability. Given values of S'_s and k', it is easy to compare excess heads at different times and depths with the aid of Fig. 7.21.

Bredehoeft and Hanshaw (1968) also described the establishment and maintenance of excess pressures in response to basin sedimentation. The solution presented is a special solution of Eq. (7.119). The initial conditions assume a layer of zero thickness, and the boundary conditions allow no drainage at the bottom of a layer increasing in thickness, but require an excess gradient across the upper boundary of the layer. The authors concluded that the establishment and maintenance of excessive-fluid pressures depend largely on the values of k' and S'_s.

We now turn attention to an identical heat-flow process in a finite solid bounded by a pair of parallel planes. The parallel planes coincide with the upper and lower surfaces of a clay unit bounded by units of higher permeability. The region of the solid $-L < z < L$ is given a zero initial temperature, and the surfaces $z = \pm L$ are assumed to be at some constant temperature T_e for $t > 0$ (Carslaw and Jaeger, 1959). Hence there is a flow of heat from the surfaces into the slab.

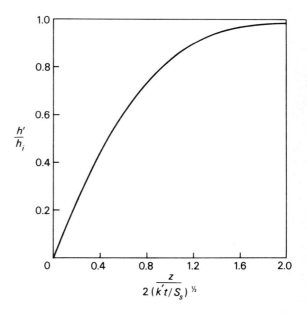

Fig. 7.21 The function erf $[z/2(k't/S'_s)^{\frac{1}{2}}]$. (*After Carslaw and Jaeger, 1959.*)

The analogous situation is shown in Fig. 7.22a. At time t equals zero, the temperature at all points in the slab is zero. After heat is applied to the surfaces, the temperature distribution changes as a function of time and depth, the temperature increasing more rapidly near the surfaces and least rapidly near the center. The curves represent the temperature distribution at any time, and are called *isochrones*. At time approaching infinity, the temperature throughout the slab is constant, and equals the applied temperature. It follows that the transient process is then completed.

Strict application of the heat-flow model to an analogous groundwater flow problem suggests that pressure levels in boundary aquifers must be lowered rapidly, and then maintained at a relatively constant level. Initially, because of constant pressure with depth, it is assumed that no flow occurs within the clay layer, Fig. 7.22b. The immediate lowering of the pressure in

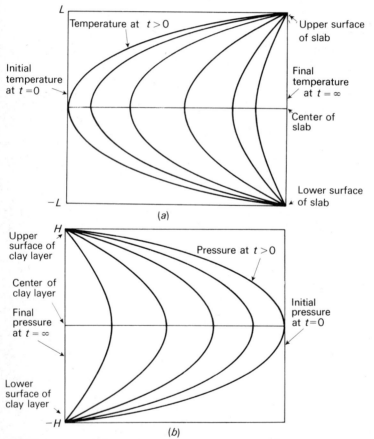

Fig. 7.22 Diagrams of (a) temperature buildup in a slab as a function of time and depth and (b) analogous pore-water pressure decline in a confining layer as a function of time and depth.

THEORETICAL MODELS OF THE UNSTEADY STATE

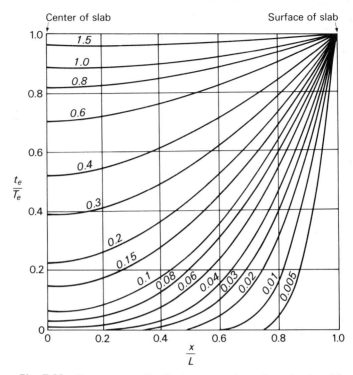

Fig. 7.23 Temperature distribution at various times in the slab $-L < x < L$, with zero initial temperature and surface temperature T_e. The numbers on the curves are the values of a time factor T^*. (*After Carslaw and Jaeger*, 1959.)

the boundary aquifers then has the same effect as an immediate application of heat to the surfaces of the slab, the only difference being that the gradient at any time is away from the center of the clay layer and toward the surfaces. It follows that Eq. (7.121) applies

$$\frac{\partial^2 \bar{P}}{\partial z^2} = \frac{S_{s'}}{k'} \frac{\partial \bar{P}}{\partial t} \tag{7.123}$$

in conjunction with the following initial and boundary conditions:

$$\bar{P}(z,0) = \bar{P}_i \quad \text{for } t = 0$$
$$\bar{P}(H,t) = 0 \quad \text{for } t \geq 0$$
$$\bar{P}(-H,t) = 0 \quad \text{for } t \geq 0$$

The initial condition corresponds to a constant value of excess pressure with depth.

The solution provided by Carslaw and Jaeger (1959) is in the form of a family of curves in two dimensions, and it demonstrates the values of t_e/T_e for various values of a dimensionless time factor T^* (Fig. 7.23). The temperature ratio t_e/T_e represents the temperature buildup in the slab, where t_e is the temperature at any time, and T_e is the constant temperature applied to the surfaces. At time t equals zero, t_e/T_e equals zero; at time t equals infinity, t_e/T_e equals 1. The ratio t_e/T_e is analogous to excess-pressure decay, stated for outward flow as $1 - \bar{P}/\bar{P}_i$, where $\bar{P}_i$ represents the value of the initial excess water pressure in the clay unit, and $\bar{P}$ is its value at any time later. At time t equals zero, $\bar{P}$ equals $\bar{P}_i$, and $1 - \bar{P}/\bar{P}_i$ equals zero; at time t equals infinity, $\bar{P}$ equals zero, and $1 - \bar{P}/\bar{P}_i$ equals 1. In other words, as the excess pressure $\bar{P}$ varies from an initial value $\bar{P}_i$ to zero, the excess-pressure decay that takes place over the same interval of time varies from 0 to 100 percent (Domenico and Mifflin, 1965).

The curves in Fig. 7.23 represent conditions in the slab after periods of time have elapsed since the first application of a constant temperature at the surfaces. The numbers on the curves are values of the time factor T^*, which is equal to $\kappa t/L^2$, where κ is the thermometric conductivity. For the clay layer,

$$T^* = \frac{k't}{H^2 S_s'} \tag{7.124}$$

where x/L in Fig. 7.23 is actually z/H.

The isochrone development in Fig. 7.22b is symmetrical about the middepth of the layer. Hence, the hydraulic gradient across the middepth is zero, and remains zero throughout the consolidation process, which allows no water to flow across the middepth. Since the clay layer in Fig. 7.22b is actually $2H$ in thickness, H in Eq. (7.124) is the thickness per drainage face; it is equal to thickness H when one surface is an impermeable boundary, and to $H/2$ where water can flow upward and downward, as in Fig. 7.22b.

The average degree of consolidation of a clay layer for a given time, represented by the average value of $1 - \bar{P}/\bar{P}_i$, is of greater interest than the values at various depths. A graphical solution for the average degree of consolidation as a function of the time factor is given in Fig. 7.24. In this figure, U represents the average percentage of consolidation.

It might be added that the assumptions of the Hantush-Jacob leaky artesian model [Eq. (7.83)] and those of the model just described differ considerably. In the former, the clay layer is merely a transmitting medium, incapable of releases from storage and attendant deformation. On the other hand, the assumptions for the modified leaky method with incorporated specific storage terms are, from a physical point of view, similar in many respects to those envisioned in consolidation theory.

Example 7.7 The problem is to determine the time rate of consolidation for the problem of Example 5.1. For this purpose, additional information is required, namely, the coefficient of vertical permeability, estimated to be 1×10^{-3} ft/day. Note that this information was not required to determine total subsidence.

THEORETICAL MODELS OF THE UNSTEADY STATE

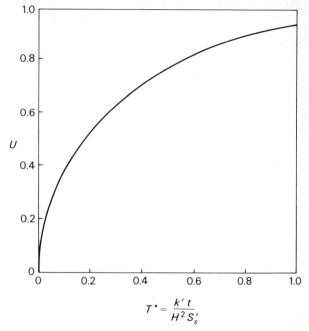

Fig. 7.24 Average degree of consolidation versus time factor. (*Modified after Carslaw and Jaeger*, 1959.)

The estimated subsidence of Example 5.1 (2 ft) is taken as 100 percent consolidation. Hence, a consolidation/time graph may be constructed by establishing the time required for various percentages of total consolidation. For U equals 50 percent, for example (1 ft of subsidence), T_{50} equals 0.196 (Fig. 7.24). From Eq. (7.124),

$$t = \frac{0.196(40/2)^2(1 \times 10^{-3})}{1 \times 10^{-3}} = 78 \text{ days}$$

Similar calculations can be made for other percentages of total consolidation, and a time/consolidation graph can be constructed. This graph will be of the same form as that in Fig. 7.24.

C. SIMULATION

Simulation, defined in broad terms, is the construction and manipulation of an operating model of a system or process. In essence, it entails experiments on a model of the system rather than on the real system itself, and so permits scientists to study problems that would otherwise be impractical to study.

Simulation models represent the essential characteristics of an actual system without superficial elegance. The techniques are carried out by a large number of scientists to investigate an equally large number of problems:

biochemical systems (Garfinkle and others, 1962), atmospheric-soil boundary problems (Estoque, 1963), operation of surface-water reservoirs (Domenico and others, 1965), wildlife and resource management strategies (Watt, 1968). Perhaps the most familiar texts dealing with simulation of water-resource systems are by Maass and others (1962) and Hufschmidt and Fiering (1966). On the other hand, one of the more familiar simulation techniques is the watershed model developed at Stanford, which was described briefly in Section 1.4.

It follows that simulation techniques vary from discipline to discipline, and are equally suited for lumped- and distributed-parameter problems. The scientist applying the simulation designs an appropriate model, observes and evaluates the results, and comes to appropriate conclusions. This procedure is generally observed whether the simulation under study is natural runoff in a watershed, a release schedule for a surface-water reservoir, or a scheme for exploring the consequences of an insect pest-control policy. The treatment of simulation in this book adheres to this general procedure, but, in common with other resource simulation studies, it is suited for one specific purpose; namely, ascertaining the response of complex aquifers to pumping stresses.

7.9 ANALOG SIMULATION

Two methods have been described by which an aquifer response to pumping wells may be obtained (Sections 7.1 and 7.5):

1. The drawdown formulas for infinite, homogeneous aquifers
2. The image well theory developed for homogeneous aquifers with highly idealized boundaries

Generally, most aquifers are relatively nonhomogeneous throughout the region of interest, with shapes and boundaries far from regular. It is not only difficult to obtain a regional picture of the hydrologic parameters for use in mathematical models, it is unrealistic to apply such idealized models to determine the effects of pumping. Faced with this dilemma, we turn to experiments which demonstrate a one-to-one correspondence in equations describing hydrologic, thermal, and electric phenomena, such as the Laplace equation, the diffusion equation, the wave equation, and other mathematical relations. When it is possible to find an analogy between one phenomenon and another that is easier to observe, information about one leads to information about the other.

The term *analogy* is defined as a relation of likeness between two things, consisting of a resemblance of their effects. Two such systems are said to be analogous if there exists this one-to-one correspondence between each element, and between the excitation and response functions of the elements as well. The term *analog* is sufficiently descriptive by definition to include the model of the hydrologic system, and the principles behind the design of the model.

Many individuals have contributed to both the theory and practical applications of electrical analogs to groundwater-flow problems. Skibitzke (1960) and members of the United States Geological Survey were instrumental in developing principles and design procedures for simulation models. In particular, Bermes (1960) has developed useful specifications for analog design. Karplus (1958) and Soroka (1954) developed much of the general theory of analog simulation. The technique has been successfully applied to a wide variety of fluid-flow problems, including flow to pumping wells (Skibitzke and da Costa, 1962; Domenico and others, 1964; Schicht, 1965; Patten, 1965; Prickett, 1967), the geometry of limestone solution (Bedinger, 1967), and the process of consolidation (Domenico and Clark, 1964).

THE HYDROLOGIC–ELECTRIC SYSTEM

The equation describing the transient hydrologic system is similar in form and principle to the electrical equation derived by utilizing Kirchhoff's current law around a resistance-capacitance network. Figure 7.25a represents the familiar differential volume, and Fig. 7.25b represents its discretized counterpart. Figure 7.25c shows the equivalent electric circuit. The resistors are the energy dissipators, similar in function to transmissivity, and the capacitors are potential-energy reservoirs, similar in function to storativity. Each resistor represents the resistance to flow in a specified direction, or "directionally oriented area," and each capacitor represents a specific directionally oriented volume of aquifer.

The operating equations can be developed for the simple, locally homogeneous case, with the more complicated nonhomogeneous development

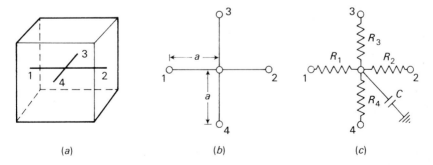

Fig. 7.25 Two-dimensional flow analogy showing (*a*) the differential flow volume, (*b*) nodal position for the finite difference approximation, and (*c*) an equivalent electric circuit.

reserved for the next section. The diffusion equation is given for two-dimensional flow

$$\frac{\partial^2 h}{\partial x^2} + \frac{\partial^2 h}{\partial y^2} = \frac{S}{T}\frac{\partial h}{\partial t} \tag{7.125}$$

which can be approximated in finite difference form (Section 6.1)

$$h_1 + h_2 + h_3 + h_4 - 4h_0 = \frac{a^2 S}{T}\frac{\partial h}{\partial t} \tag{7.126}$$

where a is the spacing. The subscripts for head refer to the nodal diagram of Fig. 7.25b. This equation is a good description of a locally homogeneous and isotropic aquifer, that is, an aquifer made up of subregions of homogeneous material, each subregion characterized by a single value of transmissivity. In practical applications, the value of the transmissivity is never known at every point, and in the absence of absolute information, most aquifers are considered to adhere to the locally homogeneous condition. The actual delineation of such subregions is often supported by preliminary geologic investigations (Domenico and Stephenson, 1964). In abbreviated form, Eq. (7.26) becomes

$$T\left(\sum_{n=1}^{4} h_i - 4h_0\right) = a^2 S \frac{\partial h}{\partial t} \tag{7.127}$$

where h_i is the head at nodes 1 through 4, and h_0 is the head at node 0.

Kirchhoff's law states that the current flowing into any node equals the current flowing out of that node. Application of this law to node 0 of the two-dimensional RC circuit of Fig. 7.25c gives

$$\frac{V_1 - V_0}{R_1} + \frac{V_2 - V_0}{R_2} + \frac{V_3 - V_0}{R_3} + \frac{V_4 - V_0}{R_4} = C\frac{\partial V_0}{\partial t} \tag{7.128}$$

If the resistors are equal in value,

$$\frac{1}{R}\left(\sum_{n=1}^{4} V_i - 4V_0\right) = C\frac{\partial V_0}{\partial t} \tag{7.129}$$

where V_i is the voltage at nodes 1 through 4, and V_0 is the voltage at node 0. Comparison of Eqs. (7.127) and (7.129) demonstrates the analogy of form for equivalent hydrologic-electric systems.

By considering another space variable, the finite difference approximation becomes (Skibitzke, 1960; Walton and Prickett, 1963)

$$T\left(\sum_{n=1}^{6} h_i - 6h_0\right) = a^2 S \frac{\partial h}{\partial t} \tag{7.130}$$

THEORETICAL MODELS OF THE UNSTEADY STATE

Electrically, the addition of a third flow direction is simulated by two more resistors (Fig. 7.26a) such that the electrical counterpart of Eq. (7.130) becomes

$$\frac{1}{R}\left(\sum_{n=1}^{6} V_i - 6V_0\right) = C\frac{\partial V_0}{\partial t} \qquad (7.131)$$

The electric circuit for two-dimensional flow in an aquifer and one-dimensional flow in an overlaying confining layer is shown in Fig. 7.26b. This is in accordance with the Hantush-Jacob developments described in Section 7.4.

In order to establish a one-to-one correspondence between the hydrologic and electric systems, it is necessary to introduce scale factors, or factors of proportionality. From observation of Eqs. (7.127) and (7.129), the following may be noted, where $\propto$ indicates proportionality,

$$T\left(\frac{L^2}{T}\right) \propto \frac{1}{R}, \text{ ohms}$$

$a^2 S(L^2) \propto C$, capacitance

$h(L) \propto V$, volts

$t_d(T) \propto t_s(T)$

From the analogy between Darcy's law and Ohm's law (Section 4.1), the rate of fluid flow is analogous to the rate-of-current flow. Further, by integrating a flow rate over time, one obtains a volume L^3 in the case of fluids,

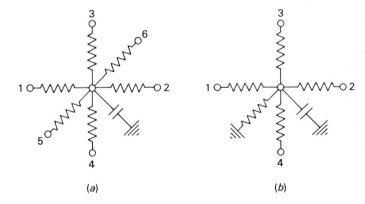

Fig. 7.26 Electric circuits for (a) three-dimensional flow and (b) two-dimensional flow in an aquifer and one-dimensional flow in an overlaying confining layer with a constant-head boundary.

and electric charge, or coulombs, Ψ in the case of current flow (current is measured in amperes, or coulombs per second). Therefore,

$$Q\left(\frac{L^3}{T}\right) \propto I, \text{ amps}$$

$$q(L^3) \propto \Psi, \text{ coulombs}$$

The following equalities are established by introducing the scale factors K:

$$\begin{aligned} q &= K_1 \Psi \\ h &= K_2 V \\ Q &= K_3 I \\ t_d &= K_4 t_s \end{aligned} \tag{7.132}$$

where the K's are identified for various systems of units:

$K_1 = \dfrac{q}{\Psi}$ L^3/coulomb $\qquad K_1 = \dfrac{q}{\Psi}$ ft^3/coulomb

$K_2 = \dfrac{h}{V}$ L/volt $\qquad K_2 = \dfrac{h}{V}$ ft/volt

$K_3 = \dfrac{L^3}{TI}$ L^3/T amp $\qquad K_3 = \dfrac{L^3}{TI}$ ft^3/(day)(amp)

$K_4 = \dfrac{t_d}{t_s}$ T/T $\qquad K_4 = \dfrac{t_d}{t_s}$ day/sec

$\qquad\qquad\qquad K_1 = \dfrac{q}{\Psi}$ gal/coulomb

$\qquad\qquad\qquad K_2 = \dfrac{h}{V}$ ft/volt

$\qquad\qquad\qquad K_3 = \dfrac{L^3}{TI}$ gal/(day)(amp)

$\qquad\qquad\qquad K_4 = \dfrac{t_d}{t_s}$ day/sec

By dimensional analysis,

$$\frac{K_3 K_4}{K_1} = 1 \tag{7.133}$$

From proportionality relations expressed above, transmissivity is in

THEORETICAL MODELS OF THE UNSTEADY STATE   373

inverse proportion to resistance, and storativity times the spacing is in direct proportion to capacitance. That is,

$$R = F' \frac{1}{T} \tag{7.134}$$

and

$$C = F'' a^2 S \tag{7.135}$$

where F' and F'' are constants to be determined from the cited scale factors. By dimensional analysis,

$$R = \frac{K_3}{K_2} \frac{1}{T} \tag{7.136}$$

and

$$C = a^2 S \frac{K_2}{K_1} \tag{7.137}$$

For the mixed so-called American practical units, Eq. (7.137) is multiplied by 7.48 gal/ft³. For leaky systems, the resistance value corresponding to vertical permeability is determined by

$$R_L = \frac{K_3}{K_2 (k'/m') a^2} \tag{7.138}$$

where m' is the thickness of the confining layer.

To summarize, descriptive equations have been presented for both the hydrologic and electric systems, their analogy of form has been shown by use of finite-difference approximations, and proportionality relations have been provided between the variables requiring expression as discrete quantities. It now remains to explore additional geological information required as model input and the electric duplication of such input, and to introduce design and operational criteria.

GEOLOGIC INPUT AND ELECTRIC DUPLICATION

As the electrical model incorporates only the essential characteristics of the prototype, the geological problem reduces to ascertaining the external geometry of the aquifer systems, their internal boundaries, if any, and the internal distribution of the transmissive and storage properties. All boundaries must be located with respect to a space coordinate system, and classified according to their function; that is, recharge or discharge. The accuracy with which the boundaries have been identified and simulated will have a one-to-one effect on the accuracy of the resulting model. Equipotential (recharge) boundaries may be simulated by grounding the network resistors along the boundary. This is valid since the potential at all nodes coinciding with the boundary is constant, and the resistors connecting these nodes are effectively shorted. Therefore, to duplicate the leaky aquifer condition, the constant-head boundary assumed to exist in the strata overlaying the confining layer is easily simulated by grounding the upper end of the resistor simulating vertical permeability (Fig. 7.26b). Similarly, constant-head boundaries represented by streams and rivers can be simulated by constructing the part of the network coinciding with the recharge boundaries in a short circuit. Equipotential boundaries may also be maintained at a variable potential by means of suitable power sources.

On the other hand, a barrier boundary to flow in the horizontal plane may be simulated by very high resistance. If the boundary is completely impermeable, the circuit is left open. Hence, simulating the exterior barrier boundaries at the margins of aquifers is accomplished merely by terminating the resistance network in the exact shape of the prototype field.

If the position of the external boundaries extends beyond the limits of the region to be studied, the general procedure is to increase the coarseness of the grid spacing in the areas of little or no concern (Karplus, 1958). The criteria for a successful approximation is that the boundaries of the model must not cause unnatural interference with the voltage or current fluctuations imposed on interior nodes.

Given the boundary conditions, the remaining geological information deals with the internal state of the system, or its stratigraphy. Ultimately, the stratigraphy must be expressed in terms of the distribution in space of the transmissive and storage properties. It is generally agreed that a sufficient number of measurements of transmissivity and storage must be made in order that interpolation between geologic variables and hydrologic parameters will provide a sufficient description of the system. Again, the accuracy of the determined hydraulic properties has a one-to-one effect on the accuracy of the resulting model.

DESIGN CRITERIA

The terms *excitation* and *response* are the electric equivalents of the general terms *cause* and *effect* or, for this particular case, *pumping* and *drawdown*. By viewing the electrical model as drawn in Fig. 7.27, excitation is the stimulus, or driving function, that produces the response in the same manner that

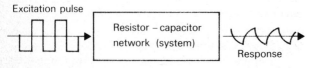

Fig. 7.27 Schematic diagram of the excitation and response functions of the analog model and operational equipment.

THEORETICAL MODELS OF THE UNSTEADY STATE

Table 7.4 Excitation-Response analogies for the hydrologic and electric systems

	Electrical	Hydrologic
Excitation, applied to either external boundaries or internal network nodes	Voltage or current	Change in potential or withdrawal or recharge of water
Response, measured at network nodes or boundaries	Voltage or current	Change in head or change in flow

pumping causes a lowering of water levels. The excitation introduces a physical quantity into the network; the amount added plus the total amount present equals the total amount in the field at any time (conservation principle). A system is termed *active* if internal energy sources are present, as in differential analyzers (or indirect analogs), and *passive* if internal energy sources are absent, as in the *RC* network under discussion. The excitation-response analogies of the hydrologic and electric systems are summarized in Table 7.4.

The selection of electric elements for the analog network is limited primarily by cost. Low voltage, fixed carbon resistors with values between 10 and 3×10^6 ohms, and ceramic capacitors of similarly low voltage ranging in value from 500 micromicrofarads (mmf) to 0.5 microfarad are economically available. These components are accurate to within ± 10 percent of their rated values. Karplus (1958) suggested that network resistors should range in value from 100 to 5×10^6 ohms in order that the interconnecting lines of low (but nonzero) resistance will not affect the resistance network.

Bermes (1960) suggested certain ranges in scale factors that would facilitate simulation of almost any range of interactions in hydrologic systems. He postulated that one must balance the value of answers provided against the cost incurred in obtaining them. The model cost can be minimized by choosing scale factors that allow the use of resistors and capacitors of reasonable cost as well as instrumentation that is commercially available. Since the most common power supply suited for exciting the model is usually limited to a maximum of 50 volts and 20 milliamperes, the following example demonstrates the logic in selecting scale factors.

Suppose it is decided that a maximum period of 20 years is sufficient for recapitulation of past cause-and-effect events, and for forecast into the future. A maximum analog excitation duration of 0.1 sec is available from the pulse generator. A candidate for scale factor K_4 becomes, from Eq. (7.132),

$$K_4 = \frac{t_d}{t_s} = \frac{20 \times 365}{0.1} = 7.3 \times 10^4 \text{ days/sec}$$

Table 7.5 Range in value for scale factors as determined by electronic equipment (*After Bermes*, 1960.)

Scale factor	Hydrological electrical relation	Assumed range in hydrological characteristics	Resultant economic range in electrical characteristics	Order of magnitude for scale factor Regional	Local
K_1	$q = K_1$	Determined by ratio $\frac{K_3 K_4}{K_1} = 1$	....	1×10^{16}	1×10^{12}
K_2	$h = K_2 V$	0.01–500 ft	0.001–50 volts	1×10^1	1×10^1
K_3	$Q = K_3 I$	0.003–3,000 ft³/sec (regionally), 25 ft³/sec (locally)	About 0.02–20 milliamperes	1×10^{11}	1×10^9
K_4	$K_4 = \frac{t_d}{t_s}$	1 day–25 + yr (regionally), 3 min–100 days (locally)	Up to 100 milliseconds	1×10^5	1×10^3

The full-scale voltage of excitation equipment is 50 volts. A maximum piezometric head change of 100 ft is assumed for any single pumping excitation, which gives a candidate for K_2,

$$K_2 = \frac{h}{V} = \frac{100}{50} = 2 \text{ ft/volt}$$

Hence, the observation along the vertical axis of the oscilloscope of a trace with an amplitude of 4 volts would be equivalent to a drawdown of 8 ft.

If the maximum amount of water to be pumped at any one time is less than 10 million gpd, and the maximum current output is 1×10^{-2} amp, a candidate for K_3 becomes

$$K_3 = \frac{Q}{I} = \frac{1 \times 10^7}{1 \times 10^{-2}} = 1 \times 10^9 \text{ gpd/amp}$$

The equality of Eq. (7.133) then requires that

$$K_1 = K_3 K_4 = (1 \times 10^9)(7.3 \times 10^4) = 7.3 \times 10^{13} \text{ gal/coulomb}$$

The areal values of transmissivity and storage of the system, coupled with the scale factors given above, permit one to determine the resistors and capacitors required. If these are not suitable in cost, the scale factors have to be changed until the level of information desired is balanced by the cost. Consider the following changes (Bermes, 1960).

THEORETICAL MODELS OF THE UNSTEADY STATE

If K_4 is reduced in order to reduce the computational speed, that is, increase t_s, the equality $K_3 K_4 = K_1$ would decrease. If K_3 is increased, the response system may not be able to record the range in currents, since K_3 is related to pumping rates. In addition, the larger K_3, the higher valued resistors required.

If, instead, K_3 is kept constant and K_1 decreased, larger capacitors are required, which allows the cost to rise rapidly. Table 7.5 relates the choice of scale factors to electronic equipment output and hydrologic characteristics.

The inherent error of the finite difference technique decreases with a decrease in network spacing. At the same time, a decrease in spacing permits the model to portray the potential field in more detail. However, for large regions and close spacing, the model is likely to get unduly large and expensive. Hence, network spacing is generally a compromise between the detail desired and the cost of the study.

OPERATIONAL CRITERIA

Conventionally, excitation sources have maximum outputs of about 50 volts and 20 milliamperes. A pulse generator is generally used to simulate the excitation (pumping). The pumping rate is approximated with step functions which are variable in duration and amplitude (Figure 7.28a). The pulse generator is activated by a waveform generator which produces varied waveforms, but most commonly sawtooths (Figure 7.28b). A dc power source is

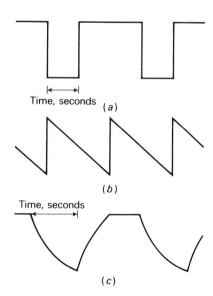

Fig. 7.28 Sequence of events in excitation and response functions of the analog: (a) output of the pulse generator, (b) sawtooth output of the waveform generator, (c) drawdown-versus-time curve as viewed by the oscilloscope at a node junction near a pumping well.

used. An arbitrary waveform generator can also be used to simulate complex pumping hydrographs, but its cost may be prohibitive.

The response of the analog is the drawdown for a given pumping rate at any time after pumping begins. Response is usually measured with a cathode ray oscilloscope, which is connected to the network node of interest. The voltage-versus-time curve observed at this point is analogous to the time/drawdown curve already familiar to the reader (Fig. 7.28c). The time of pumping is preset by adjusting the pulse width of the pulse generator in accordance with the equation encompassing scale factor K_4. The drawdown at any time is determined from the voltage curve and the equation encompassing scale factor K_2. Figure 7.29 shows typical interconnections between equipment and model.

Since the pumping rate is analogous to the current applied at a node of the network, $Q = K_3 I$, it is possible, by placing a resistor of known value in series with the pulse generator and the analog network, to compute the voltage needed to produce a desired pumping rate (that is, produce the desired current). Similarly, for any given voltage, it is possible to compute the resistance required to produce the desired current. The system may be represented by a simple circuit such as shown in Fig. 7.30, where R_p is the resistance needed to convert a known voltage (say, 50 volts from the pulse generator) to the desired current, and R_m represents the combined resistance

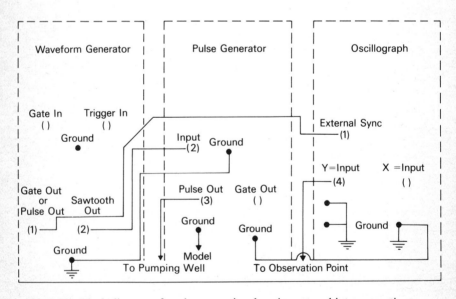

Fig. 7.29 Block diagram of analog-operational equipment and interconnections.

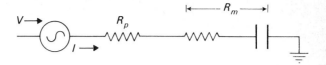

Fig. 7.30 Schematic diagram of pumping resistor in series with excitation source and model network.

offered by the resistors and capacitors of the *RC* circuit. It follows that

$$V = IR_p + IR_m = I(R_p + R_m) \tag{7.139}$$

Solving for the value of the pumping resistor,

$$R_p = \frac{V}{I} - R_m \tag{7.140}$$

If V is selected and R_m is measured in advance, I is determined for the rate of pumping desired in accordance with scale factor K_3 ($I = Q/K_3$). Making appropriate substitutions in Eq. (7.140),

$$R_p = \frac{VK_3}{Q} - R_m \tag{7.141}$$

That is, for a given value V from the pulse generator, and a known value R_m, one calculates the value of R_p to invoke any desired pumping rate Q. In essence, R_p converts a known voltage to the current desired in accordance with the equation involving K_3, I, and Q.

RECAPITULATION AND FORECAST

In the application of analog models to hydrological problems, two periods of pumpage are of interest:

1. A predetermined interval extending from some point in past time, when water levels were essentially in equilibrium with pumping, to the present
2. An arbitrarily selected interval extending from the present to some point of interest in the future

Recapitulation of historic data, as suggested in (1), provides a check on the accuracy of the modeled aquifer characteristics and boundary conditions. The historical data include pumpage and water-level response, identified in terms of time, rate, and place. The response of the model to excitation can be compared with the response of the groundwater system as measured in the field. If the comparison is favorable, that is, if the model faithfully recapitulates the past performance of the groundwater system, it may be used to predict water-level response to new pumping stresses.

The procedure of recapitulation is similar to a feedback control system (Fig. 7.31). The reference input is the desired response of the model, known in terms of historical behavior to pumping. The control element, which is the model, gives forth an output when excited, the output being a function of the properties of the model and the excitation. This output, or response to excitation, is compared with the desired output, or reference input. The properties of the control element are then changed, and the procedure repeated until the desired output is obtained. It follows that analog simulation, like all other forms of simulation, is inherently nonunique. In Karplus' terms, "To solve a problem in analysis is to determine the response due to a given excitation acting upon a known or fully specified system." The groundwater system is "specified" in this case in terms of its past behavior under pumping stresses.

Often water levels have not been in a state of equilibrium since pumping first started in an area. In order to carry out the recapitulation stated in (1), certain idealizations may be justified. One example of this is shown in Fig. 7.32, where quantities of water pumped and to be pumped in a recapitulation-forecast analysis are approximated by a step curve divided into segments along the time axis. The figure shows that prior to 1941, the rate of pumping was relatively uniform, shown by segment A. It is reasonable to assume that most of the regional water-level response to early pumping was registered on pre-1941 water levels. A condition of near equilibrium is then assumed previous to the increase in pumpage in the 1940s, and pre-1941 pumpage on post-1941 water levels is considered negligible. Thus, the extension of pre-1941 pumpage, represented by segment A', is ignored. Changes in water level over the period 1941 to 1963 are assumed to result from the increment in pumpage shown as segment B.

The regional water-level changes over the period 1963 to 1970 (the forecast period in this study) will depend on transient behavior of water levels in response to pre-1963, as well as post-1963 pumpage. The procedure is executed with the aid of step functions, which allows a continuance of pre-1963 pumpage, segment C, and the introduction of post-1963 pumpage, segment D. Total drawdown is determined by superposition.

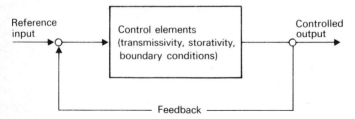

Fig. 7.31 Schematic diagram for recapitulation of past performance.

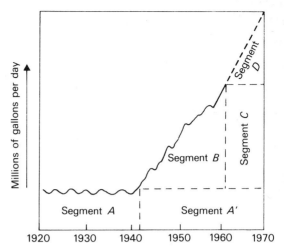

Fig. 7.32 Method of approximating functions to simulate pumpage. (*After Domenico and others*, 1964.)

7.10 DIGITAL SIMULATION

The electric model described in the previous section is, in effect, a scaled-down physical model of a prototype field that yields response data in the absence of mathematical manipulations necessary to actually solve the pertinent flow equations. A second line of progress has been to accept the mathematical equations themselves as sufficient, without using any physical model. This reasoning follows from the fact that the information contained in the electrical model represents the essential characteristics of the actual system without superficial elegance, and this same information can be formulated precisely as instructions to be given to a digital computer. Whereas the electric model yields visual information in the form of continuous curves at selected points in space, the digital model yields identical information in the form of discrete entities.

As the digital model operates on mathematical equations, it is necessary to embody within the equations the true complexity of the aquifer system. A moment's reflection would show that by employing the diffusion equation for homogeneous systems, any provisions for incorporating a varying transmissivity parameter are undesirably eliminated. Therefore, the equations to be solved must give as complete a description as possible of the nonhomogeneous prototype field.

If the principal components of the space-variable transmissivity are colinear with the coordinate axes for a two-dimensional flow problem, the

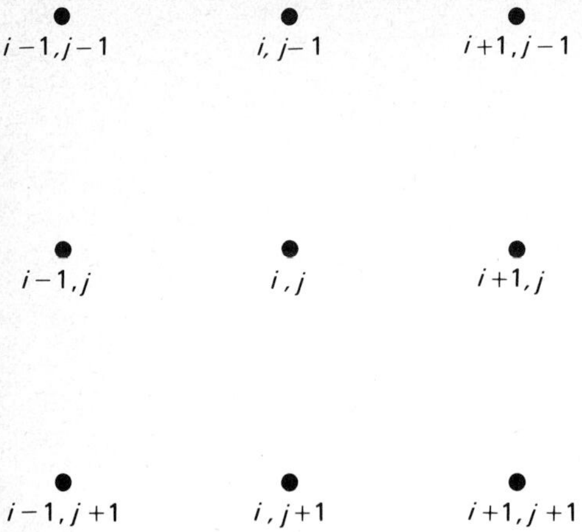

```
•               •              •
i−1,j−1       i, j−1         i+1,j−1



•               •              •
i−1,j          i,j            i+1,j



•               •              •           Fig. 7.33  Nodal diagram for
i−1,j+1       i,j+1          i+1,j+1       the finite difference approxima-
                                            tion.
```

unsteady-flow equation and its finite difference approximation become (Pinder and Bredehoeft, 1968),

$$T_{xx}\frac{\partial^2 h}{\partial x^2} + \frac{\partial T_{xx}}{\partial x}\frac{\partial h}{\partial x} + T_{yy}\frac{\partial^2 h}{\partial y^2} + \frac{\partial T_{yy}}{\partial y}\frac{\partial h}{\partial y} = S\frac{\partial h}{\partial t} + W(x,y,t) \quad (7.142)$$

and

$$T_{xx(i-\frac{1}{2},j)}\frac{h_{i-1,j}-h_{i,j}}{(\Delta x)^2} + T_{xx(i+\frac{1}{2},j)}\frac{h_{i+1,j}-h_{i,j}}{(\Delta x)^2}$$

$$+ T_{yy(i,j-\frac{1}{2})}\frac{h_{i,j-1}-h_{i,j}}{(\Delta y)^2}$$

$$+ T_{yy(i,j+\frac{1}{2})}\frac{h_{i,j+1}-h_{i,j}}{(\Delta y)^2}$$

$$= S\frac{h_{i,j,k}-h_{i,j,k-1}}{\Delta t} + W(x,y,t) \quad (7.143)$$

respectively, where T_{xx} and T_{yy} are components of the transmissivity tensor as a function of the x and y space coordinates, $W(x,y,t)$ is a source or sink term to account for the effects of pumping or recharge, and i, j, and k are indices in the x, y, and t directions, respectively. If only sink terms are considered,

$$W(x,y,t) = Q(x,y,t) \quad (7.144)$$

where Q is the pumping withdrawal rate identified in terms of the space coordinate system and duration of pumping.

Nodal points can be described with the aid of Fig. 7.33.

COMPUTATIONAL TECHNIQUES

The computational techniques employed to solve the finite-difference approximations of the unsteady-flow equation were developed originally by mathematicians for application to oil reservoirs and heat-flow problems (Peaceman and Rachford, 1955; Douglas and Peaceman, 1955; Quon and others, 1965 and 1966; Fagin and Stewart, 1966). The California Department of Water Resources made the first attempt to evaluate numerically the response of groundwater levels to pumping (Tyson and Weber, 1964), followed by the work of Eshett and Longenbaugh (1965), Bittinger and others (1967), Pinder and Bredehoeft (1968), and Prickett and Lonnquist (1968). Five various computational schemes have been described by Quon and others (1965), the two chief methods being classified by whether they are explicit or implicit.

As discussed in Section 6.1, the replacement of a continuous field by an assemblage of discrete nodal points is accompanied by the generation of n equations, one for each node of an n-node mesh, with a total of n unknown head values. Initial procedure requires that all head values at all nodal points be estimated, or guessed. The explicit method then proceeds to solve for the correct heads at various time steps in terms of the known heads at adjacent nodes. This is essentially an explicit iterative technique, in which at the kth iteration, a value of $h_{i,j}$ is determined without the simultaneous determination of a group of other values of head for other nodes (Forsythe and Wasow, 1960). The computer program presented by Prickett and Lonnquist (1968) is an example of an explicit method.

The implicit method solves for heads at nodal points in terms of the unknown heads at adjacent nodes (Fig. 7.34). An implicit method applied to groundwater problems by Pinder and Bredehoeft (1968) is the alternating direction method of Peaceman and Rachford (1955). An important feature of this method is that a given time step is divided into two equal parts, and

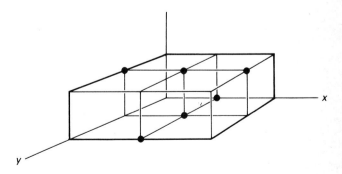

Fig. 7.34 Computational model for the alternating-direction implicit method. (*After Quon and others*, 1966.)

the equations solved implicitly, first in one coordinate direction, and second in the other coordinate direction (Douglas and Peaceman, 1955). In Fig. 7.34, for example, the head at two adjacent nodes is known from a previous time step, whereas the head at two other nodes is unknown; on the second time step, the unknowns become knowns. Hence, two equations are required, one for each half-time step. The technique, however, is noniterative in that once the equations are twice solved implicitly, the solution is complete for the given time interval under consideration. Advantages in stability and computational effort are clearly evident.

Equation (7.143) for the implicit calculations in the x direction becomes

$$\frac{T_{xx(i-\frac{1}{2},j)}}{(\Delta x)^2} h_{i-1,j} + \left[-\frac{T_{xx(i-\frac{1}{2},j)}}{(\Delta x)^2} - \frac{T_{xx(i+\frac{1}{2}),j}}{(\Delta x)^2} - \frac{S}{\Delta t} \right] h_{i,j} + \frac{T_{xx(i+\frac{1}{2},j)}}{(\Delta x)^2} h_{i+1,j}$$

$$= \frac{T_{yy(i,j-\frac{1}{2})}}{(\Delta y)^2} h_{i,j-1} + \left[\frac{T_{yy(i,j-\frac{1}{2})}}{(\Delta y)^2} + \frac{T_{yy(i,j+\frac{1}{2})}}{(\Delta y)^2} - \frac{S}{\Delta t} \right] h_{i,j}$$

$$+ \frac{T_{yy(i,j+\frac{1}{2})}}{(\Delta y)^2} h_{i,j+1} + \frac{Q}{\Delta x \, \Delta y} \quad (7.145)$$

In this formulation, the unknowns are always grouped on the left-hand side of the equation, the values on the right hand determined from a previous half-time step. For calculation in the y direction, the right-hand side of the equation is transferred to the left-hand side, and the terms assume unknown values to be solved implicitly. The resulting values at each nodal point are then used as initial conditions by which calculation over the next half-time step becomes possible, this procedure continuing until the forecast period is completed. Peaceman and Rachford (1955) provide a detailed explanation of this method, and Bittinger and others (1967) provide many general examples of digital simulation.

As with analog simulation, output may be compared with desired output, and the properties of the aquifer be subject to change until a desired output is obtained.

BOUNDARY CONDITIONS

Unlike the electrical analog, all boundary conditions must be described mathematically. Vertical leakage, for example, can be represented by a source term as provided in Eq. (7.143)

$$W(x,y,t) = K_z \frac{H_0(x,y) - h(x,y,t)}{m'(x,y)} \quad (7.146)$$

where H_0 is the constant-head water-table boundary, h is the head in the pumped aquifer, m' is the thickness of the confining layer, and K_z is the hydraulic conductivity of the confining layer. Pumpage minus vertical leakage equals the net removal of water from groundwater storage.

THEORETICAL MODELS OF THE UNSTEADY STATE

Other boundary conditions are likewise treated mathematically. No-flow boundaries are accounted for by merely assigning zero values to the nodal points outside the aquifer. Leakage from streams is easily accomplished with an equation similar to (7.146), where H_0 is the constant or time-varying head of surface water. If the position of the external boundaries does not coincide with the limits of the region to be studied, hydrologic gradients to the pertinent region may be established by assigning appropriate heads to the nodal points just beyond the study region.

7.11 SIMULATION AND OBJECTIVES FUNCTIONS

Simulation, whether by analog or digital methods, permits the ascertainment of man's effects on the natural state of the system, but does not accomplish any optimization of decisions. In general, pumping causes a change in water levels, which in turn affects the natural discharge components of the system. In addition, imbalances are established which give rise to a variety of undesirable results, such as land subsidence and seawater intrusion. The task of the hydrologist is to define the present state of the system and to describe insofar as possible (1) what stresses are to be imposed by man on the system, and (2) what the effect of these stresses will be.

Generally, the stresses to be imposed are arbitrarily selected pumping decisions, usually developed through the judgment of experienced hydrologists. The simulation model then produces the effects of these decisions, and the "best" policy is chosen from the set. Best, in this case, is most often interpreted in terms of the temporal and spatial distribution of water-level change.

Another approach is to design the simulation so as to provide a measure of the value of an objectives function for each of several alternative decisions. This idea was fostered for surface-water studies by the Harvard water group (Maass and others, 1962; Hufschmidt and Fiering, 1966), and first applied to a groundwater problem by Chun and others (1964). By incorporating an objectives function in simulation studies, one combines one of the stronger points of decision theory with a detailed distributed-parameter groundwater model. Alternative courses of action, such as imposing a use tax or quota system, may be tested and quantitatively evaluated as to their influence on long-term costs and benefits. Hence, the tasks cited in Section 3.1 are equally valid here:

1. The listing of alternative courses of action
2. The determination of the consequences that follow from each of the alternatives
3. The comparative evaluation of these sets of consequences in terms of the value or values to be maximized.

The comparative evaluation in 3 represents a novel modern-day departure from conventional simulation studies in groundwater hydrology.

At least two approaches are possible. The first follows closely in the path of the Harvard water program by establishing an objectives function that is evaluated directly for alternative plans of operation without recourse to the techniques of mathematical programming. The net benefit is generally evaluated over an economic life of T years. For example, Maass and others (1962) designated the net benefit from a multiple-purpose water-resource system as

$$\sum_{t=1}^{T} \frac{B_t(Y_t) - C_t(X)}{(1 + \gamma)^t} - K(X)$$

where $B_t(Y_t)$ equals gross benefits in year t from a given management policy as a function of variables Y_t which influences gross benefits, $C_t(X)$ equals operation, maintenance, and replacement costs in year t as a function of variables X, γ is the interest rate, and $K(X)$ is the initial capital construction costs. The value for this expression is calculated for each of several different management policies, by using digital simulation.

Chun and others (1964) addressed the problem of finding the present worth of future costs for a large number of alternatives in the conjunctive operation of the water resources of the Coastal Plain of Los Angeles County. All plans were formulated to provide identical water services so that the benefits, which are not measured, are equal in all cases. The heart of the simulation was the digital model described in previous sections. Each alternative was expressed in terms of groundwater basin operation, and was made up of four variables: a method of preventing sea-water intrusion, a pumping pattern, a water-spreading recharge schedule, and an extraction schedule. The most economical set of spreading and extraction schedules was determined for a specified pumping pattern and for a specified method of preventing sea-water intrusion. A total of 58 plans of operation were analyzed (California Department of Water Resources, 1966).

A cost equation of the following form was used in this study,

$$C_t = C_p + C_b + C_s \tag{7.147}$$

in which C_t indicates total cost of pumping, boosting, and storage of water. Each of the terms on the right-hand side is a function of several variables. Pumping costs included cost of pump and well units, and electrical energy; boosting costs included the cost of boosting units and electric energy; and storage costs included the costs of the storage facilities. In the economic comparison, the alternative with the least total present worth of future costs was selected as best meeting the objectives of the project. This study is the most comprehensive conjunctive-use investigation attempted thus far.

A second approach entails incorporation of the optimization techniques of mathematical programming. As discussed in Chapter 3, the pumping response surface in optimization studies is a lumped basin average described completely by the simple hydrological equation. Where this assumption is not altogether appropriate, the simulation model may provide the response surface and mathematical programming used to optimize allocation decisions.

The logic of this dual operation is closely linked to the idea of lumped- and distributed-parameter systems in groundwater hydrology. It has been mentioned in Section 6.1 that the finite difference approach with its characteristic nodal pattern involves the replacement of a continuous field with an assemblage of discrete (or lumped) elements. Each node in digital simulation (or each resistor in electrical simulation) represents a specified area in space within which permeability variations may be ignored. It is reasonable, therefore, to consider the possibility of linking a lumped-parameter allocation model to each node, or lump, making up the distributed-parameter model. In practice, such a linking is only necessary at those nodes where optimizing decisions must be made, such as pumping wells. The change in the pumping surface in response to these decisions is then recorded at the same node. Included in these observations is the effect of interactions between allocations at spatially distributed pumping centers, a point that is totally disregarded in conventional optimization studies.

Linear programming inputs to simulation models have been described for both analog and digital simulation studies (Martin and others, 1969; Bredehoeft and Young, 1970). The applications thus far have concentrated on real or hypothetical areas of competitive pumping, where it is desirable (1) to maximize benefits for each of a long series of irrigation seasons, or (2) to examine the ramifications of policy decisions, such as a use tax or quota system, designed to extend the life of the resource. These policy decisions constitute man-made constraints, and are the source of conflicts of interest in groundwater basins.

The linear programming model developed by Bredehoeft and Young (1970) is designed to forecast the quantity of water demanded in any subarea for any year, and is of the form

$$\max \sum_{i,j} A_{i,j} R_{i,j} - CX$$

where $A_{i,j}$ is the acreage allocated to the ith crop at the jth level of water input for a given year, $R_{i,j}$ is the net revenue over variable operation costs (other than the cost of developing water) for the ith crop with the jth level of water use, C is the variable cost of pumping 1 unit of water, and X is the quantity pumped. The flow diagram of Fig. 7.35 illustrates the linking of the allocation and simulation models. Initial conditions include the net return over variable costs for each potential use of irrigation water, depth to water, and constraints on water use (Martin and others, 1969). These conditions are stipulated as parameters in the linear programming model, which yields two fundamental outputs: the activities required to maximize net

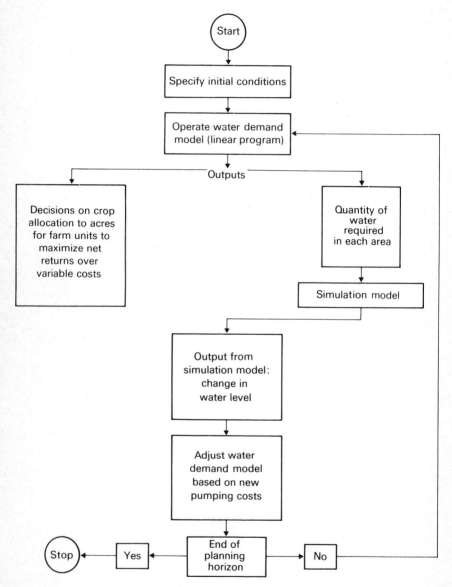

Fig. 7.35 Flow diagram for a combined linear programming–simulation study. (*Modified after Martin and others*, 1969.)

THEORETICAL MODELS OF THE UNSTEADY STATE

returns, and the amount of water required in the maximization process. The latter serves as an input to the simulation model. Simulation output is change in water levels in response to this pumping, which gives rise to changes in water demand. The feedback loop serves to operate the allocation model for the next stage. Hence, this procedure couples N single stages of groundwater development for a problem similar to that developed for linear programming only (Section 3.5) by Stults (1966).

The objectives function utilized by Bredehoeft and Young (1970) is of the form

$$Z = \sum_{t=1}^{T} [(B_t - C_{vt} - C_{ft} - C_{at})(1 + \gamma)^{-t}] \tag{7.148}$$

where Z is net economic benefits, B_t is gross value of irrigated crops in year t, variable costs of production are given by C_{vt} including water costs, C_{ft} is fixed costs, C_{at} is administrative costs associated with imposition of a management scheme, and T is the length of the planning horizon.

The linking of optimal resource allocation models and simulation models need not be restricted to problems of the type described above. Logical extensions of this approach are certain to include the more difficult problems associated with integrated use, as well as those dealing with natural constraints, such as sea-water intrusion, land subsidence, or groundwater–surface-water development in irrigated areas.

PROBLEMS AND DISCUSSION QUESTIONS

7.1 After 24 hours of pumping a confined aquifer, the drawdown in an observation well at a distance of 320 ft is 1.8 ft, and the drawdown in an observation well at a distance of 110 ft is 3.5 ft. The pumping rate is 192×10^3 ft^3/day. Find the transmissivity. (Answer: $T = 19.3 \times 10^3$ ft^2/day.)

7.2 Time/drawdown data collected at a distance of 100 ft from a well pumping at a rate of 192×10^3 ft^3/day are as follows:

t, min	s, ft
1	3.8
2	5.2
3	6.2
4	7.0
5	7.6
6	8.3
7	8.8
10	10.0
20	12.2
40	14.0
80	15.8
100	16.4
300	19.0
500	20.2
1,000	21.6

Calculate the transmissivity and storativity. (Answer: $T = 5 \times 10^3$ ft²/day and $S = 2.7 \times 10^{-4}$.)

7.3 A well 250 ft deep is planned in an aquifer with a transmissivity of 1,340 ft²/day and a storativity of 1×10^{-2}. The well is expected to yield 960×10^2 ft³/day, and will be 12 in. in diameter. If the nonpumping water level is 50 ft below land surface, estimate the depth to water after 1 year's operation. After 3 years' operation. (Answer: 1 year, 162 ft; 3 years, 168 ft.)

7.4 An 18-in.-diameter well within an aquifer with a transmissivity of 1,070 ft²/day and a storativity of 7×10^{-2} is to be pumped continuously. What pumping rate should be used so that the maximum drawdown after 2 years will not exceed 20 ft? (Answer: 15.5×10^3 ft³/day.)

7.5 A 24-in.-diameter well is in an aquifer with a transmissivity of 1,340 ft²/day and a storativity of 5×10^{-2}. If a fault (barrier) is located 1,000 ft from the well, calculate the drawdown at the fault and at the midpoint between the fault and the well after 1 year of pumping at a rate of 96×10^3 ft³/day. (Answer: 35 ft at the fault, 39 ft at the midpoint.)

7.6 Replace the fault in Problem 7.5 with a fully penetrating stream, and calculate the drawdown at the same two points for the same pumping rate and duration of pumping. (Answer: 0 ft at the stream, 12.4 ft at the midpoint.)

7.7 An aquifer has a transmissivity of 1,340 ft²/day and a storativity of 1×10^{-4}. A 24-in.-diameter well in this aquifer has a drawdown of 300 ft after 24 hours of pumping at a rate of 192×10^3 ft³/day. What is the efficiency of the well? (Answer: $E = 65$ percent.)

7.8 A step drawdown test is conducted at pumping rates of 1 ft³/sec, 3 ft³/sec, and 5 ft³/sec. Incremental drawdowns in the pumping well for a common pumping duration for each step are 10 ft, 26 ft, and 34 ft, which gives a total of 70 ft.
  a. Determine the constants B and C. (Answer: $B = 9$ sec/ft², $C = 1$ sec²/ft⁵.)
  b. Make a plot of drawdown versus yield for the well.
  c. On the same plot for question b, plot drawdown versus yield for a 100 percent efficient well.
  d. What is the efficiency of the well at rates of 1 ft³/sec, 3 ft³/sec, and 5 ft³/sec?

7.9 An aquifer has a transmissivity of 1,000 ft²/day and a storativity of 1×10^{-4}. A 24-in. pumping well has a drawdown of 235 ft at the end of 1 day's pumping at a rate of 100×10^3 ft³/day. The efficiency of the well is determined as 58 percent. What is the efficiency when the well is pumped at 200×10^3 ft³/day for a 1-day pumping period? (Answer: approximately 40 percent.)

7.10
  a. List three reasons which might explain an upward inflection of a semilogarithmic time/drawdown plot.
  b. List three reasons which might explain a downward inflection of a semilogarithmic time/drawdown plot.

7.11 Suppose the only type curve available to you was a plot of $W(u)$ versus u. How would you plot time drawdown data obtained from the field in order to use this curve in the matching procedure?

7.12 Consider the pumping-test result shown in Fig. P7.12. For this case, $s_1 = s_2$ and
$$u_1 = r_1^2 S/4Tt_1 \quad \text{and} \quad u_2 = r_2^2 S/4Tt_2$$
where r_1 is the distance from the pumping well to the observation well, which is known, and r_2 is the distance from the imaginary pumping well causing the inflection to the observation well, which is not known. If $u_1 = u_2$,
$$r_2 = r_1(t_2/t_1)^{\frac{1}{2}}$$
  a. Can you satisfy yourself that u_1 actually equals u_2, as assumed above?
  b. The distance r_2 in the above formulation can be easily calculated, but it

merely defines a set of all points which form a circle of radius r_2 with the observation well at the center. Explain.

c. What kind of pumping-test arrangement would be required to determine exactly the location of the barrier boundary?

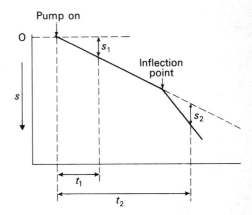

7.13 Examine Fig. 7.13, and explain why drawdown decreases with increasing sources of leakage for a given time.

7.14 Use the dimensionless graph of Fig. 7.21 to answer the following questions, which treat of the decay of excess pressure in a semiinfinite medium.

a. How long does it take to dissipate 50 and 75 percent of the original excess head at a depth of 6,000 ft? Assume that the permeability is 1×10^{-3} ft/year and that the specific storage is 1×10^{-4} ft^{-1}. (Answer: 3.6×10^6 years and 1.44×10^7 years.)

b. How long does it take to dissipate 50 and 75 percent of the excess head at a depth of 6,000 ft if the permeability is 1×10^{-1} ft/year and the specific storage is as above? (Answer: 3.6×10^4 years and 1.44×10^5 years.)

c. What is the percentage of the excess head that is dissipated at depths of 3,000 and 9,000 ft at time equal to 3.6×10^6 years? Assume that the permeability and specific storage are as given in question *a*. (Answer: approximately 75 and 28 percent.)

d. At what depth is the excess pressure unaffected after 3.6×10^6 years have elapsed? Assume that the permeability and specific storage are as given in question *a*. (Answer: 2.4×10^4 ft.)

e. From your answers and the information given in questions *a*, *c*, and *d*, make a plot of the ratio h'/h_i versus depth for time equal to 3.6×10^6 years.

7.15 Draw a time/consolidation graph for the data given in Example 7.7.

7.16 Summarize the geological and hydrological information required to simulate successfully a groundwater basin subject to extensive pumping.

7.17 From the material given in Sections 5.6 and 7.9, any process described by the diffusion equation can be simulated with an appropriate RC network. Consider the one-dimensional consolidation equation [Eq. (7.121)] and the equivalent expression for flow in a one-dimensional RC network [Eq. (5.117)]. A voltage applied to the uppermost node in the RC network can be made equivalent to the initial excess pressure (assumed known). Measurements at the other nodes by means of the oscilloscope would give the excess pressure decay versus time at various depths (corresponding to the nodes).

a. Draw a schematic of the appropriate RC network that can be applied to the one-dimensional consolidation process.

b. Develop the finite-difference equivalents for both equations cited above.

c. From the equations derived in part *b*, relate analogous terms by means of four scale factors. These factors should relate hydrologic quantities, such as permeability, storativity, and excess pressure with electric quantities, such as voltage, resistance, capacitance, and time.

d. Develop the following relation by dimensional analysis

$$\frac{F' \times F''}{F'''} = 1$$

where the F's represent three of the four developed scale factors.

7.18 Select any one of the water-resource problems dealing with the integrated use of surface water and groundwater for which a linear or dynamic programming solution has been discussed (Section 3.5). Discuss the possibility of reexamining this problem, or some modified version, by combining linear programming with an aquifer simulation model. Include in your discussion your ideas on an appropriate objective function; the water allocation scheme from sources to uses, or users, and from sources to sources, as in the case of artificial recharge; the possibility of a finite or infinite planning horizon; the type of data that would be required to put your program actually into action; the various problems you might expect in attempting such a project. Prepare a flow diagram along the lines of that in Fig. 7.35, which conveys the main message of the operations to be carried out.

REFERENCES

Bedinger, M. S.: An electrical analog study of the geometry of limestone solution, *Groundwater*, vol. 5, pp. 24–28, 1967.

Bermes, B. J.: An electric analog model for use in quantitative hydrologic studies, *U.S. Geol. Surv., Open-File Rept.*, Groundwater Branch, Phoenix, Arizona, 1960.

Bittinger, M. W., H. R. Duke, and R. A. Longenbaugh: Mathematical simulations for better aquifer management, *Intern. Assoc. Sci. Hydrol., Symp. Haifa*, Publ. 72, pp. 509–519, 1967.

Boulton, H. S.: Unsteady radial flow to a pumped well allowing for delayed yield from storage, *Intern. Assoc. Sci. Hydrol.*, Publ. 37, pp. 472–477, 1954.

————: Analysis of data from nonequilibrium pumping tests allowing for delayed yield from storage, *Proc. Inst. of Civil Engrs. (London)*, vol. 26, pp. 469–482, 1963.

Bredehoeft, J., and B. Hanshaw: On the maintenance of anomalous pressures, I: Thick sedimentary sequences, *Bull. Geol. Soc. Amer.*, vol. 79, pp. 1097–1106, 1968.

———— and R. Young: The temporal allocation of groundwater—A simulation approach, *Water Resources Res.*, vol. 6, pp. 3–21, 1970.

Brown, R. H.: Estimating the transmissibility of an artesian aquifer from the specific capacity of a well, in "Methods of Determining Permeability, Transmissibility, and Drawdown," *U.S. Geol. Surv., Water Supply Papers*, 1536-I, pp. 336–338, 1963.

Bruin, J., and H. Hudson: Selected methods for pumping test analysis, *Illinois State Water Surv., Rept. Invest.* 25, 1958.

California Department of Water Resources: Planned utilization of groundwater basins, Coastal Plain of Los Angeles County, Appendix C: Operations and Economics, 1966.

Carslaw, H., and J. Jaeger: "Conduction of Heat in Solids," The Clarendon Press, Oxford, 1959.

Chun, R. Y. D., L. R. Mitchell, and K. W. Mido: Groundwater management for the nation's future—optimum conjunctive operation of groundwater basins, *J. Hydraulics Div., Amer. Soc. Civil Engrs.*, HY4, vol. 90, pp. 79–105, 1964.

Cooper, and C. E. Jacob: A generalized graphical method for evaluating formation constants and summarizing well field history, *Trans. Amer. Geophys. Union*, vol. 27, pp. 526–534, 1946.

———, J. D. Bredehoeft, and I. S. Papadopulos: Response of a finite diameter well to an instantaneous charge of water, *Water Resources Res.*, vol. 3, no. 1, pp. 263–269, 1967.

Csallany, S., and W. Walton: Yields of shallow dolomite wells in northern Illinois, *Illinois State Water Surv., Rept. of Invest.*, 46, 1963.

Domenico, P., and G. Clark: Electric analogs in time-settlement problems, *J. Soil Mech. Found. Div. Amer. Soc. Civil Engrs.*, SM3, vol. 90, pp. 31–51, 1964.

——— and D. Stephenson: Application of quantitative mapping techniques to aid in hydrologic systems analysis of alluvial aquifers, *J. Hydrol.*, vol. 2, pp. 164–181, 1964.

———, D. Stephenson, and G. Maxey: Groundwater in Las Vegas Valley, *Desert Res. Inst., Tech. Rept.* 7, Reno, Nevada, 1964.

——— and M. Mifflin: Water from low permeability sediments and land subsidence, *Water Resources Res.*, vol. 1, pp. 563–576, 1965.

——— and others: Feasibility study of the operational plan, in "Operation Plan for the Humboldt River," Div. of Water Resources, Carson City, Nevada, 1965.

Douglas, J., and D. Peaceman: Numerical solution of two-dimensional heat flow problems, *J. Amer. Inst. Chem. Eng.*, vol. 1, pp. 505–512, 1955.

Eshett, A., and R. Longenbaugh: Mathematical model for transient flow in porous media, *Progr. Rept.*, Civil Eng. Dept., Colorado State Univ., Fort Collins, 1965.

Estoque, M.: A numerical model of the atmospheric boundary layer, *J. Geophys. Res.*, vol. 68, pp. 1103–1113, 1963.

Fagin, R. G., and C. H. Stewart: A new approach to the two-dimensional multiphase reservoir simulator, *J. Petrol. Technol., Soc. Petrol. Engrs., AIME,* pp. 175–182, June, 1966.

Ferris, J. G.: Groundwater, Chap. 7, in C. O. Wisler and E. F. Brater, "Hydrology," John Wiley & Sons, Inc., New York, pp. 198–272, 1959.

——— and others: Theory of aquifer tests, *U.S. Geol. Surv., Water Supply Papers,* 1536-E, pp. 69–174, 1962.

Forsythe, G. E., and W. R. Wasow: "Finite Difference Methods for Partial Differential Equations," John Wiley & Sons, Inc., New York, 1960.

Garfinkle, D., and others: Simulation and analysis of biochemical systems, *Commun. Assoc. Computing Machinery,* pp. 115–118, 1962.

Hantush, M. S.: Analysis of data from pumping tests in leaky aquifers, *Trans. Amer. Geophys. Union,* vol. 37, pp. 702–714, 1956.

———: Nonsteady flow to a well partially penetrating an infinite leaky aquifer, *Proc. Iraqi Sci. Soc.,* vol. 1, pp. 10–19, 1957.

———: Modification of the theory of leaky aquifers, *J. Geophys. Res.,* vol. 65, pp. 3713–3725, 1960.

———: Hydraulics of Wells, in V. T. Chow (ed.), "Advances in Hydroscience," Academic Press, Inc., New York, pp. 281–432, 1964.

——— and C. E. Jacob: Nonsteady radial flow in an infinite leaky aquifer, *Trans. Amer. Geophys. Union,* vol. 36, pp. 95–100, 1955.

Hufschmidt, M. M., and M. B. Fiering: "Simulation Techniques for Design of Water Resource Systems," Harvard University Press, Cambridge, Mass., 1966.

Israelson, O. W.: "Irrigation Principles and Practices," John Wiley & Sons, Inc., New York, 1950.

Jacob, C. E.: Flow of groundwater, in H. Rouse (ed.), "Engineering Hydraulics," John Wiley & Sons, Inc., New York, pp. 321–386, 1950.

────── and S. W. Lohman: Nonsteady flow to a well of constant drawdown in an extensive aquifer, *Trans. Amer. Geophys. Union,* vol. 33, pp. 559–569, 1952.

Johnson, E. E., Inc.: The principles and practical methods of developing water wells, Bull. 1033, Saint Paul, Minn., 1959.

Karplus, W. J.: "Analog Simulation," McGraw Hill Book Company, New York, 1958.

Kazmann, R. G.: Notes on determining the effective distance to a line of recharge, *Trans. Amer. Geophys. Union,* vol. 27, pp. 854–859, 1946.

LeGrand, H. E.: Geology and groundwater in the Statesville area, North Carolina, *N. Carolina Dept. Conserv. Develop., Div. Mineral Resources, Bull.* 68, 1954.

Maass, A., and others: "Design of Water Resource Systems," Harvard University Press, Cambridge, Mass., 1962.

Mansur, C. I., and R. I. Kaufman: Dewatering, in G. A. Leonards (ed.), "Foundation Engineering," McGraw-Hill Book Company, New York, pp. 241–350, 1962.

Martin, W. E., T. G. Burdak, and R. W. Young: Projecting hydrologic and economic interrelationships in groundwater basin management, paper presented at International Conference on Arid Lands in a Changing World, Tucson, Ariz., June, 1969.

Muskat, M.: "The Flow of Homogeneous Fluids through Porous Media," McGraw-Hill Book Company, New York, 1937.

Papadopulos, I. S., and H. H. Cooper: Drawdown in a well of large diameter, *Water Resources Res.,* vol. 3, no. 1, pp. 241–244, 1967.

Patten, E. D.: Design, construction, and use of electric analog models, in L. A. Wood, Analog model study of groundwater in Houston district, *Texas Board Water Engrs., Bull.,* 6508, pp. 41–103, 1965.

Peaceman, D. W., and H. H. Rachford: The numerical solution of parabolic and elliptical difference equations, *J. Soc. Ind. Appl. Math.,* vol. 3, pp. 28–41, 1955.

Pinder, G. F., and J. D. Bredehoeft: Application of digital computer for aquifer evaluation, *Water Resources Res.,* vol. 4, pp. 1069–1093, 1968.

Prickett, T. A.: Type curve solution to aquifer tests under water table conditions, *Groundwater,* vol. 3, pp. 5–14, 1965.

──────, Designing pumped well characteristics into electric analog models, *Groundwater,* vol. 5, pp. 38–46, 1967.

────── and C. G. Lonnquist: Comparison between analog and digital simulation techniques for aquifer evaluation, *Use of Analog and Digital Computers in Hydrology Symp.* Tucson, Arizona, pp. 625–634, December, 1968.

Quon, D., and others: A stable explicit computationally efficient method for solving two dimensional mathematical models of petroleum reservoirs, *J. Can. Petrol. Tech.,* vol. 4, pp. 53–58, 1965.

────── and others: Application of alternating direction explicit procedure to two dimensional natural gas reservoirs: *J. Petrol. Technol., Soc. Petrol. Engrs., AIME,* pp. 137–142, 1966.

Rubey, W. W., and M. K. Hubbert: Role of fluid pressure in mechanics of overthrust faulting, II: Overthrust belt in geosyncline area of western Wyoming in light of fluid pressure hypothesis, *Bull. Geol. Soc. Amer.,* vol. 20, pp. 168–205, 1959.

Schicht, R. J.: Groundwater development in East St. Louis area, Illinois, *Illinois State Water Surv., Rept. Invest.,* 51, 1965.

Skibitzke, H. E.: Electronic computers as an aid to the analysis of hydrologic problems, *Intern. Assoc. Sci. Hydrol., Publ.* 52, pp. 347–358, 1960.

——— and J. A. da Costa, The groundwater flow system in the Snake River plain, Idaho—An idealized analysis, *U.S. Geol. Surv., Water Supply Papers,* 1536-D, 1962.

———: Determination of the coefficient of transmissibility from measurements of residual drawdown in a bailed well, in Methods of determining permeability, transmissibility, and drawdown, *U.S. Geol. Surv., Water Supply Papers,* 1536-I, pp. 293–298, 1963.

Slichter, C. S.: Theoretical investigation of the motion of groundwaters, *U.S. Geol. Surv., 19th Annual Rept.,* II, Washington, D.C., pp. 295–384, 1899.

Soroka, W. W.: "Analog Methods in Computation and Simulation," McGraw Hill Book Company, New York, 1954.

Stallman, R. W.: Nonequilibrium type curves modified for two well systems, *U.S. Geol. Surv., Groundwater Notes, Open-File Rept.* no. 3, 1952.

Stults, J. M.: Predicting farmer response to a falling water table—an Arizona case study, *Proc. Econ. Water Resource Develop. Western Agri. Econ. Res. Council, Las Vegas, Nevada, Rept.* 5, pp. 127–141, December, 1966.

Terzaghi, K.: "Erdbaumechanic Auf Bodenphysikalischer Grundlage," Franz Deuticke, 1925.

Theis, C. V.: The relation between the lowering of the piezometric surface and the rate and duration of discharge of a well using groundwater storage, *Trans. Amer. Geophys. Union,* vol. 2, pp. 519–524, 1935.

———: The source of water derived from wells—Essential factors controlling the response of an aquifer to development, *Civil Eng.,* Amer. Soc. Civil Engrs., pp. 277–280, May, 1940.

———: Drawdowns resulting from cyclic pumping, in Methods of determining permeability, transmissibility, and drawdown, *U.S. Geol. Surv., Water Supply Papers,* 1536-I, pp. 319–323, 1963.

———, R. H. Brown, and R. R. Meyer: Estimating the transmissibility of aquifers from the specific capacity of wells, in Methods of determining permeability, transmissibility, and drawdown, *U.S. Geol. Surv., Water Supply Papers,* 1536-I, pp. 331–340, 1963.

Thiem, G.: "Hydrologische Methode," Gebhardt, Leipzig, 1906.

Turneaure, F. E., and H. L. Russel: "Public Water Supplies," John Wiley & Sons, Inc., New York, 1901.

Tyson, N. H., and E. M. Weber: Groundwater management for the nation's future—computer simulation of groundwater basins, *J. Hydraulics Div., Amer. Soc. Civil Engrs.,* HY 4, vol. 90, pp. 59–77, 1964.

Walton, W. C.: Selected analytical methods for well and aquifer evaluation, *Illinois State Water Surv., Bull.* 49, 1962.

——— and J. C. Neill: Statistical analysis of specific capacity data for a dolomite aquifer, *J. Geophys. Res.,* vol. 68, pp. 2251–2262, 1963.

——— and T. A. Prickett: Hydrogeologic electric analog computers, *J. Hydraulics Div., Amer. Soc. Civil Engrs.,* HY 6, vol. 89, pp. 67–91, 1963.

Watt, K. E. F.: "Ecology and Resource Management," McGraw-Hill Book Company, New York, 1968.

Wenzel, L. K.: Methods for determining permeability of water-bearing materials with special reference to discharging well methods, *U.S. Geol. Surv., Water Supply Papers,* 887, 1942.

Zeizel, A. J., and others: Groundwater resources of Dupage County, Illinois, *Illinois State Water Surv., State Geol. Surv. Coop. Rept.* 2, 1962.

Name Index

Ackermann, W. C., 16, 36
Allen, D. N. G., 267, 310
American Society Civil Engineers, 79-80, 139
Amorocho, J., 23-25, 27, 30, 36
Anderson, A. B. C., 182, 206
Anderson, D., 140
Aron, G., 115, 129, 139
Ashby, W. R., 4, 36

Baas-Becking, L. G. M., 200, 205
Bachmat, Y., 183, 206
Back, W., 180, 183, 193, 195-196, 200-201, 205-206, 288-291, 310
Bagley, E. S., 21, 36, 86, 139
Banks, H. O., 31, 43, 76, 183, 206
Barksdale, H. C., 183, 206
Barnes, B. S., 48, 76
Barnes, I., 193, 200-201, 205-206
Bauer, J. W., 77
Baumann, P., 21, 36
Bear, J., 87-88, 126-127, 139-140, 183, 206
Bedinger, M. S., 369, 392
Behnke, J., 21, 36
Bellman, R., 124, 130, 139
Bennett, R. R., 183, 206
Benson, M. A., 55, 76
Bermes, B. J., 369, 375-376, 392
Bernard, G. G., 298, 311
Berry, F., 180, 206
Bianchi, W., 21, 36
Birch, F., 299, 311
Bittinger, M. W., 383-384, 392
Blaney, H. F., 15, 36
Bloomenthal, H. S., 142
Boas, M., 163, 206
Boato, G., 311
Botset, H. G., 154, 207
Boulton, H. S., 327, 392
Brashears, M. L., 21, 36
Brater, E. F., 10, 38
Bredehoeft, J., 180, 206, 238, 249, 280, 311, 359, 362-363, 382-383, 387, 389, 392-394
Brown, I. C., 295, 311
Brown, R. H., 301, 311, 338, 392, 395
Brown, R. J., 308, 311
Bruin, J., 353, 392
Buchanan, J. M., 82, 139
Buras, N., 21, 36, 89, 126-127, 129, 131, 139, 140
Burdak, T. G., 394
Burt, O., 103, 106, 125, 140
Butler, S. S., 49, 76

Cady, R. C., 301, 311
California Department of Water Resources, 380, 392
Candler, W., 119, 141
Carlston, C. W., 272, 311
Carman, P. C., 154, 206
Carslaw, H., 317, 327, 362-363, 365, 367, 393
Cartwright, K., 300, 311
Casagrande, A., 175, 206
Case, C., 140
Castle, E., 119, 140
Chamberlin, T. C., 174, 206
Chapman, T. G., 47, 76
Chebotarev, I. I., 283, 285, 291-293, 311
Chen, C. L., 15, 36
Cherry, R., 193, 196, 206
Chorley, R. J., 6, 29, 36
Chow, V. T., 15, 36
Christ, C. L., 198, 206
Chun, R. Y. D., 385-386, 393
Ciriacy-Wantrup, S. V., 91, 140
Clark, G., 369, 393
Cochran, G., 128, 131, 140
Cohen, P. O., 21, 37
Colburn, W. A., 312
Conkling, H., 43-44, 73, 76
Cooper, H. H., 216, 243, 249, 327, 332, 338, 393-394
Counts, H. B., 249
Craig, H., 281, 311
Crawford, N. H., 24, 37
Cropper, W. H., 201, 206
Cross, W. P., 51, 54, 76
Cross Section, The, 107, 140
Cssallany, S., 356, 393
Cuevas, J. A., 228, 249

DaCosta, J. A., 169, 207, 369, 395
Darcy, H. P. G., 149, 206
Davis, G. H., 21, 31, 37, 228, 249-250, 284, 311
Davis, I., 91, 140
DeGrys, A., 281, 311
DeHaven, J. C., 141
DeSitter, L. U., 180, 206
DeWeist, R. J. M., 216, 249
DiStefano, J. J., 9, 37
Domenico, P. A., 84, 98, 100-102, 140, 211, 228, 231-232, 234, 237, 249, 359, 366, 368-370, 381, 393
Dooge, J. C. I., 5, 12-13, 37
Dorn, W. S., 267, 312
Douglas, J., 383-384, 393
Dracup, J., 20, 37, 121, 140

NAME INDEX

Drescher, W. J., 298-299, 311
Drew, L. J., 77
Dreyfus, S., 124, 139
Duckstein, L., 8, 37, 97, 140
Duke, H. R., 392
Dusinberre, G. M., 267, 311

Eakin, T. A., 279-280, 311
Eckstein, O., 81-82, 141
Edlefsen, N. E., 182, 206
Elder, J. W., 282-283, 311
Eshett, A., 383, 393
Estoque, M., 368, 393

Fagin, R. G., 383, 393
Farvolden, R. N., 54, 76, 312
Fayers, F. J., 267, 311
Ferris, J. G., 154, 206, 221, 223, 239, 249, 326, 332, 347, 349-351, 393
Feth, J. H., 290, 308, 311
Fiering, M. B., 368, 385, 394
Fischel, V. C., 45, 76, 151, 206, 207
Forsythe, G. E., 383, 393
Foxworthy, B., 37
Frank, O., 37
Freeze, R. A., 262-264, 266-267, 270, 311
Friedman, L., 141

Garfinkle, D., 368, 393
Garrels, R. M., 198, 206, 277, 312
Garza, S., 279, 312
Geraghty, J. J., 207
Gerand, J. R., 142
Ghyben, W. B., 183, 206
Gilluly, J., 228, 233, 249
Gleason, G. B., 45, 76
Glover, R. E., 190, 206
Goldschmidt, V. M., 287, 312
Gordon, H. S., 83, 140
Gorrel, H. A., 286, 312
Grace, H., 228, 250
Grant, E. L., 91, 140
Grant, U. S., 228, 233, 249
Grubb, H. W., 92, 140

Hall, W. A., 20, 37, 124, 131, 140
Hanshaw, B., 179-180, 183, 193, 206, 210, 238, 249, 288, 291, 310, 359, 362-363, 392
Hantush, M. S., 46-47, 76, 196, 220, 228, 249, 327, 338-340, 343-346, 393
Hard, H. H., 209, 213, 249
Harr, M. E., 257, 312
Hart, W. E., 23-25, 27, 30, 36
Hartman, L. M., 81, 119, 141
Heady, E., 119, 141
Hely, A. G., 51, 54, 76
Hem, J. D., 201, 206, 286-287, 312, 314
Henningsen, E. R., 295, 312
Herzberg, B., 184, 207
Hill, G. A., 303-304, 312

Hillier, F., 119, 141
Hirshleifer, J., 84, 141
Hitchon, B., 303-304, 312
Hodgson, G. W., 302, 304, 312
Hsu, K. J., 193, 207
Hubbert, M. K., 147, 150-152, 154, 161, 166, 172, 175-176, 186-187, 190, 202, 207, 211, 235, 246, 249-250, 254, 256, 301-306, 312, 359, 394
Hudson, A., 353, 392
Hufschmidt, M. M., 81, 141, 368, 385, 394
Hughes, G., 300, 312
Hutchins, W. A., 85, 141

Israelson, O. W., 332, 394

Jacob, C. E., 45-47, 51, 76, 210-211, 216, 224, 228, 241, 249, 327, 330, 332, 338, 343, 353, 362, 393, 394
Jaeger, J., 317, 327, 362-363, 365, 367, 393
Johnson, A. H., 21, 37
Johnson, E. E., 211, 249, 352, 394
Jones, J. F., 52-53, 77
Jones, P. H., 279, 283, 312

Kantrowitz, I. H., 281, 313
Kaplan, I. R., 205
Karplus, W. J., 4, 37, 242, 244, 249, 267, 269, 312, 369, 375, 394
Kaufman, R. I., 332, 394
Kazmann, R. G., 15, 18, 37, 44, 76, 348, 394
Kelso, M. M., 95-96, 141
King, F. H., 151, 207
Kisiel, C. C., 8, 37, 97, 140
Klein, H., 31, 183-185, 207
Knight, J. W., 312
Kögler, F., 231, 249
Kohler, M. A., 37
Kohout, F. A., 31, 183-185, 207
Kozeney, J., 154, 207
Krauskopf, K. B., 201, 207
Krumbein, W. C., 27, 37, 65, 76
Krutilla, J., 81-82, 141

Lamone, R., 119, 141
Landon, R., 312
Lane, E. W., 55, 77
Langbein, W. R., 6, 37, 62, 77, 307, 312
Lasdon, L., 129, 141
Latimer, W. M., 195, 207
Lee, C. H., 43, 77
Leggette, R. M., 45, 77, 183, 207
LeGrand, H. E., 308-309, 314, 356, 394
Lei, K., 55, 77
Leonards, G. A., 154, 207
Leopold, L. B., 6, 35, 37, 62, 77, 254, 314
Levin, O., 87-88, 126-127, 139
Levin, R., 119, 141
Lieberman, G., 119, 141

NAME INDEX

Liefrinck, F. A., 183, 207
Lindeborg, K., 119, 140
Linsley, R. K., 15, 24, 26-27, 37
Livingstone, D. A., 277, 312
Lohman, S. W., 210, 234, 249, 338, 394
Longenbaugh, R. A., 383, 392-393
Lonnquist, C. G., 383, 394
Lusczynski, N. J., 189, 207

Maass, A., 368, 385-386, 394
MacArthur, R., 61, 77
McCracken, D. D., 267, 312
McGinnis, L. D., 308, 313
McGuinness, C. L., 18, 35, 37
MacKenzie, F. T., 277, 312
Makower, M. S., 119, 141
Malmberg, G. T., 46, 77
Mansur, C. I., 332, 394
Marshall, H., 88, 141
Martin, W. E., 387-388, 394
Matalas, N. C., 62, 77
Maxey, G. B., 249, 295-296, 312, 393
Meinzer, O. E., 17-18, 35, 37, 42-43, 77,
    151, 166, 173, 207, 209-210, 213,
    249, 254, 277-279, 313
Meyboom, P., 9, 37, 49, 50, 77, 246, 250,
    254, 273-276, 301, 313
Meyer, O. H., 48, 77
Meyer, R. R., 183, 206, 395
Mido, K. W., 393
Mifflin, M. D., 172, 207, 211, 231-232, 249,
    254, 279, 282-283, 295-296, 308,
    312-313, 359, 366, 393
Miller, J. P., 54, 77
Milliman, J. W., 86, 141
Mindling, A. L., 249
Mitchell, L. R., 393
Moore, D., 205
Murray, C. R., 22, 37
Muskat, M., 154, 207, 257, 313, 327, 394

Neill, J. C., 356, 395

Olmstead, F. H., 51-54, 76

Paige, S., 301, 313
Papadopulos, I. S., 280, 311, 327, 393-394
Parizek, R. R., 56-57, 77
Parker, G. G., 228, 250, 301, 311
Patten, E. D., 369, 394
Paulhus, J. L. H., 37
Peaceman, D. W., 383-384, 393-394
Peck, R. B., 211, 250
Pelto, C. R., 61, 77
Penman, H. L., 15, 37
Perlmutter, N. M., 188, 207
Pinder, G. F., 52-53, 77, 382-383, 394
Piper, A. M., 288, 313
Plane, R. A., 200, 207
Pluhowski, E. J., 281, 313

Poland, J. F., 31, 210, 228, 233, 250
Polubarinova-Kochina, P. Ya., 257, 313
Prickett, T. A., 327, 369-370, 383, 394-395
Prigogine, I., 6, 37, 159-160, 193, 207

Quon, D., 383, 394

Rachford, H. H., 383-384, 394
Renshaw, E. F., 83, 94, 141
Reynolds, O., 147, 207
Rhodehamel, E. C., 311
Richter, R. C., 31, 183, 206
Robinson, T. W., 225, 228, 250
Rorabaugh, M. I., 281, 313
Rubey, W. W., 211, 235, 249-250, 359, 394
Russell, H. L., 332, 395

Sasieni, M. A., 119, 141
Scheidig, A., 231, 249
Schicht, R. J., 369, 394
Schiff, L., 21, 37
Schmorak, S., 183, 207
Schneider, R., 281-282, 313
Schneider, W. J., 55, 77
Schoeffler, J., 129, 141, 313
Schoeller, H., 283, 291, 293
Schwindiman, L. C., 298, 313
Scott, A., 81, 83-84, 141
Scott, R. F., 266, 268, 313
Seaber, P. B., 290, 314
Senio, K., 183, 207
Shannon, C. E., 55, 65, 77
Sheldon, J. W., 267, 311
Sherman, F. B., 300, 311
Shinners, S. M., 119, 141
Sienko, M. J., 200, 207
Skibitzke, H. E., 169, 207, 338, 369-370,
    395
Slichter, C. S., 151, 207, 332, 395
Small, J. B., 249
Snyder, J. H., 80, 141
Soroka, W. W., 369, 395
Southwell, R. V., 267, 314
Stallman, R. W., 280-281, 314, 347, 395
Stearns, N. D., 42, 77, 151, 207
Stephenson, D. A., 249, 370, 393
Stewart, C. H., 383, 393
Streeter, V. L., 4, 37, 148, 207
Stringfield, V. T., 228, 250, 308-309, 314
Stubberud, A. R., 9, 37
Stults, J. M., 121-122, 141, 389, 395
Swenson, F. A., 293-294, 314

Taguchi, K., 312
Terzaghi, K., 210-211, 213, 222, 235, 250,
    301, 314, 359, 362, 395
Thatcher, L. C., 311
Theis, C. V., 6-7, 31, 38, 90, 141, 210, 246,
    250, 316, 318, 338, 351, 395
Thiem, G., 330, 395

NAME INDEX

Thomas, H. E., 18-19, 38, 44, 77, 86, 142, 254, 314
Thomas, R. O., 90, 142
Thornthwaite, C. W., 15, 38
Todd, D. K., 20-21, 38, 43-44, 77, 148, 207
Tolman, C. F., 18, 35, 38
Toth, J., 173, 207, 257-262, 268, 271-272, 277-278, 290, 295, 314
Trelease, F. J., 85-86, 142
Turneaure, F. E., 332, 395
Tyson, N. H., 383, 395

Upson, J. E., 207

Von Bertalanffy, L., 6, 38
Von Foerster, H., 66, 77

Walker, T. R., 307, 314
Walton, W. C., 55-56, 77, 353, 355-356, 370, 393, 395
Waring, G. A., 314
Warner, D. L., 298, 314

Wasow, W. R., 383, 393
Watt, K. E. F., 124, 142, 368, 395
Watts, E. V., 238, 250
Weaver, W., 59, 65, 77
Weber, E. M., 383, 395
Wenzel, L. K., 45-46, 77, 320, 395
Weschler, L. F., 112-113, 142
White, D. E., 281, 286, 311, 314
Whittelsey, N., 119, 140
Williams, I. J., 9, 37
Williamson, E., 119, 141
Wilson, G., 228, 250
Winograd, I. H., 279, 281, 314
Winslow, A. G., 228, 250
Wisler, C. O., 10, 38
Witherspoon, P. A., 262-264, 266-267, 270, 311
Wood, L. A., 228, 250

Yaspan, A., 141
Young, R., 387, 389, 392, 394

Zeizel, A. J., 355, 395
Zen, E., 179, 206

Subject Index

Active analog models, 375
Activities of concentrated solutions, 194
Activity coefficient, 194, 198
Alternative yield, 79-80
Alternative-yield policy, 97-98
American hydrologic system of units, 153
　for pumping test results, 323-325, 334-335
Analog, definition of, 368-369
Analog models:
　steady-state flow, 267-270
　transient flow, 368-380
Analytical solutions of Laplace's equation, 256-263
Angle of internal friction, 235
Anions:
　common to groundwater, 285
　as hydrochemical facies, 288-290
Anisotropic medium:
　effect on liquid waste, 299
　to fluid flow, 151, 165-166
Appropriation doctrine, 85
Aquifer:
　classification of, 17-20
　definition of, 17
　factors controlling response, 6-7
　major types, 18-19
Aquifers:
　elastic compression of, 210-211, 215-216, 233-235
　as storage reservoirs, 20
　water withdrawals, by states, 21-22
Artesian, 214
Artificial recharge, 20-21

Barometric efficiency, 227-228
Barrier boundary, 255-256, 347-348
　electrical duplication of, 374
Base exchange, 295
　(*See also* Ion exchange)
Baseflow, 48
Baseflow recession (*see* Recession)
Benefit function, 104
Bernoulli equation, 155-160
Big Smokey Valley, Nevada, 278-279
Black box, 8, 52
Boundary conditions:
　in consolidation theory, 362-367
　for electric analog models, 373-374
　in numerical analysis, 384-385
　for radial flow in finite aquifers, 347-348
　for solution of Laplace's equation, 255-256

Boundary conditions:
　for steady flow in homogeneous, isotropic layers, 263
　for steady radial flow in an infinite aquifer, 329-331
　for transient radial flow in an infinite aquifer, 318-319
Boundary-value problem, 255-256
Brines, occurrence in deep sedimentary basins, 180
Bulk modulus of compression, 217, 219
　of aquifer matrix, 217
　of fluids, 219
　of low permeability sediments, 230
　range in values, 231

Capillary rise, 181-182
Capillary water, 14
Carbonate flow systems, 279-280, 283, 295
Carbonate rocks:
　chemical character of groundwater, 295-297
　in flow variability studies, 56-57
　as groundwater flow systems, 279-281
　in information theory, 71-72
Centralized decision making in common pool problems, 84
Chebotarev sequence, 292
Chemical classification of groundwater, 285-288
Chemical equilibrium, 193-199
Chemical mass balance for hydrograph separation, 52-53
Chemical osmosis, 177-180
　as a mechanism for hydrocarbon entrapment, 303-304
Classification of flow, 147-149
Closed system, 6
Coastal plain of Los Angeles County, 386
Coefficient of compressibility, 231, 357-358
Coefficient of consolidation, 231-232
Coefficient of permeability (*see* Permeability)
Coefficient of storage (*see* Storativity)
Coefficient of transmissibility (*see* Transmissivity)
Common-pool problems, 83-85
Competitive pumping, 84, 115-116
　effect of: by linear programming models, 121-122
　by simulation models, 387-388
Compressibility:
　of an aquifer matrix, 217

400

SUBJECT INDEX 401

Compressibility:
    of aquifers, 210-211, 215-216, 233-235
    coefficient of, 231, 357-358
    of confining layers, 215-216, 230-233
    definition of, 209
    of fluids, 219
Conductive liquids, 267
Conductive sheets and solids, 267-269
Cone of depression in an ideal aquifer, 319-321
Confined flow, 173-176
Confining bed:
    definition of, 17
    subsidence of, 229, 232
    vertical permeability of, 344-345
Conflicts of interest in water resource management, 90-91
Connate water, 286
Consolidation:
    coefficient of, 231-232
    definition of, 356
    one-dimensional, 356-367
    theory of, 215-216
Constant head boundary, 255, 348
    electrical duplication of, 374
Continuity equation, 241
Continuity principle, 146
Control, definition of, 2
Control volume, 6, 147
Correlative rights doctrine, 86
Coulomb's equation, 235

Dakota sandstone:
    chemical character of groundwater, 293-295
    compressibility of, 209-210
    recharge area, 293
Darcy's law, 149-151, 161-162
    in consolidation theory, 361
    in the derivation of flow equations, 241-243
    in fluid to fluid interface problems, 189
    in potentiometric surface interpretations, 167-168
Decentralization principle, 129
Decision variable, 2, 116
    in information theory, 69
Depletion deduction, 107-108
Depth-pressure diagram, 232
Diffusion equation, 164, 243-245, 318
    discussion of, 315-316
    finite difference approximation, 370
    in polar coordinate form, 318, 319
Diffusion zone, 183
    in Biscayne aquifer, 183-185
    off Long Island, 188-189
Diffusivity, 245, 317-318
Dimensionless variables, 362
Discharge area:
    chemical character of groundwater, 290-292, 295, 296
    definition of, 259
    field observations, 273-279

Dispersion, 298, 299
Distance-drawdown curves, 324
Distance-drawdown method, well hydraulics, 336-338
Drawdown, 319
Drawdown-yield curve, 354
Dupuit-Forchheimer assumptions, 332
Dynamic equilibrium:
    of a fluid to fluid interface, 186-187, 189-190
Dynamic programming, 122-135
    for coastal aquifers, 124-125
    for importation schemes, 127-128
    for isolated basins, 124-125

Earthquakes caused by fluid injections, 299
Economic efficiency, 81-83, 113
Effective average rate of precipitation, 46, 47
Effective pressure area, 233
Effective radius of wells, 352
Effective stress, 210-216
    in consolidation theory, 357-362
    importance in shear, 235, 238-239
    in relation to storativity, 219
    in relation to water level fluctuations, 223-228
Eh-pH diagram, 201-202
Elasticity of aquifers (see Aquifers)
Electrical analog models (see Analog models)
Electrical resistance, 162
Electromotive series, 199-200
Electroosmosis, 182-183
Elements, mobility of, 285
Endothermic reaction, 193
Englishtown formation, 290
Entropy:
    in chemical equilibrium, 191-193
    in closed systems, 6
    in information theory, 60-61
    in irreversible flow, 160
Ephemeral water bodies, conditions for, 275-277
Equilibrium concepts (see Chemical equilibrium)
Equilibrium storage, 105, 106
Equipotential line, 157, 164-166
Equivalents per million, 288, 289
Evaporite, 287, 307-308
Evapotranspiration, 15
Excess-fluid pressure:
    in association with geothermal gradients, 283
    in consolidation theory, 357-362
    as a mechanism for hydrocarbon migration, 303
Excess-pressure oil pools, 238
Excitation-response apparatus, 377-379
Exothermic reaction, 193
Expected variable, 4
Explicit methods of numerical analysis, 383

Feedback control system, 380
Finite aquifers, 347-352
Finite differences, 263-266
  (*See also* Numerical methods)
Flow-line refraction, 174-176
Flow pattern, 253, 271-273
Flow systems:
  conditions for development, 259-262
  definition of, 259
  delineation by chemical means, 293-297
  in environmental considerations, 297-302
  geologic and hydrologic controls, 270-271
  in hydrocarbon migration, 304
  study of, 253-254
  types, 259
Formation loss, 352
Free energy:
  in capillary phenomena, 182
  in chemical equilibrium concepts, 191-196
  in chemical osmosis, 178-179
  definition of, 177
  in relation to oxidation potential, 200
Free surface, 256

Gas storage structures, 299
Genetic chemical classifications, 286
Geochemical cycle, 283-285
Ghyben-Herzberg theory, 184-186
Gradient, hydraulic, 149
Groundwater hydrology, definition of, 1
Groundwater law, 85-87
Groundwater mining, major states, 21
Groundwater movement, 146-149
Groundwater outcrops, 273-279
Groundwater temperature (*see* Temperature of groundwater)

Hagan-Poiseuille equation, 152
Hanford atomic energy plant, 298
Heat flow:
  from earth's interior, 281-282
  in an infinite slab, 317, 318
  in a semiinfinite medium, 362-366
High plains, 94, 107-108
Homogeneous medium to fluid flow, 166, 168, 256
Humboldt River, Nevada, 68
Hydraulic conductivity, 151-155
  in relation to flow patterns, 264
  in relation to a fluid to fluid interface, 191
Hydraulic head, gradient of, 172
  (*See also* Potential energy; Total head)
Hydrocarbons, migration and accumulation, 302-307
Hydrochemical facies, 288-293
Hydrogen ion concentration, 201
Hydrogeology, research activities of, 254
Hydrographs, surface and groudwater, 45-49

Hydrograph separation, 48
  by chemical mass balance, 52-53
Hydrologic cycle, 10-17
Hydrologic equation:
  applications of, 42-43
  in inventory calculations, 4
  in optimization models, 88, 93
  for subsystems of the hydrologic cycle, 12-15
Hydrology, definition of, 10
Hydrolysate, definition of, 287
Hydrothermal systems, 282, 283

Ideal fluid, 148, 155, 156
Ignatovitch-Souline sequence, 293
Image well theory, 347-351
Impermeable boundary, 256
Implicit methods of numerical analysis, 383-384
Infiltration, 10, 27
Infinite reservoir, 127
Information systems:
  combination of variables, 67-72
  self-information, 62-66
Initial conditions, 318, 362-365
Input-output analysis:
  for hydrologic models, 42-52
  for optimization models, 87-89
Interest rate:
  effect on optimal allocation of groundwater, 102, 106
  effect on optimal timing of importation, 110
Interflow, 48
Intermediate zone in the water profile, 13-14
Inventory:
  definition of, 42
  for flow-system delineation, 279, 280
  of Pomperaug basin, Connecticut, 42
Ion exchange, 287, 295
Ionic activity product, 194
Ionic strength, 197-198
Iron in groundwater, 201-202
Isochrone, 229, 364, 366
Isolated basin, definition of, 93-94
Isotropic medium to fluid flow, 151, 166, 168, 256

Joint probability, 72
Juvenile water, 280

Kinetic energy, 156, 157
Kirchhoff's equation, 269, 369
Kozeny-Carman equation, 154

Lake Bonneville, Utah, 290
Laminar flow, 147-148
Land-ownership doctrine, 85-86
Landslides, 235, 301
Laplace equation, 164, 241-243

SUBJECT INDEX

Laplace equation:
  discussion of, 255
  finite difference approximation, 264-265
  solutions of, 256-270, 328-329
Laws of thermodynamics, 146
Leachate, 299, 300
Leakage, vertical, 344-345
Leaky aquifers, 338-346
Legal aspects, groundwater, 85-87
Linear programming, 117-120
  in multistage decision problems, 120-122
  in simulation studies, 378-389
Liquid waste disposal, 298-299
Lumped element analogs, 269-270, 368-381

Magmatic water, 286
Management problem, definition of, 2
Marginal cost of pumping, definition of, 95
Marine water, 286
Markov process:
  definition of, 27
  example of, 28
  in information theory, 64-65
Mathematical programming, 114
Metamorphic water, 286
Meteoric water, 286
Mining yield, 79-80
Moapa Valley, Nevada, 279, 280
Mobility of ions, 285
Model, definition of, 2
Model Water Use Act, 85
Models in hydrology, 22-30
Modified nonequilibrium method, well hydraulics, 333-335
Monte Carlo method, 23, 27
Multiple-boundary aquifers, 349-351
Multistage process, 116

Navier-Stokes equation, 147
Nernst equation, 200
Neutral stress, 211-212, 214-215
Nevada test site, flow system of, 279, 281
No-flow boundary, 255
  (*See also* Barrier boundary)
Nonlinear programming, 119
Normal pressure, 170, 359, 361
Numerical methods:
  in steady-state flow, 263-267
  in transient flow, 381-385

Objective function:
  discussion of, 88-89
  for dynamic programming models, 123
  for isolated basins, 93
  for linear programming models, 118-119
  for simulation models, 385-389
Open system, 6-7
Optimal groundwater mining, 97-102
Optimal temporal allocation of groundwater, 102-106
Optimal yield, 45, 87-89, 102

Optimization, 8, 78, 93
Orange County Water District, California, 112-113
Osmotic pressure, definition of, 179
  (*See also* Chemical osmosis)
Overdraft, types of, 80
Oxidation-reduction potential, 199-202, 307

Pahranagat Valley, Nevada, 279, 280
Paleohydrology, 307-309
Parametric hydrology, 24-27
Parts per million, 285-286
Passive analog models, 375
Permanent water bodies, conditions for, 275-277
Permeability, 151, 153-154
  control on potentiometric configuration, 167-168
  effect on flow-line refraction, 175-176
  of semipervious confining layers, 344-345
  variations in limestone terrain, 308-309
Phreatic zone in water profile, 13-14
Phreatophytes, 273-275, 279
Piezometric surface (*see* Potentiometric surface)
Pleistocene lakes, 277-278
Plutonic water, 286
Pluvial, 308
Pore-water pressure (*see* Excess-fluid pressure)
Porosity, 150
Potential, 161
  electrical, 182
  oxidation-reduction, 199
Potential energy:
  in closed systems, 6
  field interpretation of, 156-157
  gradient of, 162-163
  in relation to Darcy's law, 161-162
  in relation to hydrocarbon migration and accumulation, 304-307
  in relation to irreversible flow, 159-160
  in relation to total hydraulic head, 161
  of two homogeneous fluids, 187
Potential gradient, divergence of, 163-164
Potentiometric surface, 166-169
  for confined flow, 174, 176
  modified by osmotic withdrawals, 304
  in relation to sea water intrusion, 183-184
Prairie profile, 273-275
Precipitate, definition of, 287
Precipitation-recharge relations, 45-48
Present worth, 91-92
  calculations for isolated groundwater basins, 94-97
  calculations for optimal importation schemes, 108-110
Principle of optimality, 124, 131
Probability paper, 55, 56
Programming, 114

SUBJECT INDEX

Pumping by states, 21-22
Pumping tax, 113

Quicksand, conditions for, 238-239

Radial flow, 148
Radioactive wastes, disposal of, 298
Radius of influence, pumping wells, 320
Rainfall-runoff relations, 23-27, 30
Real fluid, 148, 155, 156
Rebound in consolidation theory, 358, 359
Recession, baseflow and spring discharge, 48-53
Recession characteristics in relation to geology, 51, 55-57
Recharge:
    analyzed from departure curves, 45-46
    artificial, 20-21
    from baseflow calculations, 48-51
    effect on flow patterns, 172
    methods of determining, 46-47
    in optimization models, 89-91
    in safe-yield studies, 146
Recharge area:
    chemical character of water, 290-292, 295-296
    of Dakota sandstone, 293-294
    definition of, 259
    in precipitation-recharge models, 46-47
    in relation to field observations, 273-279
Recovery method, well hydraulics, 335-336
Redundancy:
    calculation of, 64
    of constraints, 118
    of the hydrologic cycle, 58-59
    rate of change of, 66
Refraction (*see* Flow-line refraction)
Regional-flow system, definition of, 254
Regression equations in flow variability studies, 53-57
Replenishment assessment, 113
Resistate, definition of, 287
Reversible flow, 158-159
Reynold's number, 147-148, 150, 155
Rights in groundwater law (*see* Legal aspects, groundwater)
Riparian doctrine, 85-86
Roswell Basin, New Mexico, 47
Runoff, surface (*see* Surface-water hydrology)

Safe yield, 43-45, 79-80
    optimization of, 89-90
Safe-yield policy, 94, 97
Saline soils, 273-277, 279
Salt filtering, 180
Salt water-fresh water interface, 184-191
Scalar field, 162-163
Scale factors, electric analog models, 371-373, 375-377
Sea-level canal, 301-302

Sea-water intrusion, 183-184
Self-information, 61-66
Sensitivity analysis, optimization models, 137
Separation of variables, 258
Simplex method, 119
Simulation:
    definition of, 367
    by electrical analog models, 368-381
    by numerical methods, 381-385
    of surface water systems, 24-27
Single-stage process, 114-115
Single-stage processes, coupling of, for simulation, 389
Slope stability, 235, 300-301
Social rate of interest, 110
Social welfare as objective in water-resource design, 80-83
Sole ownership in common-pool problems, 84
Solid-waste disposal, 299-301
Soluble salts common to groundwater, 285
Sorption, 299
Specific capacity, 351-352
    geologic controls on, 355-356
Specific discharge, 149-150
    of compressible confining layers, 232
Specific heat, 317
Specific storage, 216
    in aquifer loading problems, 228
    in consolidation theory, 361-367
    definition of, 220
    of low permeability materials, 231-233
    range in values, 231
Specific subsidence, 234, 237
Specific yield, 95, 136
Stage, 130
State variable:
    definition of, 2
    effect on dimensionality, 129
    in information theory, 68-69
    in multistage decision models, 116
Steady confined flow, 330-331
Steady flow, 147-148, 241-243
Steady state of open systems, 6
Steady unconfined flow, 331-332
Step-drawdown test, 353-354
Stochastic hydrology, methods of, 23
Stochastic variable, 4
    in precipitation-recharge studies, 47, 48
Storage equation (*see* Hydrologic equation)
Storage-flow ratio, 47, 48
Storativity:
    definition of, 220
    derivation of, 216-220
    determined from pumping tests (*see* Well hydraulics)
    in diffusion equation, 243-245
    in land subsidence calculations, 231-233
    in recharge calculations, 51
Stream hydrograph (*see* Hydrograph)
Streamflow parameters in relation to geology, 53-57
Streamline flow, 147

SUBJECT INDEX

Suboptimization, 91, 107, 131
Subsidence of the land surface, 228-235
Sulfate reduction, 287, 288
Superficial velocity, 149-150
Surface features of groundwater flow, 273-279
Surface tension, 180-182
Surface water, optimal importation of, 108-110
Surface water hydrology, methods in, 22-28
Sustained yield, 79-80
Sustained yield policy, 98
System:
    black-box, 8
    deterministic, 3
    distributed parameter, 8, 9, 145-146
    linear, 9
    lumped parameter, 8, 145
    in model formation, 4-9
    nonlinear, 9
    open versus closed, 6-7
    probabilistic, 4
    schematic representation of, 5
    in science, engineering and management, 1-2

Tangent refraction law, 174-176
    in boundary value problems, 262-263
Temperature of groundwater, 279-283
Temperature potential, 162
Terzaghi consolidation equation, 359, 361-362
Theis equation, 321
Thermal conductivity, 162
Thermodynamic potential, 177, 191-192
Thermoosmosis, 182-183
Thiem equation, 330
Tidal efficiency, 226-228
Time constant in resistor-capacitor networks, 245
Time-drawdown curves, 324
Time factor in consolidation theory, 365-367
Total head, 156
Total potential groundwater discharge, 50
Transmissivity, 153, 154
    determined from pumping tests (*see* Well hydraulics)

Transmissivity:
    in recharge calculations, 51
Transmissivity tensor, 382
Tritium, 272
Turbulent flow, 147-148
Type curves for well functions, 340, 343

Ultrafiltration, 180
Unconfined flow, 172-176
Unsteady flow, 147-148, 243-245

Vadose zone in water profile, 13-14
Vector field, 163
Vertical permeability, 344-345
Void ratio, 230-231
Volcanic water, 286
Voltage potential, 162

Water budget, 279-280
Water-level fluctuations, 45-46, 223-228
Water profile, 13-14
Water-resource systems, 15-17
Water table, 14
    mathematical approximations, 256-258
Water withdrawals by states, 21-22
Welfare economics, 81-83
    in common-pool problems, 84-85
Well capacity, 351
Well-development techniques, 352
Well functions, 320, 344
Well hydraulics, 317-354
    distance-drawdown method, 336-338
    equilibrium method, 327-332
    leaky aquifer method, 342-345
    modified nonequilibrium method, 333-335
    nonequilibrium method, 321-327
    recovery method, 335-336
Well loss, 353-354
White River, Nevada, 279, 280

Yield, 43-45
    of wells, 351-356